Advances in
Physical Organic Chemistry

Advances in Physical Organic Chemistry

Volume 26

Edited by

D. BETHELL

The Robert Robinson Laboratories
Department of Chemistry
University of Liverpool
P.O. Box 147
Liverpool L69 3BX

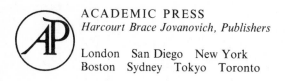

ACADEMIC PRESS
Harcourt Brace Jovanovich, Publishers

London   San Diego   New York
Boston   Sydney   Tokyo   Toronto

ACADEMIC PRESS LIMITED
24/28 Oval Road
London NW1 7DX

United States Edition published by
ACADEMIC PRESS INC.
San Diego, CA 92101

QD476
A4
Vol. 26
1990

British Library Cataloguing in Publication Data

Advances in physical organic chemistry
  Vol. 26
  1. Physical organic chemistry
  547.1'3'05

ISBN 0-12-033526-3
ISSN 0065-3160

FILMSET BY BATH TYPESETTING LTD, BATH, UK
AND PRINTED IN GREAT BRITAIN BY T. J. PRESS, PADSTOW, CORNWALL

Contents

Hydrogen Bonding and Chemical Reactivity   **255**

FRANK HIBBERT AND JOHN EMSLEY

Preface

This series of volumes, established by Victor Gold in 1963, aims to bring before a wide readership among the chemical community substantial, authoritative and considered reviews of areas of chemistry in which quantitative methods are used in the study of the structures of organic compounds and their relation to physical and chemical properties.

Physical organic chemistry is to be viewed as a particular approach to scientific enquiry rather than a further intellectual specialization. Thus organic compounds are taken to include organometallic compounds, and relevant aspects of physical, theoretical, inorganic and biological chemistry are incorporated in reviews where appropriate. Contributors are encouraged to provide sufficient introductory material to permit non-specialists to appreciate fully current problems and the most recent advances.

The series has been extremely fortunate in the quality of the contributors, who have allowed the editors to persuade them to devote much time and effort in order to expound their specialist interests for the benefit of a wider audience. The Editor would welcome feedback from readers. This might merely take the form of criticism. It might also contain suggestions of developing areas of chemistry that merit a forward-looking exposition or of the need for a new appraisal of better established topics that have escaped the notice of the Editor and his distinguished Advisory Board.

D. BETHELL

Contributors to Volume 26

John Emsley   Department of Chemistry, King's College London, Strand, London, WC2R 2LS, UK

Frank Hibbert   Department of Chemistry, King's College London, Strand, London, WC2R 2LS, UK

Hiizu Iwamura   Department of Chemistry, Faculty of Science, The University of Tokyo, Bunkyo-Ku, Tokyo 113, Japan

Hans-Gert Korth   Institut für Organische Chemie der Universität Gesamthochschule Essen, Postfach 103764, D-4300 Essen 1, FRG

Jean-Michel Savéant   Laboratoire d'Electrochimie Moleculaire, Université Paris 7, 2 Place Jussieu, 75251 Paris Cedex 05, France

Reiner Sustmann   Institut für Organische Chemie der Universität Gesamthochschule Essen, Postfach 103764, D-4300 Essen 1, FRG

Single Electron Transfer and Nucleophilic Substitution

JEAN-MICHEL SAVÉANT

Laboratoire d'Electrochimie Moléculaire, Université de Paris 7, 2 Place Jussieu, 75251 Paris Cedex 05, France

ADVANCES IN PHYSICAL ORGANIC CHEMISTRY
VOLUME 26   ISBN 0-12-033526-3

1 Introduction

The possible role of single electron transfer in organic reactions, as opposed to the classical notion of electron-pair transfer, has attracted continuous and active attention during the past twenty-five years. An important step in this connection, experimentally exemplifying such reaction pathways, has been the discovery of nucleophilic substitution reactions proceeding via anion radical intermediates and taking place at benzylic carbon centres (Kornblum *et al.*, 1966; Russell and Danen, 1966) or at aromatic carbon centres (Kim and Bunnett, 1970). The term "$S_{RN}1$" used to designate these reactions (Kim and Bunnett, 1970; Bunnett, 1978) underlines that, while belonging to the general class of nucleophilic (N) substitution (S) reactions, they involve radical intermediates (R), the first of which in the reaction sequence, i.e. the anion radical of the substrate, undergoes a monomolecular cleavage of the nucleofugic group (1). On the other hand, the continuous development of organic electrochemistry, particularly in its mechanistic and kinetic aspects (Andrieux and Savéant, 1986a; Baizer and Lund, 1983; Savéant, 1986, 1988), has been another source of interest and information for reactions triggered by single electron transfer.

The $S_{RN}1$ mechanism is now reasonably well understood (Bunnett, 1978; Kornblum, 1971, 1975; Rossi and Rossi, 1983; Russell, 1970; Savéant, 1980a). In the case of $S_{RN}1$ aromatic nucleophilic substitution, the electrochemical approach to the problem has allowed a detailed and quantitative description of the various steps of the substitution process and of the side-reactions (Savéant, 1980a, 1986, 1988). As discussed in more detail in the following, the reaction proceeds by a chain mechanism in which the electron supplied during the (chemical, electrochemical, photochemical) initiation process plays, *stricto sensu*, the role of a catalyst. The species that reacts with the nucleophile is not the substrate but the aryl radical deriving from its anion radical by cleavage of the nucleofugic group. Single electron transfer is thus involved in the generation of the key intermediate of the reaction, viz., the aryl radical. It is also involved in the reoxidation of the anion radical of the product which closes up the propagation loop. One of the most important side-reactions (termination step), namely the reduction of the aryl radical, also proceeds by means of single electron transfer from the various reducing species present in the reaction medium (initiator, anion radical of the substrate and of the product).

The $S_{RN}1$ reaction thus appears as a reaction in which single electron transfer plays a pre-eminent role but is by no means a single elementary step. A different problem is that of the possible involvement of single electron transfer in reactions that are not catalysed by electron injection (or removal). A typical example of such processes is another substitution reaction, namely,

$$RX + D^- \rightleftharpoons RX\cdot^- + D\cdot \qquad (1)$$

$$RX\cdot^- \longrightarrow R\cdot + X^- \qquad (2)$$

$$R\cdot + D\cdot \longrightarrow RD \qquad (3)$$

the S_N2 reaction. The question which then arises is whether or not these reactions, that are classically viewed as proceeding via electron-pair transfer, in fact involve single electron transfer. The latter would then be associated with bond breaking and bond formation. The problem thus amounts to distinguishing between processes in which electron transfer, bond breaking and bond formation are stepwise [as in (1)–(3) where X^- = nucleofugic group, D^- = single electron donor or nucleophile] and processes where they are concerted. Although the outer sphere/inner sphere terminology was coined originally for electron-transfer reactions involving coordination complexes (Espenson, 1986; Taube, 1970), it can be used profitably for organic processes of the type under discussion after some extension of the definitions (Lexa *et al.*, 1988). In outer sphere electron-transfer reactions, either no bond is cleaved or formed within the time scale of the experiment, or in the opposite case, bond breaking and bond formation take place in separated steps, distinct from the electron-transfer step. Conversely, if all three steps are concerted one will deal with an inner sphere electron transfer. In this context, an S_N2 reaction may be considered as being formally equivalent to an inner sphere electron-transfer reaction, or even close to being truly equivalent in many instances, as discussed in the following. An intermediate situation exists, however, in between the two extreme cases of outer sphere and inner sphere electron transfers, just defined. The reduction of alkyl halides by inert electrodes or by aromatic anion radicals in solution offers an example of such a situation. Electron transfer is dissociative there in the sense that the carbon–halogen bond is cleaved in concert with electron transfer. On the other hand, the electron donor, while transferring one electron, is not the object of any bond formation or cleavage. The reaction has thus an outer sphere character from the point of view of the electron donor and an inner sphere character from the point of view of the electron acceptor. Such a reaction may be termed an outer sphere dissociative electron transfer. The outer sphere/inner sphere terminology may also be used to characterize the way in which the reactants react rather than to characterize the overall reaction. In the preceding case, the alkyl halide behaves as an inner sphere electron acceptor whereas the inert electrode or the aromatic anion radical behaves as an outer sphere electron donor.

Alkyl and aryl halides are among the most commonly investigated substrates in S_N2 and $S_{RN}1$ reactions. They will thus serve as the main experimental examples in the general discussion presented in the first part of this chapter, which aims to answer the following questions. How does a frangible substrate react with outer sphere electron-transfer donors? Do electron transfer and bond breaking then occur concertedly or in a stepwise manner? What kind of activation *vs* driving force relationships are applicable in each case? The second part will be devoted to the $S_{RN}1$ reaction and the third to the S_N2 reaction. In all cases, some emphasis will be laid on electrochemical approaches. This is not only a matter of personal inclination but also because it offers some distinct advantages over strictly chemical approaches. One is that, rather obviously, electrochemical methods may be of help for gathering thermodynamic information, such as standard potentials, required for the analysis of reactivity patterns in reactions where electron transfer is involved. On the other hand, electrochemical generation of homogeneous electron donors (nucleophiles) belonging to a reversible redox couple has several advantages. One is that the species thus generated need not be very stable. It suffices that its reaction with the desired substrate be faster than all the reactions that cause its disappearance in the reaction medium. The standard potential characterizing thermodynamic properties of the electron donor can be easily measured by current electrochemical techniques such as cyclic voltammetry. The same techniques can be used conveniently for determining, upon addition of the substrate, the kinetics and mechanism of its reaction with the electrogenerated homogeneous electron donor (nucleophile). Much faster reactions can thus be characterized than with conventional chemical techniques. On the other hand, as far as outer sphere electron donors are concerned, it is of interest to examine whether or not the same activation–driving force relationships are applicable to heterogeneous (inert electrodes) and homogeneous outer sphere electron donors as is, for example, predicted by Marcus–Hush theory in the particular case of purely outer sphere electron transfer (Hush, 1958, 1961; Marcus, 1956, 1963, 1964, 1965, 1977, 1982; Marcus and Sutin, 1985; Waisman *et al.*, 1977).

2   Single electron transfer and bond breaking

A substitution reaction involves the breaking of one bond and the formation of another one. We discuss in this section reactions in which electron transfer is associated with the breaking of a bond but in which either no bond formation occurs or, if it does, takes place in a further separated step. For such processes, a series of fundamental questions arise. Are electron transfer and bond breaking concerted or stepwise? Can experimental data concerning elementary steps (in the first case the dissociative electron transfer, in the second the electron transfer and the bond breaking steps) be organized under the form of activation versus driving force relationships (rate constant versus equilibrium constant or synonymously "Brønsted plots", "Marcus plots" and, for electrochemical reactions, "Tafel plots")? If so, what models are available to analyse these plots in terms of shape and reorganization barriers?

In the case of stepwise electron-transfer bond-breaking processes, the kinetics of the electron transfer can be analysed according to the Marcus–Hush theory of outer sphere electron transfer. This is a first reason why we will start by recalling the bases and main outcomes of this theory. It will also serve as a starting point for attempting to analyse inner sphere processes. Alkyl and aryl halides will serve as the main experimental examples because they are common reactants in substitution reactions and because, at the same time, a large body of rate data, both electrochemical and chemical, are available. A few additional experimental examples will also be discussed.

MODELLING OF OUTER SPHERE ELECTRON TRANSFER

The Marcus–Hush theory of outer sphere electron transfer (Hush, 1958, 1961; Marcus, 1956, 1963, 1964, 1965, 1977, 1982; Marcus and Sutin, 1985) is based on the Born–Oppenheimer approximation and thus relates the activation barriers to the nuclear reorganization that accompanies electron transfer. In its currently most often used version, it deals with adiabatic processes: at the intersection between potential energy surfaces of the reactants and products (Fig. 1), the resonance energy between the reactant and product states (crossing avoidance) is sufficient for the probability for the system to cross the barrier to be unity. Note, however, that, in current applications, this energy is, at the same time, considered small enough so as not to affect significantly the height of the barrier.

The nuclear reorganization energy is assumed to be the sum of two independent terms, one representing the solvent reorganization (external reorganization, outer sphere reorganization) and the other, the internal (inner sphere) reorganization, i.e. the changes in bond lengths and angles in the reactants occurring upon electron transfer. Both terms are evaluated within a harmonic approximation. This appears quite natural for the internal reorganization term, provided the transition state is not too far from the reactant and product equilibrium states in term of bond distances and angles (otherwise the stretching of the bond might fall in the anharmonic region). Insofar as the various vibrational modes involved in the internal

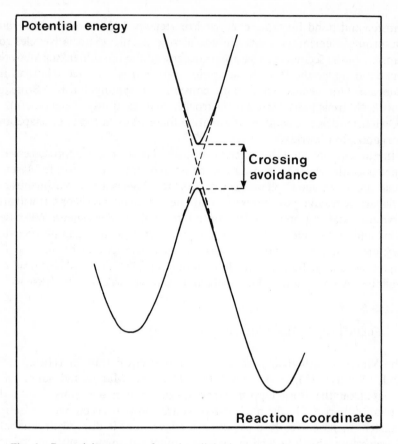

Fig. 1 Potential energy surfaces in adiabatic outer sphere electron transfer.

$$G_{i,R} = \tfrac{1}{2} \sum f_R (y - y_R)^2 \tag{4}$$

$$G_{i,P} = \tfrac{1}{2} \sum f_P (y - y_P)^2 \tag{5}$$

reorganization can be considered as independent of each other, their contributions are additive. So also are the internal reorganization terms of the two reactants in the case of a homogeneous cross-exchange electron-transfer reaction. Thus, the dependence of the free energy of the reactant and product systems upon the variations of the vibration coordinates can be expressed as in (4) for the reactant system and in (5) for the product system. Here, y designates the various vibration coordinates involved in the electron transfer process, y_R and y_P are their values for the reactant and product, respectively, and f_R and f_P are the corresponding force constants in the reactants and products. The summations are extended to all vibration coordinates involved and to the two reactants and products in the homogeneous case.

The contribution of solvent fluctuational reorganization to the free energy of the reactant and product systems can also be expressed in a quadratic manner. In the framework of the Born model of solvation, a fictitious electric charge, x, borne by the reactant or product is taken as the reaction coordinate (in the homogeneous case, the increase of the fictitious charge borne by one reactant occurs at the expense of that borne by the other reactant and the same applies to the products). The contribution of solvent reorganization to the free energy of the reactant and product systems can thus be expressed (Marcus, 1965, 1977) as in (6a) and (6b), where x_R and x_P are the charges borne by the reactant and product at their equilibrium state.

$$G_{O,R} = \tfrac{1}{2}\lambda_0(x - x_R)^2 \tag{6a}$$

$$G_{O,P} = \tfrac{1}{2}\lambda_0(x - x_P)^2 \tag{6b}$$

$$\lambda_0^{\text{hom}} = e_0^2\left(\frac{1}{D_{op}} - \frac{1}{D_S}\right)\left(\frac{1}{2a_1} + \frac{1}{2a_2} - \frac{1}{a_1 + a_2}\right) \tag{7}$$

In the homogeneous case, λ_0 is given by (7), where e_0 is the electron charge, D_{op} and D_S the optical and static dielectric constants of the solvent respectively, and a_1 and a_2 are equivalent hard sphere radii of the two reactants (and products). For the electrochemical case, there are two versions for the expression of λ_0. In Marcus's treatment (Marcus, 1965) the reaction site is assumed to be located at a distance from the electrode equal to its radius, a, and the effect of image forces in the electrode is taken into account (8). In Hush's treatment (Hush, 1961) the reaction site is assumed to be located farther from the electrode surface and the effect of image forces is neglected (9).

$$\lambda_0^{\text{het}} = e_0^2\left(\frac{1}{D_{op}} - \frac{1}{D_S}\right)\frac{1}{4a} \tag{8}$$

$$\lambda_0^{\text{het}} = e_0^2\left(\frac{1}{D_{op}} - \frac{1}{D_S}\right)\frac{1}{2a} \tag{9}$$

In total, the free energy surfaces for the reactant and product systems are (as shown in Fig. 1, but in terms of free energy rather than energy, and with more than one reaction coordinate, namely, the fictitious charge representing solvent reorganization and all the bond lengths and bond angles of the reactants that vary significantly upon electron transfer) given by (10) and (11). G_R^0 and G_P^0 are the standard free energies of the reactants and products

$$G_R = G_R^0 + w_R + \tfrac{1}{2}\lambda_0(x - x_R)^2 + \tfrac{1}{2}\sum f_R(y - y_R)^2 \tag{10}$$

$$G_P = G_P^0 + w_P + \tfrac{1}{2}\lambda_0(x - x_P)^2 + \tfrac{1}{2}\sum f_P(y - y_P)^2 \tag{11}$$

respectively when they are at infinite distance one from each other in the homogeneous case, or of the reactant and product when they are at infinite distance from the electrode surface in the electrochemical case. The term w_R is the work required to bring the reactants from infinity to the reacting distance, i.e. to form the "precursor complex" whereas w_P is the work required to form the "successor complex" from infinity to the reacting distance. In the electrochemical case, it is often assumed that the reaction site is located at the external boundary of the compact double layer, i.e. in the outer Helmholtz plane (Delahay, 1965). The introduction of the w_R and w_P terms in the above equations then amounts to performing the "Frumkin correction" of the double layer effect on the kinetics of the electrochemical electron-transfer reaction (Delahay, 1965). Minimizing then the free energy of the pathways going from the reactant to the product systems leads to the expressions (12) and (13) for the activation barriers, in terms of free energy,

$$\Delta G_+^{\ddagger} = w_R + \Delta G_0^{\ddagger}\left(1 + \frac{\Delta G^0 - w_R + w_P}{4\Delta G_0^{\ddagger}}\right)^2 \tag{12}$$

$$\Delta G_-^{\ddagger} = w_P + \Delta G_0^{\ddagger}\left(1 - \frac{\Delta G^0 - w_R + w_P}{4\Delta G_0^{\ddagger}}\right)^2 \tag{13}$$

for the forward and backward reactions. Here, $\Delta G^0 = G_R^0 - G_P^0$ is the standard free energy of the reaction (the opposite of the driving force in terms of free energy) and $\Delta G_0^{\ddagger}$ the standard free energy of activation, i.e. the free energy of activation of the forward and backward reactions at zero driving force, in other words, the intrinsic barrier. $\Delta G_0^{\ddagger}$ is related to the internal and external reorganization factors by (14) where λ_0 is given by (7)–

$$\Delta G_0^{\ddagger} = \frac{\lambda_0 + \lambda_i}{4} \tag{14}$$

(9) and λ_i by (15), in which the summation involves all relevant vibration coordinates and the two reactants in the homogeneous case f being given by (16). The derivation of (12)–(16) implies that, for each vibration coordinate,

$$\lambda_i = \tfrac{1}{2}\sum f(y_P - y_R)^2 \tag{15}$$

$$f = \frac{2f_R f_P}{f_R + f_P} \tag{16}$$

the force constant does not vary considerably upon passing from the reactant to the product. In other words, one introduces symmetrical functions of the force constants as given by (16) and antisymmetrical functions (17) are neglected.

$$f = \frac{f_R - f_P}{f_R + f_P} \tag{17}$$

The rate constants, k_+ and k_- of the forward and backward reactions are finally derived from (12) and (13) according to the transition-state theory, i.e. assuming that the transition and the initial states, on the one hand, and the transition and final states, on the other, are in equilibrium (Glasstone *et al.*, 1941). Thus, estimating the partition function of these three states in the classical way gives (18) and (19), where μ is the reduced mass of the two reactants in the homogeneous case and m the mass of the reactant in the electrochemical case.

$$k_+ = Z\exp\left(\frac{\Delta G_+^{\ddagger}}{RT}\right) \quad ; \quad k_- = Z\exp\left(\frac{\Delta G_-^{\ddagger}}{RT}\right) \tag{18}$$

$$Z_{\text{hom}} = (a_1 + a_2)^2 \left(\frac{8\pi kT}{\mu}\right)^{1/2} \quad ; \quad Z_{\text{het}} = \left(\frac{kT}{2\pi m}\right)^{1/2} \tag{19}$$

One of the most valuable outcomes of the Marcus theory is that it provides a quantitative, albeit simple, description of how the activation barrier varies with the driving force of the reaction. The factors constituting the intrinsic barrier are clearly identified and can be approximately estimated from available data characterizing the structure of the reactants of the products and of the reaction medium.

The experimental construction of the activation–driving force relationship is particularly simple in the *electrochemical case* since an easy and accurate way of making the driving force vary is to change the electrode potential. The theory predicts that the activation–driving force relationships should be

$$\ln\left(\frac{k_+}{Z_{het}}\right) = -\frac{zF\varphi_r}{RT} - \frac{\Delta G_0^{\ddagger}}{RT}\left(1 + F\frac{E - E^0 - \varphi_r}{4\Delta G_0^{\ddagger}}\right)^2 \tag{20}$$

$$\ln\left(\frac{k_-}{Z_{het}}\right) = -\frac{(z-1)F\varphi_r}{RT} - \frac{\Delta G_0^{\ddagger}}{RT}\left(1 - F\frac{E - E^0 - \varphi_r}{4\Delta G_0^{\ddagger}}\right)^2 \tag{21}$$

(20) and (21) once the work terms w_R and w_P have been estimated as functions of the electrical potential at the reaction site, φ_r: $w_R = zF\varphi_r/RT$, $w_P = (z - 1)F\varphi_r/RT$. E is the electrode potential, E^0 is the standard potential, or more exactly the formal potential when activity effects cannot be neglected, and z is the charge number of the reactant. Thus, the current–electrode potential relationship characterizing the kinetics of an outer sphere electron-transfer reaction is given by (22) (i is the current flowing through

$$\frac{i}{FS} = Z_{het}\exp\left[-\frac{zF\varphi_r}{RT} - \frac{\Delta G_0^{\ddagger}}{RT}\left(1 + F\frac{E - E^0 - \varphi_r}{4\Delta G_0^{\ddagger}}\right)^2\right]$$

$$\times \left\{[C_R]_0 - [C_P]_0 \exp\left(\frac{F}{RT}(E - E^0)\right)\right\} \tag{22}$$

the electrode, S is the electrode surface area, and $[C_R]_0$ and $[C_P]_0$ are the concentrations of the reactant and product at the electrode surface, respectively). This equation replaces the classical Butler–Volmer law (Butler, 1924a,b; Erdey-Gruz and Volmer, 1930) shown in (23) and (24), which was

$$\frac{i}{FS} = k_S^{ap}\exp\left[-\frac{\alpha F}{RT}(E - E^0)\right]\left\{[C_R]_0 - [C_P]_0 \exp\left(\frac{F}{RT}(E - E^0)\right)\right\} \tag{23}$$

$$k_S^{ap} = k_S\exp\left[-\frac{(\alpha - z)F\varphi_r}{RT}\right] \tag{24}$$

based on the phenomenological *assumption* that the effect of a variation of potential on the activation free energy is a constant fraction, α, of this variation. In the framework of the Marcus theory, α, the transfer coefficient, i.e. the electrochemical symmetry factor, is no longer a constant but varies linearly with the electrode potential as in (25) if it can be assumed that the

$$\alpha = \frac{\partial \Delta G_+^{\ddagger}}{\partial \Delta G_0} = \frac{\partial \Delta G_+^{\ddagger}}{\partial F(E - E^0)} = \frac{1}{2}\left(1 + F\frac{E - E^0 - \varphi_r}{4\Delta G_0^{\ddagger}}\right) \tag{25}$$

potential of the reaction site, φ_r, does not vary appreciably in the range of electrode potentials investigated. If not, the factor $\partial \varphi_r / \partial E$ appears in the expression for α (26).

$$\alpha = \frac{\partial \Delta G_+^{\ddagger}}{\partial \Delta G_0} = \frac{\partial \Delta G_+^{\ddagger}}{\partial F(E - E^0)} = \frac{1}{2}\left(1 + F\frac{E - E^0 - \varphi_r}{4\Delta G_0^{\ddagger}}\right)\left(1 - \frac{\partial \varphi_r}{\partial E}\right) + z\frac{\partial \varphi_r}{\partial E}$$

(26)

The electrochemical electron-transfer kinetics can be investigated by means of a large variety of techniques that are all based on the same principle, namely the competition between the rate of electron transfer and the rate of diffusion of the reacting species to and from the electrode surface. The rate of diffusion, and hence the competition with electron transfer, can be varied in a way which is specific to each electrochemical technique. For example, in steady-state techniques such as rotating disc electrode voltammetry (Andrieux and Savéant, 1986a; Bard and Faulkner, 1980) the rotation rate of the electrode, which imposes the thickness of the diffusion layer, is used for this purpose. In a transient technique, such as cyclic voltammetry, where the electrode potential is scanned linearly with time, this role is played by the scan rate. There are thus two ways of changing the driving force of the reaction. Given the rotation rate or the scan rate, the available variation of the driving force corresponds to the variation of the electrode potential along the "wave" (i.e. the current–potential curve). On the other hand, the wave shifts as the rotation rate or the scan rate are varied (negatively as they are raised for a reduction process and vice versa for an oxidation process). This is thus an additional way for varying the driving force. At a given rotation rate or scan rate, the range of electrode potentials in which the wave is controlled, at least partially, by the electron transfer kinetics, being comprised between its foot and its peak, is usually quite narrow. This is the reason why detection of the usually small variations of the transfer coefficient with the electrode potential requires not only the analysis of the wave at a given rotation rate or scan rate, but also that these parameters be varied to the largest possible extent. Another consequence is that, in current practice of electrochemical kinetic investigations, the variation of the transfer coefficient with the electrode potential can often be neglected at each rotation rate or scan rate, the Butler–Volmer law (23, 24) being then applicable, whereas it should be taken into account when one varies the rotation rate or the scan rate to a large extent (Andrieux *et al.*, 1989a; Andrieux and Savéant, 1989; Lexa *et al.*, 1987).

It is thus, in principle, possible to derive from the potential location and from its shape all the parameters contained in the Marcus–Hush model, namely, the standard potential, E^0, and the intrinsic barrier, $\Delta G_0^{\ddagger}$ (Klinger and Kochi, 1981, 1982; Lexa et al., 1987). In cyclic voltammetry, for example, these two parameters can be derived from the potential, E_m, located midway between the peak and the half-peak and from the width of the wave as defined by the difference between the half-peak and peak potentials, $E_{p/2} - E_p$ using (27–30), where v represents the scan rate and D the diffusion

$$\Delta G_0^{\ddagger} = \frac{1}{4\alpha_m^2} \frac{RT}{F} \ln\left(\frac{Z_{het}}{k_{+,m}}\right) \tag{27}$$

$$E^0 = E_m + 4\Delta G_0^{\ddagger}(1 - 2\alpha_m) - \varphi_r \tag{28}$$

$$\alpha_m = \frac{RT}{F} \frac{1.85}{E_{p/2} - E_p} \tag{29}$$

$$\frac{RT}{F} \ln\left(\frac{Z_{het}}{k_{+,m}}\right) = \frac{RT}{F} \ln\left[\frac{Z_{het}}{[\alpha_m v D(F/RT)]^{1/2}}\right] + 0.145\frac{RT}{F} \tag{30}$$

coefficient. This procedure should, however, be applied with caution. A first precaution is to check that the voltammetric wave is actually controlled by the electron-transfer kinetics. This is not always done (Fukusumi et al., 1987) and, in such a case, the procedure is not applicable (Hapiot et al., 1989). On the other hand, even if the reaction is controlled by electron-transfer kinetics and if reasonable evidence exists as to the applicability of the Marcus–Hush model, the determination of α from the wave shape is not very precise and rather large errors on E^0 and $\Delta G_0^{\ddagger}$ may ensue (Andrieux and Savéant, 1989; Andrieux et al., 1989b). This approach remains nevertheless useful for semi-quantitative purposes.

What precedes is true for any other electrochemical technique, using in each case the appropriate experimental parameter for varying the diffusion rate (the frequency in impedance methods, the measurement time in potential-step techniques, and so on).

In the *homogeneous case*, the derivation of an activation–driving force relationship and of an expression of the intrinsic barrier is less straightforward. It is interesting in this connection to relate the kinetics of cross-exchange reactions (31) to those of the two self-exchange reactions (also called identity or isotopic reactions) [(32) and (33)].

$$Ox_1 + Red_2 \rightleftharpoons Red_1 + Ox_2 \tag{31}$$

$$Ox_1 + Red_1 \rightleftharpoons Red_1 + Ox_1 \tag{32}$$

$$Ox_2 + Red_2 \rightleftharpoons Red_2 + Ox_2 \tag{33}$$

From this, it is clear that the internal reorganization factor $\lambda_{i(12)}$ is given by (34).

$$\lambda_{i(12)} = \frac{\lambda_{i(11)} + \lambda_{i(22)}}{2} \tag{34}$$

This is also approximately true for solvent reorganization (Marcus, 1965). Thus, the standard fuel energy of activation, defined by (35) leads to expression (36) for k_{12}, in which f_{12} is given by (37) and K_{12} is the equilibrium constant of the cross-exchange reaction.

$$\Delta G^{\ddagger}_{0(12)} \simeq \frac{\Delta G^{\ddagger}_{0(11)} + \Delta G^{\ddagger}_{0(22)}}{2} \tag{35}$$

$$k_{12} \simeq (k_{11} k_{22} K_{12} f_{12})^{1/2} \tag{36}$$

$$f_{12} = \frac{(\ln K_{12})^2}{4 \ln (k_{11} k_{22} / Z^2_{\text{hom}})} \tag{37}$$

These equations can be used to analyse any set of cross-exchange electron transfer reactions. A particularly simple situation is met when, given the acceptor (or vice versa the donor), a series of donors (or vice versa acceptors) of variable standard potentials, but having approximately the same intrinsic barriers for self-exchange, are used. In terms of relative accuracy, the situation is improved if the intrinsic barriers in the series of donors (or acceptors) are small as compared to that of the investigated acceptor (or donor). Then a Brønsted-type diagram is obtained upon plotting the log of the rate constant against the difference between the standard potential of the acceptor (or vice versa the donor) and those of the donors (or vice versa acceptors). The plot is, however, predicted to be parabolic instead of linear. In other words, the symmetry factor is predicted to vary linearly with the standard free energy of the reaction according to (38) in the case where the work terms do not vary in the series.

$$\alpha = \frac{\partial \Delta G^{\ddagger}_+}{\partial \Delta G_0} = \frac{1}{2} \left(1 + \frac{\Delta G^0 - w_R + w_P}{4 \Delta G^{\ddagger}_0} \right) \tag{38}$$

Let us again emphasize the connection between the Marcus–Hush model, on the one hand, and Brønsted-type activation–driving force relationships and Brønsted slopes, on the other. In the electrochemical case, the Marcus–Hush model predicts a definite free energy relationship of the Brønsted type, provided the driving force is varied by means of the electrode potential. There is another way of varying the driving force, namely the standard potential when one passses from one redox couple to another. However, the reorganization factors, both internal and external, and hence the intrinsic barrier, then also vary. There is thus no reason, in the general case, why a Brønsted-type relationship should be obtained when the driving force is varied in this way. The same is true when a common acceptor (or donor) is opposed to a series of donors (acceptors), provided the self-exchange intrinsic barriers of the latter are approximately the same in the series. Such a series constitutes a *family* of donors (acceptors) for which a Brønsted-type relationship is anticipated. If one changes family with the same acceptor (donor) or if one changes acceptor keeping the same (acceptor) family, different Brønsted plots are anticipated. If two arbitrary series of acceptors and donors are opposed no Brønsted relationship is predicted to hold. Instead, the cross-exchange relationship (37, 38) should then apply.

Under these restrictive conditions, the Brønsted plot is expected to be slightly parabolic rather than linear. As represented schematically in Fig. 2, a Brønsted slope smaller than, close to, and bigger than 0.5 indicates that the driving force is, respectively, big, close to zero and small.

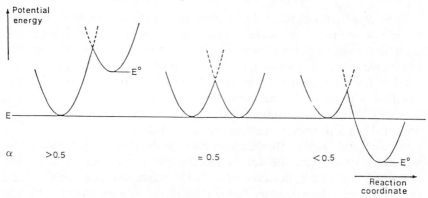

Fig. 2 Variation of the Brønsted slope (transfer coefficient, symmetry factor) with the driving force. (Adapted from Savéant, 1986.)

In other words, under these restrictive conditions, outer sphere electron-transfer reactions obeying the Marcus–Hush model are typical examples where the Hammond–Leffler postulate and the reactivity–selectivity principle (see, for example, Pross, 1977, and references cited therein, for the definition of these notions) are expected to apply.

APPLICATION TO ORGANIC REDOX SYSTEMS

What about the experimental verification and application of the Marcus theory of outer sphere electron transfer in the organic field? They are not very numerous, and are definitely less than in coordination chemistry. It is remarkable, in this connection, that in a recent book devoted to electron transfer in organic chemistry (Eberson, 1987), where the Marcus theory of outer sphere electron transfer is described in detail, one does not find a single purely organic experimental example of its application to a reaction that would unequivocally be of the outer sphere type from the point of view of both the donor and the acceptor. Thanks to electrochemistry and line broadening in esr spectroscopy, the situation is not as dull as this would suggest, since there are in fact several examples where various aspects of the Marcus theory have been tested quantitatively or semi-quantitatively. The reason why these tests are less numerous and/or less precise than in coordination chemistry is that injection or removal of one electron from an organic molecule is very often accompanied by bond breaking or bond formation. Even if these reactions are not concerted with electron transfer and are slow enough for the primary intermediate to be detected, this is often too unstable for an accurate analysis of its structural properties to be possible. The same type of reactions (ligand exchange) may also be associated with electron transfer in coordination chemistry. There are, however, many complexes in which the coordination number is the same in both members of the redox couple. Detailed structural information concerning the vibrations in both states can then be gathered, allowing a precise testing of the theory (Marcus and Sutin, 1985).

   The standard activation energies of the electrochemical reduction of a series of nitro-derivatives were measured in N,N-dimethylformamide (Peover and Powell, 1969). In this solvent, the corresponding anion radicals are stable enough for a ready determination of the kinetics to be possible by the impedance technique then used. It was shown that the standard activation energies correlate linearly with the electron density on the nitrogen of the nitro-group in the anion radical as measured by esr spectroscopy. The electron density on the nitrogen being assumed to be proportional to the negative charge on the nitro group, the observed correlation was interpreted as showing that the internal reorganization term does not vary significantly in the series and that solvent reorganization is correctly represented by the Marcus–Hush model. A similar, although more qualitative, study of the electrochemical reduction in the same solvent of a series of organic thiocarbonates:

$$
\begin{array}{cccc}
\text{RS—C—SR} & \text{RS—C—SR} & \text{RS—C—OR} & \text{RS—C—OR} \\
\parallel & \parallel & \parallel & \parallel \\
\text{S} & \text{O} & \text{S} & \text{O}
\end{array}
$$

has been carried out, using cyclic voltammetry, as a function of the nature of the functional group and the substituents (Falsig *et al.*, 1980). The variations of the standard activation free energy in the series was found to correlate qualitatively with the concentration of charge in the radical anion as predicted by the Marcus theory, again, in the context of a predominance of solvent reorganization over internal reorganization.

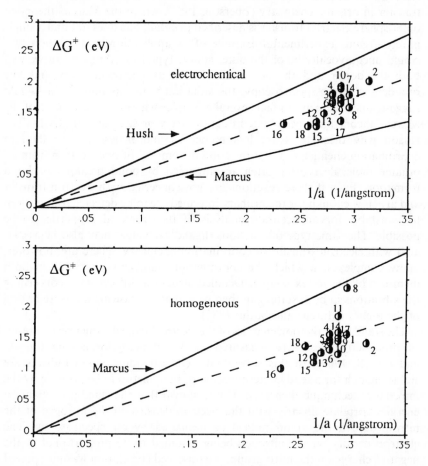

Fig. 3 Electrochemical and homogeneous standard free energies of activation for self-exchange in the reduction of aromatic hydrocarbons in *N,N*-dimethylformamide as a function of their equivalent hard sphere radius, *a*. 1, Benzonitrile; 2, 4-cyanopyridine; 3, *o*-toluonitrile; 4, *m*-toluonitrile; 5, *p*-toluonitrile; 6, phthalonitrile; 7, terephthalonitrile; 8, nitrobenzene; 9, *m*-dinitrobenzene; 10, *p*-dinitrobenzene; 11, *m*-nitrobenzonitrile; 12, dibenzofuran; 13, dibenzothiophene; 14, *p*-naphthoquinone; 15, anthracene; 16, perylene; 17, naphthalene; 18, *trans*-stilbene. Solid lines denote theoretical predictions. (Adapted from Kojima and Bard, 1975.)

A more systematic and quantitative investigation has been carried out with an extended series of aromatic anion radicals in N,N-dimethylformamide (Kojima and Bard, 1975). Both electrochemical and homogeneous standard free energies of activation for self-exchange were determined using the impedance technique and line broadening in esr spectra, respectively. It was again assumed that solvent reorganization is the dominant factor, which seems reasonable since the strength of the aromatic rings makes improbable a significant change in bond lengths or angles upon passing from the hydrocarbon to its anion radical. The variations of the standard free energies of activation with the inverse of the radius of the equivalent hard sphere are shown in Fig. 3. In both cases, the standard free energies of activation vary in a roughly linear fashion with $1/a$ as predicted by theory. In the homogeneous case, it appears that the theory overestimates solvent reorganization. This is not too surprising since it is based on the Born model, which itself overestimates the free energies of solvation. In the electrochemical case, the experimental points fall in between the lines representing the Hush and Marcus models. If it is considered that the model overestimates the contribution of solvent reorganization, as it does in the homogeneous case, then the Hush model appears to fit the experimental data better. This falls in line with the notion that the reaction site, being probably located in the outer Helmholtz plane (Delahay, 1965), is not, in the case of tetrabutylammonium supporting cations, in close contact with the electrode, thus contributing to decrease the image force effects. This study is important not only because it provides an experimental test of the Marcus–Hush theory but also because it makes available the intrinsic barriers characterizing the heterogeneous and homogeneous generation of large series of aromatic anion radicals which, as discussed later, function as outer sphere electron donors in most cases. Thus, one disposes of an extended series of reference redox couples that can be used for investigating, on a comparative basis, whether other electron donors behave in an inner sphere or outer sphere manner.

In the same work, encumbered stilbenes were found to have much higher intrinsic barriers than the hydrocarbons listed in the caption of Fig. 3. This is due to a strong internal reorganization around the central C—C bond (Dietz and Peover, 1968). Similar effects have been found in the reduction of cyclooctatetraene and related systems (see Evans and O'Connell, 1986, and references cited therein).

Another important aspect of the Marcus theory has also been systematically investigated with organic molecules, namely the quadratic, or at least the non-linear, character of the activation–driving force relationship for outer sphere electron transfer. In other words, does the transfer coefficient (symmetry factor) vary with the driving force, i.e. with the electrode potential in the electrochemical case? This question has attracted the attention of electrochemists for a long time. However, the first attempts in this connection (Angel and Dickinson, 1972; Anson et al., 1970; Bindra et al., 1975; Mohilner, 1969; Momot and Bronoel, 1974; Parsons and Passeron, 1966; Suga et al., 1973) were not very conclusive. One reason for this is that the predicted variation is small and that the accuracy of the measurements was not sufficient for an unambiguous answer to be obtained. Another reason deals with the selection of the systems investigated: with highly charged reactants the terms in $\partial \varphi_r / \partial E$ (see equation 26) may become important. The determination of $\partial \varphi_r / \partial E$ over the entire range of potential is not free from uncertainties because it derives from a rather approximate model of the double layer (Delahay, 1965) and because of experimental uncertainty in the determination of the double layer capacitance data from which φ_r is calculated. Under these conditions, it is difficult to prove unambiguously whether the transfer coefficient does or does not vary with the electrode potential. The simplest way of avoiding these difficulties is to start from an uncharged reactant. Another condition to be fulfilled is that the electrode reaction must not be significantly affected by adsorption, even non-specific adsorption, of the reactants on to the electrode surface. (Chemisorption, if important, would make the electron transfer inner sphere rather than outer sphere since a sort of chemical bond is then formed between the reactant and the electrode surface.)

The above conditions have been met in the investigation of the reduction of a series of organic molecules, mostly, but not exclusively (benzaldehyde belongs to the series) nitro-compounds in N,N-dimethylformamide and acetonitrile at a mercury electrode (Savéant and Tessier, 1975, 1977, 1978, 1982). The kinetics of the reduction were investigated by cyclic voltammetry (treated by the convolution method) and by the impedance technique. The investigated molecules were selected because they give rise to stable anion radicals, at least within the time scale of the experiments, and also because electron transfer to these molecules is not too fast. Thus, in the available range of scan rates (cyclic voltammetry) or of frequencies (impedance), a sufficiently large potential range could be investigated. The main result of this study is that, in all cases, the transfer coefficient was found to vary with the electrode potential, approximately in a linear manner, with a rate of the same order of magnitude as predicted by the theory. These conclusions have been confirmed by a study of the reduction of one member of the series, 2-nitropropane, on a different electrode material, namely platinum (Corrigan and Evans, 1980).

These findings contrast with the results of one investigation of the electrochemical kinetics of the III/II couple series of chromium complexes (Weaver and Anson, 1976) where the transfer coefficient was found not to vary appreciably with the potential, although the accuracy of the measurements was sufficient for detecting the variation predicted by the theory. The electron transfer is, in this case, much slower than with the organic couples invoked above. There is accordingly a dominant contribution of the reorganization of the coordination sphere of the complex, whereas, in the organic case, solvent reorganization was prevalent. The explanation offered to interpret the lack of variation of α with the potential was that the inner sphere reorganization barrier is so high that the vibration of the metal ligand bond is likely to be in an anharmonic region.

Another example where α does not appear to vary with potential is the hydrogen evolution reaction (39) in water (Conway, 1985). It should,

$$H^+ + e^- \rightleftharpoons H \text{ (chemisorbed)} \tag{39}$$

however, be noted that this reaction, which, at first sight, could appear as one of the simplest electrochemical processes, is in fact more difficult to model than electron transfer to a large organic molecule yielding a stable anion radical. In an H-bonded and H-bonding solvent such as water, solvent reorganization around H^+ is quite unlikely to obey such a simple model as the Born model used in the Marcus theory. In this respect, if one views the reactant as the solvated H_3O^+, the reaction is rather an inner sphere than an outer sphere process, since a bond has to be broken upon electron transfer. The chemisorption of the immediate product of this reaction, the H atom, at the electrode surface is another factor conferring an inner sphere character on the reaction. It is thus not much of a surprise that this inner sphere electron-transfer reaction does not follow the predictions of a model devised, strictly speaking, for outer sphere processes.

In the homogeneous case, it is much more difficult than in the electrochemical case to detect a variation of the symmetry factor with the driving force. If a common acceptor (donor) is opposed to a set of donors (acceptors) having different standard potentials so as to produce the desired driving force variation, one deals with a set of cross-exchange reactions, the intrinsic barriers of which are combinations of the intrinsic barriers of the corresponding acceptor and donor self-exchange reactions (29). In the donor (acceptor) series, the intrinsic barrier may well vary. In such a case, these variations have to be independently known and the cross-reaction (31) used rather than a Brønsted relationship. Cases exist in which the intrinsic barrier for self-exchange does not vary very much in the donor (acceptor) family opposed to the common acceptor (donor) and is small compared to the intrinsic barrier for self-exchange of the latter; a single Brønsted plot is then anticipated. However, this is only approximate and the resulting inaccuracy may prevent the detection of small variations of the symmetry factor. On the other hand, unlike the electrochemical case, the variation of the driving force cannot be made continuous and thus the donor (acceptor) family should contain a very large number of members to reach a comparable accuracy, as required for detecting minute variations of the symmetry factor. In the best experimental examples (Schlesener et al., 1986; Wong and Kochi, 1979), quite a large number of points were indeed gathered and were found to be compatible with a Marcus-type quadratic law. As in the electrochemical case, the standard potential and the intrinsic barrier of the investigated reaction could in principle be derived from the Brønsted plot. However, even more caution than in the electrochemical case should be exerted in the application of such procedures. As discussed in the following, there are other ways for determining these two parameters in the homogeneous case when the system lends itself to the observation of both activation and diffusion controls of the electron transfer.

Another, and even more striking aspect of the variation of the symmetry factor with the driving force is the prediction, from (9) and (10), that an "inverted region" should exist at large driving forces, i.e. when the inequality (40) applies. The activation free energy is then predicted to increase with the

$$-\Delta G^0 > 4\Delta G_0^{\ddagger} \tag{40}$$

driving force and thus the symmetry factor is predicted to become negative. This phenomenon is expected to be opposed by nuclear tunnelling which should then be more significant than in the "normal region" (because the barrier is thinner). Nevertheless, examples of "inverted region" behaviour have been reported for several outer sphere electron-transfer reactions (Closs and Miller, 1988; Fox and Gray, 1990; Gould et al., 1987; Gunner, et al., 1986; McLendon, 1988; Miller et al., 1986; Ohno et al., 1986, 1987; Snadrini et al., 1985; Wasielevski et al., 1985).

As a general conclusion of the above discussion, the Marcus–Hush theory of outer sphere electron transfer can be considered to fit reasonably well the experimental data available for organic redox couples provided they are actually of the outer sphere type. The fit could probably be improved by taking into account additional factors such as electronic factors (Newton and Sutin, 1984) and also solvent dynamic effects (Calef and Wolynes, 1983a,b; Fraunfelder and Wolynes, 1985; Gennet et al., 1985; Hupp et al., 1983; Hupp and Weaver, 1984, 1985; Marcus and Sumi, 1986a,b,c, 1987; McManis et al., 1986; Nadler and Marcus, 1987; Newton and Sutin, 1984; Nielson et al., 1988; Van der Zawn and Hynes, 1982, 1985; Weaver and Genneth, 1985; Zhang, et al., 1985, 1987; Zusman, 1980). Although approximate, in the form discussed above, it is a quite valuable tool for investigating structure–reactivity patterns in the field of outer sphere electron-transfer reactions.

MODELLING OF DISSOCIATIVE ELECTRON TRANSFER

If "dissociative" is intended to mean not only that electron transfer is accompanied by bond breaking but that the two steps are concerted, then the reaction loses its purely outer sphere character and, therefore, the Marcus theory is not, strictly speaking, applicable. It is true that the Marcus theory has been applied successfully to several reactions involving concerted bond breaking and bond formation, which are not, therefore, of the outer sphere type, such as proton transfer (Keefe and Kresge, 1986; Kreevoy and Truhler, 1986; Kresge, 1975; Marcus, 1982; Schlesener et al., 1986) and methyl cation transfer (Albery and Kreevoy, 1978; Lewis, 1986). There is some theoretical justification for this, based on a bond energy–bond order description of the reaction (Marcus, 1968). However, essential features, which make the Marcus theory of outer sphere electron transfer so valuable, are lost in this generalization, namely, the clear definition of intrinsic barriers independent from the driving force (Marcus, 1969) as well as their expression as functions of the structural characteristics of the reacting molecules and of their interactions with the reaction medium. If these valuable features are to be restored in a model of dissociative electron transfer, the Marcus theory of outer sphere electron transfer obviously needs an extension that is able to model the contribution of bond breaking to the activation barrier.

The harmonic oscillator approximation used to describe the reactants and products in terms of internal reorganization is certainly not suited for describing the broken bond in the product system in the case of dissociative electron transfer. For reactions of the type (41) and (42), where RX does

Electrochemical: $RX + e^- \rightleftharpoons R\cdot + X^-$ (41)

Homogeneous: $RX + D\cdot^- \rightleftharpoons R\cdot + X^- + D$ (42)

not represent necessarily an alkyl halide but, more generally, any electron acceptor undergoing a dissociative electron transfer, the following model has been proposed (Savéant, 1987). As in the Marcus theory, the Born–Oppenheimer approximation is assumed to hold and the reaction is assumed to be adiabatic. The potential energy surfaces for the reactants and products depend upon three types of reaction coordinate: the stretching of the R—X bond from equilibrium, z, the same fictitious charge, x, describing the

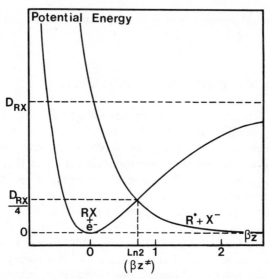

Fig. 4 Morse curves of the reactants and products at zero driving force (z is the elongation of the RX distance from equilibrium, $\beta = v_0(2\pi^2\mu/D_{RX})^{1/2}$, with v_0 representing the vibration frequency, μ the reduced mass of the two atoms of the R—X bond, and D_{RX} the RX bond dissociation energy).

solvent fluctuational configurations as in the Marcus theory of outer sphere electron transfer, and the vibration coordinate, y, of the bonds that are not broken during the reaction. The corresponding contributions to the free energy of the reactant and product systems are regarded as independent one from the other and additive. The two last contributions are treated exactly in the same way as in the Marcus theory. As concerns bond breaking, the potential energy of the reactants is assumed to depend upon the R—X distance according to the RX Morse curve, and the potential energy curve for the products is assumed to be purely dissociative and to be the same as the repulsive part of the *reactant* Morse curve (Fig. 4). The assumption that the repulsive part of the potential energy curves of the reactants and products are the same has previously been made in the interpretation of the kinetics of gas-phase thermal electron attachment to alkyl halides (Wentworth *et al.*, 1969). It was based on the notion that the repulsive term arises from interactions involving the core electrons and the nuclei and should, therefore, not be dramatically affected by the presence of an additional peripheral electron. Possible attractive interactions in the product system, such as induced dipole-charge and quadrupole-charge interactions between R· and X⁻ are regarded as small, smaller than in the gas phase, owing to the presence of a surrounding polar medium. Under these conditions, the free energies are expressed as (43) for the reactants and (44) for the products. G_R^0 and G_P^0 are the equilibrium free energies of the reactants and product

$$G_R = G_R^0 + w_R + \tfrac{1}{2}\lambda_0(x - x_R)^2$$

$$+ \tfrac{1}{2}\sum f_R(y - y_R)^2 + D_{RX}[1 + \exp(-2\beta z) - 2\exp(-\beta z)] \tag{43}$$

$$G_P = G_P^0 + w_P + \tfrac{1}{2}\lambda_0(x - x_P)^2 + \tfrac{1}{2}\sum f_P(y - y_P)^2 + D_{RX}\exp(-2\beta z) \tag{44}$$

systems, respectively, and the other symbols are the same as defined previously, or are in the caption of Fig. 4. The resulting activation–driving force relationship has the same quadratic form as in the Marcus–Hush theory of outer sphere electron transfer (12, 13), the standard free energy of

$$\Delta G_+^{\ddagger} = w_R + \Delta G_0^{\ddagger}\left(1 + \frac{\Delta G^0 - w_R + w_P}{4\Delta G_0^{\ddagger}}\right)^2 \tag{12}$$

$$\Delta G_-^{\ddagger} = w_P + \Delta G_0^{\ddagger}\left(1 - \frac{\Delta G^0 - w_R + w_P}{4\Delta G_0^{\ddagger}}\right)^2 \tag{13}$$

activation (activation barrier) being given by (45) in which λ_0 and λ_i are the usual Marcus–Hush solvent and internal reorganization factors (8, 9, 15, 16). According to this model *the contribution of bond breaking to the intrinsic barrier is simply one-quarter of the dissociation energy of the bond being broken.*

$$\Delta G_0^{\ddagger} = \frac{D_{RX}}{4} + \frac{\lambda_0}{4} + \frac{\lambda_i}{4} \tag{45}$$

The activation–driving force relationships (12, 13) are exactly the same as in the Marcus–Hush theory of outer sphere electron transfer. Thus, the preceding conclusions about the possibility of obtaining Brønsted plots and the variation of the transfer coefficient (symmetry factor) with the driving force in the electrochemical and homogeneous cases are, likewise, the same in the framework of the present model of dissociative electron transfer. Of particular relevance for the discussion below is the conclusion that values of α significantly smaller than 0.5 are associated with large driving forces and vice versa for values larger than 0.5. An illustration of the application of the model to experimental systems is given below in the section devoted to the reduction of alkyl halides (p. 54).

STEPWISE AND CONCERTED PROCESSES

When a bond is broken upon electron transfer, a first question arises: are electron transfer and bond breaking stepwise (46, 47) or concerted (48).

$$RX + e^- \rightleftharpoons RX\cdot^- \qquad (E^0, \alpha, k_s^{ap}) \tag{46}$$

$$RX\cdot^- \xrightarrow{k} R\cdot + X^- \tag{47}$$

$$RX + e^- \longrightarrow R\cdot + X^- \tag{48}$$

In the *electrochemical case*, i.e. when e^- designates the uptake of one electron from an inert electrode, the first of these mechanisms belongs to the general family of "EC" mechanisms (Andrieux and Savéant, 1986a; Bard and Faulkner, 1980), in which an electrode electron transfer, "E", is followed by a chemical reaction, "C", taking place in the solution or at the electrode surface. The kinetics of the overall process is governed jointly by those of the electron-transfer step (46), of the follow-up reaction (47) and of the diffusion of the species RX and $RX\cdot^-$ to and from the electrode surface. When the problem is to distinguish between concerted and stepwise electron-transfer–bond-breaking reactions, the follow-up reaction (47), in the context of the stepwise mechanism, is certainly fast. "Pure kinetic conditions" (Andrieux and Savéant, 1986a; Nadjo and Savéant, 1973) are then achieved in the case of a solution follow-up reaction, meaning that a steady state is established from mutual compensation of the chemical and diffusional processes. Thus, the decay of the $RX\cdot^-$ species from the electrode surface takes place entirely within a thin reaction layer (as compared to the diffusion layer) adjacent to the electrode surface. The thickness of the reaction layer is equal to $(D/k)^{1/2}$ (Andrieux and Savéant, 1986a). Thus, "pure kinetic conditions" are typically achieved when k is larger than $10^3 \, \text{s}^{-1}$ (the thickness of the diffusion layer is typically $10^{-3} \, \text{cm}$ and the diffusion coefficient $10^{-5} \, \text{cm}^2 \, \text{s}^{-1}$). As long as the reaction layer is larger than the double layer, the overall kinetics results from a combination of electron transfer (46), subject to the normal double layer effects (Frumkin correction), the follow-up reaction (47) and diffusion to and from the electrode. Under these conditions, the competition between electron transfer (46) and the follow-up reaction (47) is a function of a single parameter, p, given by (49), where α is the transfer coefficient and where the expressions of Λ and λ,

$$p = \alpha^{(\alpha - 1)/2\alpha} \Lambda^{1/\alpha} \lambda^{1/2} \tag{49}$$

which represent the competition between electron transfer and the follow-up reaction on the one hand and diffusion on the other, depend on the particular electrochemical technique employed. For example, in steady-state techniques, such as rotating disc electrode voltammetry (Andrieux and Savéant, 1986a; Bard and Faulkner, 1980), they are given by (50), where D is

$$\Lambda = \frac{k_s^{ap}\delta}{D} \tag{50a}$$

$$\lambda = \frac{k\delta^2}{D} \tag{50b}$$

the diffusion coefficient of RX and RX·⁻ and δ is the thickness of the diffusion layer which can be varied by changing the rotation rate of the electrode. Then the half-wave potential of the RX reduction "wave" (i.e. the current-potential curve), $E_{1/2}$, is given by (51) in which the value of E^* is given by (52).

$$\exp\left[\frac{\alpha F}{RT}(E_{1/2} - E^*)\right] + p\exp\left[\frac{F}{RT}(E_{1/2} - E^*)\right] = 1 \tag{51}$$

$$E^* = E^0 + \frac{RT}{\alpha F}\ln\left(\frac{k_s^{ap}\delta}{D}\right) \tag{52}$$

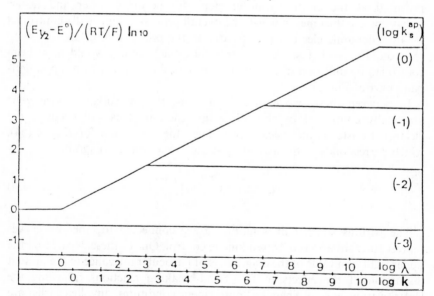

Fig. 5 Electrochemical stepwise electron-transfer–bond-breaking reactions. Competition between electron transfer, bond breaking and diffusion. $E_{1/2}$, Half-wave potential; E^0, RX/RX·⁻ standard potential. The horizontal scale is given both in terms of λ and k. The number on each curve is the value of Λ, and the value of $\log k_s^{ap}$ is given in parentheses. δ is taken as 10^{-3} cm and D as 10^{-5} cm^2 s^{-1}. (Adapted from Andrieux et al., 1978.)

The outcome of the competition is represented in Fig. 5 in terms of the location of the half-wave potential of the RX reduction "wave" (i.e. the current-potential curve), $E_{1/2}$, relative to the standard potential of the RX/RX·⁻ couple, E^0 (Andrieux et al., 1978). As concerns the competition, three main regions of interest appear in the diagram. On the left-hand side, the follow-up reaction is so slow (as compared to diffusion) that the overall process is kinetically controlled by the parameter Λ, i.e. by electron transfer and diffusion. Then, going upward, the kinetic control passes from electron transfer to diffusion. In the upper section $E_{1/2} = E^0$ and in the lower section $E_{1/2}$ is given by (53).

$$E_{1/2} = E^0 + \frac{RT}{\alpha F} \ln \left(\frac{k_s^{ap} \delta}{D} \right) \tag{53}$$

On the extreme right-hand side of the diagram, the follow-up reaction has become so fast that it prevents the back electron transfer. Kinetic control is then by the forward electron transfer and the half-wave potential is then, once more, given by (53). It becomes more and more positive of the standard potential as the electron-transfer step (46) becomes faster and faster. Situations are thus met in which the overall process is kinetically controlled by an endergonic electron transfer due to the presence of a fast follow-up reaction. For such fast electron transfers, the reaction would have been controlled by diffusion in the absence of the follow-up reaction (upper left-hand part of Fig. 5).

In between these two extreme situations, the overall process may be kinetically controlled by the follow-up reaction. This will occur if the electron transfer is intrinsically fast. Then the expression for $E_{1/2}$ is (54), which corresponds to the ascending straight-lines on the diagram.

$$E_{1/2} = E^0 + \frac{RT}{2F} \ln \left(\frac{k \delta^2}{D} \right) \tag{54}$$

The various straight lines in Fig. 5 are connected together by curves representing situations of mixed kinetic control. One of these, located on the upper right-hand side of the diagram, is of particular relevance to the present discussion of the effects of fast bond-breaking reactions following the electron-transfer step. Then the "pure kinetic conditions" invoked earlier are met and the kinetic competition is governed by the parameter p defined in (49) and (50). In all of the above analysis, the validity of the Butler–Volmer law for representing the kinetics of the electron-transfer step has been implicitly assumed. As discussed earlier, this is perfectly acceptable at a given rotation rate. However, when passing from one rotation rate to another, the value of the transfer coefficient used in the analysis should be changed.

When the follow-up reaction becomes so fast that the thickness of the reaction layer comes close to molecular dimensions, the above analysis breaks down because the diffusion of $RX\cdot^-$ ceases to obey Fick's law. An extreme situation in this connection is when the reaction is so fast that $RX\cdot^-$ has no time to diffuse away from the electrode and collapses instead at the surface. The follow-up reaction should then be viewed as a surface reaction and the half-wave potential is given (Savéant, 1980b, 1983) by (55), where

$$\exp\left[\frac{\alpha F}{RT}(E_{1/2} - E^*)\right] + p_s \exp\left[\frac{F}{RT}(E_{1/2} - E^*)\right] = 1 \qquad (55)$$

the competition parameter, p_s, is now defined by (56) and E^* by (52). For a moderately fast follow-up reaction (47) and a fast electron transfer (46), the

$$p_s = (k_s^{ap})^{1/\alpha} \exp\left[\frac{(z - 1)F\varphi_r}{RT}\right](ka)^{-1} \delta^{1/\alpha - 1} D^{-1/\alpha + 1} \qquad (56)$$

$$E^* = E^0 + \frac{RT}{\alpha F} \ln\left(\frac{k_s^{ap}\delta}{D}\right) \qquad (52)$$

overall kinetics are controlled by the follow-up reaction (47). Then (57) applies, in which a is, as earlier, the radius of the hard sphere equivalent to the reactant. In most cases, however, the follow-up reaction is so fast, since it has become a surface reaction, that kinetic control is by the forward electron transfer (46). Then, $E_{1/2}$ is given again by (53).

$$E_{1/2} = E^0 + (1 - z)\varphi_r + \frac{RT}{F} \ln\left(\frac{ka\delta}{D}\right) \qquad (57)$$

$$E_{1/2} = E^0 + \frac{RT}{\alpha F} \ln\left(\frac{k_s^{ap}\delta}{D}\right) \qquad (53)$$

In the case of a solution follow-up reaction (47) so fast that the thickness of the reaction layer is of the same order of magnitude as the double layer, the competition between the electron transfer step (46) and the follow-up reaction (47) can be described in a similar way with a somewhat different effect of the double layer on the electron-transfer kinetics (Savéant, 1980b, 1983).

A very similar analysis applies to other electrochemical techniques, for example to cyclic voltammetry. The scan rate would then play the role of the rotation rate of the electrode in rotating disc electrode voltammetry, the half-wave potential leading to the definitions (58) of Λ and λ (Nadjo and Savéant, 1973).

$$\Lambda = \frac{RT}{F} \frac{k_S^{ap}}{(Dv)^{1/2}} \qquad (58a)$$

$$\lambda = \frac{RT}{F} \frac{k}{v} \qquad (58b)$$

What about the values of the transfer coefficient in the various kinetic control regions, if it is assumed that the electron-transfer step obeys the Marcus–Hush model and that the follow-up reaction takes place in the solution? In the case where the electron-transfer step (46) is intrinsically slow, i.e. in the lower part of Fig. 5, we deal with an exergonic electron transfer and, therefore, the transfer coefficient should be smaller than 0.5 (see Fig. 3). In the upper right-hand part of Fig. 5, we deal with an endergonic electron transfer and, therefore, the transfer coefficient should be larger than 0.5. In the intermediate oblique straight-line region, where the overall kinetics are controlled by the follow-up bond-breaking reaction, the di/dE slope is not a reflection of the transfer coefficient of the electron-transfer step since this is not the rate determining step as can be easily recognized on the experimental curves. If we deal with a very fast follow-up surface reaction resulting in the electron-transfer step (46) being rate determining, the transfer coefficient should be larger than 0.5.

In the case, now, of a concerted electron-transfer–bond-breaking reaction (48), assumed to obey the model described in the preceding section, the current–potential curves will have the characteristics of waves jointly controlled by electron transfer and diffusion. At each rotation rate or scan rate, the Butler–Volmer law is usually applicable in an approximate fashion for the same reasons as already discussed for purely outer sphere electron-transfer reactions (Andrieux *et al.*, 1989a; Andrieux and Savéant, 1989; Lexa *et al.*, 1987). As seen earlier, the contribution of bond breaking to the intrinsic barrier is approximately one-quarter of the bond-dissociation energy. Taking this and the solvent reorganization factor into account indicates that the intrinsic barriers for dissociative electron transfer should be large in most cases and, therefore, that the reduction potential (as measured for example by the half-wave potential in rotating disc electrode voltammetry or the peak potential in cyclic voltammetry) should be considerably negative of the standard potential of the reaction, i.e. this should be largely exergonic. The standard potential we are referring to is not that of

RX/RX·⁻ couple but that of the RX/R· + X⁻ couple. Under such conditions, the transfer coefficient should thus be smaller than 0.5 in the framework of the quadratic activation–driving force relationship predicted to hold for dissociative electron transfer by the model developed in the preceding section.

It follows that the value of the electrochemical transfer coefficient may allow the distinction between stepwise and concerted electron-transfer–bond-breaking reactions when a chemical bond of normal strength is involved (Andrieux and Savéant, 1986b; Andrieux et al., 1990b). If the reduction wave possesses the characteristics of a process controlled by slow electron transfer rather than controlled by a follow-up reaction, and if α is significantly larger than 0.5, then one can conclude that the reaction proceeds in a stepwise manner. The same is true when the wave exhibits the characteristics of a process controlled by a follow-up reaction, electron transfer remaining at equilibrium.

The case where the transfer coefficient is significantly smaller than 0.5 is less unambiguous since it may correspond either to a concerted process or to a stepwise process involving an intrinsically slow electron-transfer step (lower part of Fig. 5). Outer sphere electron transfer to organic molecules is often regarded as fast, which would lead to the conclusion that a value of α smaller than 0.5 excludes the possibility of a stepwise process (Perrin, 1984). This belief derives from the fact that electron transfer to large aromatic molecules is actually fast for the reasons discussed in a preceding section. This is not general, however, as outer sphere electron transfer to aliphatic molecules may well be slow both because of vigorous internal reorganization and also because of a large solvent reorganization caused by a high concentration of electric charge (Andrieux et al., 1989; Falsig et al., 1980; Peover and Powell, 1969; Savéant and Tessier, 1982). In spite of the remaining general ambiguity, a distinction between the two possibilities can nevertheless be made if the values of the standard potential and of the intrinsic barrier derived from the experimental data are compatible or not with what may be known independently concerning the properties of RX·⁻. In this connection, a particularly favourable situation is met when it is independently known that the standard potential of the presumed RX/RX·⁻ couple is negative to the reduction potential of RX (as measured, for example, by the half-wave potential in rotating disc electrode voltammetry or the peak potential in cyclic voltammetry). Then a value of α smaller than 0.5 clearly points to the conclusion that the reaction follows a concerted pathway. The reduction of alkyl and aryl halides, described in the following sections, will provide experimental examples of the distinction between concerted and stepwise pathways.

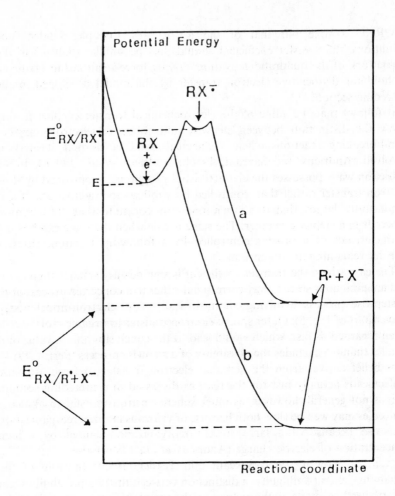

Fig. 6 Stepwise and concerted electron transfer and bond breaking. Schematic representation of the potential energy surface. (a) Stepwise process, $\alpha > 0.5$. (b) Concerted process, $\alpha < 0.5$. (Adapted from Andrieux *et al.*, 1985.)

Another question of interest is that of the transition between stepwise and concerted pathways. A schematic representation of the problem is given in Fig. 6; this is schematic in the sense that the system is represented as depending on a single reaction coordinate, viz., the distance between the atoms forming the bond to be broken during the reaction. This is obviously the main reaction coordinate as far as the concerted pathway (48) and the decomposition of $RX\cdot^-$ (47) are concerned. It is also an important reaction coordinate for the outer sphere electron-transfer step (46) in the representation of internal reorganization. Strictly speaking, one should also consider solvent reorganization and the other internal vibration modes besides R—X stretching. The schematic representation of Fig. 6 is, however, sufficient for the following qualitative discussion. The case represented is that invoked above where the reduction potential is positive of the standard potential of the RX/RX·$^-$ couple. It is seen that whether the reaction will follow the stepwise rather than the concerted pathways primarily depends on the structural properties of the RX reactant under examination. However, the preference for the reacting system of the stepwise versus the concerted pathway is also a function of the driving force, here represented by the electrode potential: the more positive the latter, i.e. the smaller the driving force, the stronger the tendency for the system to follow the concerted pathway and vice versa (Andrieux *et al.*, 1985; Andrieux and Savéant, 1986b). In most cases, of course, the reacting system will follow one or other pathway in the whole available driving force range, but borderline cases are conceivable where the reaction would go through RX·$^-$ at high driving forces and would proceed concertedly at low driving forces. In this sense, the course of the reaction does not strictly depend on the "existence" of the anion radical taken in an absolute sense: the anion radical may exist but the barrier for going through this intermediate might be higher than that for the concerted pathway, which would then be followed instead.

The same problem has been previously addressed in terms of two parameters representing the "extent of bond breaking" and the "extent of electron transfer" (Perrin, 1984). This model is, however, confusing since the notion of "extent of electron transfer" is difficult to master: the "extent of bond breaking", i.e. the length of the R—X bond, is itself an essential coordinate in governing the occurrence of electron transfer, subject to Frank–Condon restrictions, in the concerted electron-transfer–bond-breaking pathway (48) and in the outer sphere electron transfer (46) in the framework of the Born–Oppenheimer approximation. This is also the case for the bond-breaking reaction (47), which can be viewed as an intramolecular concerted electron-transfer–bond-breaking process in which the odd electron is transferred from a first residence orbital to one orbital of the anionic leaving group (Savéant, 1988). It is true that the occurrence of electron transfer depends upon additional coordinates describing solvent reorganization and vibrational modes other than R—X stretching, but it does not seem appropriate to view electron transfer and bond breaking as two independent phenomena. If a more exact representation than that given above (Fig. 6) is sought, one should rather investigate the occurrence of the stepwise versus concerted pathways on a hypersurface involving all the reaction coordinates. Thus, besides the R—X distance, solvent reorganization, which may well be different for reactions (46), (47) and (48), should be considered as well as other relevant internal vibrational modes.

Similar, although not exactly identical, analyses apply to the *homogeneous case* [(59), (47), 60)]. If the reaction proceeds in a stepwise fashion, there is again competition between the activation-controlled electron transfer (59),

$$RX + D^{\cdot -} \underset{k_+}{\overset{k_-}{\rightleftharpoons}} RX^{\cdot -} + D \tag{59}$$

$$RX^{\cdot -} \xrightarrow{k} R^{\cdot} + X^- \tag{47}$$

$$RX + D^{\cdot -} \longrightarrow R^{\cdot} + X^- + D \tag{60}$$

the follow-up reaction (47) and diffusion (Andrieux *et al.* 1979, 1980; Andrieux and Savéant, 1986b; Marcus and Sutin, 1985). This last phenomenon occurs in conditions that are quite different from the planar diffusion dealt with in the electrochemical case, since what is now involved is the diffusion of the reactants one towards the other, leading to the possibility that the forward and backward electron transfer reach the diffusion limit rather than being under activation control.

If the electron donor is so efficient a reductant as to react with the acceptor with a rate constant equal to the diffusion limit, then not much information can be derived from the experiments, except the knowledge of the diffusion limit itself. The opposite situation, where an endergonic electron transfer is followed by a fast bond-breaking step, is of more interest. There is then competition between the follow-up reaction and the backward electron-transfer step. If the latter is faster than the former, kinetic control is by the bond-breaking step, the electron-transfer step acting as a pre-equilibrium. Under these conditions, there is no difficulty to conclude from the adherence to the rate law (61) that the overall reaction is stepwise rather than concerted, since, in the concerted case, the rate law would be (62). If, in

$$\frac{d[RX]}{dt} = -\frac{k_+}{k_-} k[RX] \frac{[D^{\cdot -}]}{[D]} \tag{61}$$

$$\frac{d[RX]}{dt} = -k[RX][D^{\cdot -}] \tag{62}$$

the stepwise case, the electron-transfer equilibrium constant, i.e. the difference in standard potentials between the electron donor and the $RX/RX^{\cdot -}$ couple, is known independently, the rate constant of the follow-up reaction is readily derived. When $k \gg k_-[D]$, the forward electron transfer is the rate-determining step, the rate law is again of the form of (62), allowing easy recognition of the rate control by the forward electron-transfer step. Distinction between this situation and a concerted pathway is not possible on simple kinetic grounds since the rate law is the same in both cases. It can be achieved, however, under two types of favourable circumstances.

One such set of circumstances is when it is possible, upon raising the electron donor concentration, to change the rate law from (62) to (61), indicating that kinetic control passes from forward electron transfer to bond breaking according to (63). The passage across the mixed kinetics region

$$\frac{d[RX]}{dt} = -k_+ k \frac{[RX][D^{\cdot -}]}{k_-[D] + k} \tag{63}$$

may not be complete; provided that it is possible to go from one of the limiting situations to the mixed kinetics region upon changing the electron donor concentration, it is readily possible to conclude that the reaction is stepwise rather than concerted. In the latter case, indeed, the reaction order in electron donor would be unity whatever its concentration. In such circumstances it is also possible to derive both the rate constant of the follow-up reaction and the standard potential of the $RX/RX^{\cdot -}$ couple (the standard potential of the outer sphere electron donor is readily obtained from electrochemical measurements since its reduced form is chemically stable).

When the follow-up reaction is so fast as to overcome the backward electron transfer whatever the electron donor concentration, a possibility of distinguishing between stepwise and concerted pathways still remains if the forward electron transfer is so endergonic that the backward electron transfer is under diffusion control. This situation is schematically represented in Fig. 7 in the form of a plot of the overall rate constant (which is now the rate constant of the forward electron transfer) against the standard potential of the outer sphere electron donor (Andrieux et al., 1979; Andrieux and Savéant, 1986b). The rate constant of the forward electron transfer may fall into one of three regions, corresponding, from left to right, to diffusion control, activation control and "counter-diffusion control", the last term meaning that the backward electron transfer is under diffusion control. In total, the rate constant of the forward electron transfer can be expressed by (64). The rate constant of the backward electron transfer shows the converse behaviour (65), with the ratio k_+/k_- given by (66). Here k_D is the bimolecular diffusion limit (67) according to the Smoluchovski–Debye hard sphere model (N_A is Avogadro's constant, D is twice the average diffusion coefficient of the two reactants and R is the distance of closest approach between their centres) (Debye, 1942; Smoluchovski, 1916a,b, 1917).

$$\frac{1}{k_+} = \frac{1}{k_{+\,\text{act}}} + \frac{1}{k_\text{D}} + \frac{1}{k_\text{D}\exp\left[\dfrac{F}{RT}(E^0_{\text{RX/RX}^{\cdot-}} - E^0_{\text{D/D}^{\cdot-}})\right]} \tag{64}$$

$$\frac{1}{k_-} = \frac{1}{k_{-\,\text{act}}} + \frac{1}{k_\text{D}} + \frac{1}{k_\text{D}\exp\left[\dfrac{F}{RT}(E^0_{\text{RX/RX}^{\cdot-}} - E^0_{\text{D/D}^{\cdot-}})\right]} \tag{65}$$

$$\frac{k_+}{k_-} = \frac{k_{+\,\text{act}}}{k_{-\,\text{act}}} = \exp\left[\frac{F}{RT}(E^0_{\text{RX/RX}^{\cdot-}} - E^0_{\text{D/D}^{\cdot-}})\right] \tag{66}$$

$$k_\text{D} = 4\pi N_\text{A} D R \tag{67}$$

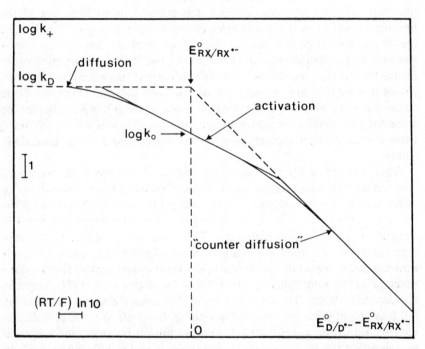

Fig. 7 Representation of (64) showing the procedure for determining the standard potential, $E^0_{\text{RX/RX}^{\cdot-}}$, and the intrinsic barrier, $\Delta G_0^{\ddagger}$ ($\Delta G_0^{\ddagger} = RT \ln k_0$). (Adapted from Andrieux and Savéant, 1986b.)

The three successive portions of the plot have slopes equal to zero, $\alpha F/RT$ and F/RT respectively, in terms of a $\ln k_+$ versus E^0 diagram. The (small) variation of α with the driving force has been omitted for clarity. Thus, if experimental points arising from a family of electron donors are available both in the second and third regions, the standard potential and the intrinsic barrier of the investigated redox couple can be obtained as shown on Fig. 7. It should again be emphasized that, in the activation region, the electron donors in the family should have about the same intrinsic barrier for self-exchange if a meaningful activation–driving force relationship (Brønsted plot) is to be obtained, as discussed earlier for outer sphere and dissociative electron transfers.

The same relationships apply, in principle, to the concerted pathway (60). However, the intrinsic barrier is now so high, because of the contribution of bond breaking, that the possibility of observing a region that is "counter-diffusion" controlled is quite unlikely since the rate constant of the forward reaction would then be immeasurably small in most cases. At any rate, even if one conceives that such a situation might occur, E^0 and $\Delta G_0^{\ddagger}$ then determined would be so different from that of the outer sphere electron transfer (59) in the stepwise pathway that the confusion would hardly be possible.

An explicit assumption in the preceding discussion was that, in the case of a stepwise process, the follow-up reaction is so fast that the electron-transfer step (59) is rate determining. A tacit additional assumption was that it is, however, not so fast that the decay of $RX^{\cdot -}$ does not take place solely outside the diffusion layer surrounding the electron donor molecule. In other words, the reaction scheme (68) was assumed, where the parentheses represent a solvent cage.

$$RX + D^{\cdot -} \underset{k_D}{\overset{k_D}{\rightleftharpoons}} (RX + D^{\cdot -}) \underset{k_{+act}}{\overset{k_{-act}}{\rightleftharpoons}} (RX^{\cdot -} + D) \underset{k_D}{\overset{k_D}{\rightleftharpoons}} RX^{\cdot -} + D \qquad (68)$$

$$\downarrow k$$

$$R^{\cdot} + X^- + D$$

For faster reactions, the decay of $RX^{\cdot -}$ within the diffusion layer, i.e. within the solvent cage, should be taken into account (Andrieux and Savéant, 1986b; Grimshaw and Thompson, 1986). An extreme situation, opposite to that just discussed, is when the follow-up reaction is so fast that $RX^{\cdot -}$ collapses before having time to diffuse away from the donor molecule. This is the homogeneous equivalent of the surface follow-up reaction case in electrochemistry. Equation (64) is then replaced by (69) (Andrieux and

$$\frac{1}{k_+} = \frac{1}{k_{+act}} + \frac{1}{k_D} + \frac{1}{(kN_A V) \exp\left[\dfrac{F}{RT}(E^0_{RX/RX^{\cdot -}} - E^0_{D/D^{\cdot -}})\right]} \qquad (69)$$

Savéant, 1986b), where V is the maximal volume occupied by the RX molecules when they are in contact with the electron donor. In between these two extremes, $RX\cdot^-$ diffuses within the molecular diffusion layer while being converted into $R\cdot$ and X^-. Then (70) applies. Figure 8(a) represents the variations of the forward rate constant with the standard potential difference resulting from the combination of (64), (69) and (70) for typical values

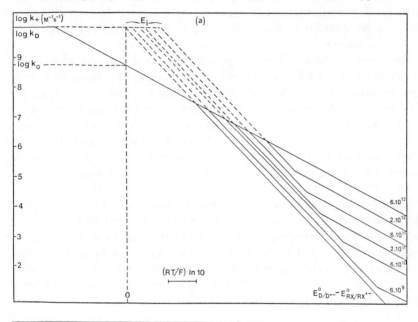

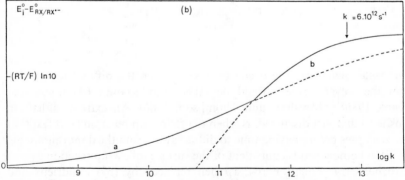

Fig. 8 (a) Variations of the forward rate constant, k_+, with the standard potential difference, $E^0_{D/D\cdot^-} - E^0_{RX/RX\cdot^-}$, as a function of the rate constant of the follow-up reaction, k (values on each curve in s^{-1}) for typical values of k_D ($10^{10}\,M^{-1}\,s^{-1}$), D ($2 \times 10^{-5}\,cm^{-2}\,s^{-1}$) and R (3 Å). (b) Shift of the intersection potential, E_i [see Fig. 8(a)] as a function of the follow-up reaction rate constant, k. $E_i - E^0_{RX/RX\cdot^-}$ is a measure of the error in the determination of $E^0_{RX/RX\cdot^-}$ made when the conversion of $RX\cdot^-$ into $R\cdot + X^-$ within the solvent cage is ignored. (Adapted from Andrieux and Savéant, 1986b.)

$$\frac{1}{k_+} = \frac{1}{k_{+\,\text{act}}} + \frac{1}{k_D} + \cfrac{1}{k_D\left(1 + \dfrac{Rk^{1/2}}{D^{1/2}}\right)\exp\left[\dfrac{F}{RT}(E^0_{RX/RX^{\cdot-}} - E^0_{D/D^{\cdot-}})\right]} \qquad (70)$$

of the various parameters. The diagram breaks down into three asymptotes with, from left to right, slopes of zero, $\alpha F/RT$ and F/RT. They correspond to different controls of the forward reaction, diffusion control, activation control and "counter-diffusion" control, respectively, combined with control by the follow-up reaction taking place within the solvent cage. The first two asymptotes are the same whatever the value of k. The third is a function of k. Its location is represented in Fig. 8(b) in the form of the variation of $E_i - E^0_{RX/RX^{\cdot-}}$, with k, E_i being the intersection of the F/RT slope asymptote and the zero slope asymptote. It is seen in Fig. 8(b) that the treatments on which (69) and (70) are based represent two limiting types of behaviour, one in which $RX^{\cdot-}$ does not diffuse and one in which it diffuses within the solvent cage according to Fick's law. The behaviour in between would be a diffusion process occurring over too short a distance, as compared to molecular sizes, for Fick's law to be obeyed. On the other hand, $E_i - E^0_{RX/RX^{\cdot-}}$ is a measure of the error in the determination of $E^0_{RX/RX^{\cdot-}}$ made when the conversion of $RX^{\cdot-}$ into $R^{\cdot} + X^-$ within the solvent cage is ignored.

Upon going to smaller and smaller driving forces, a fourth asymptote may be met which corresponds to the passage from a stepwise to a concerted process for the same reasons as already discussed in the electrochemical case. Its slope is again $\alpha F/RT$, but with a value of α corresponding to the concerted process rather than to the stepwise process. It is expected that the former will be significantly smaller than the latter, since the electron transfer is exergonic ($\alpha < 0.5$) in the concerted case and endergonic ($\alpha > 0.5$) in the stepwise case. The passage from the stepwise to the concerted case will be the more rapid the faster the follow-up reaction, as represented in Fig 8(a). Taking 6×10^{12} $M^{-1}\,s^{-1}$ as an estimate of the limit over which a concerted process prevails, whatever the driving force, it can be seen (Fig. 8) that the error on the determination of $E^0_{RX/RX^{\cdot-}}$ made when the conversion of $RX^{\cdot-}$ into $R^{\cdot} + X^-$ within the solvent cage is ignored is small, at most of the order of 80 mV (Andrieux and Savéant, 1986b).

ARYL HALIDES

The electron-transfer reduction of a large number of aromatic molecules involving an aryl carbon–heteroatom σ-bond produces a frangible anion radical which decomposes to the corresponding aryl radical and an anion containing the heteroatom. The most widely investigated compounds in this connection have been the aryl halides. Mechanistic and kinetic information for the reduction of these molecules is available (for reviews see Becker, 1983; Hawley, 1980) from direct and indirect electrochemistry in organic polar solvents (Aalstaad and Parker, 1982; Alwair and Grimshaw, 1973a,b; Andrieux *et al.*, 1980, 1984a, 1987, 1988a; Danen *et al.*, 1969; Grimshaw and Trocha-Grimshaw, 1974; Heinze and Schwart, 1981; Houser *et al.*, 1973; Lawless and Hawley, 1969; M'Halla *et al.*, 1978, 1980; Nadjo and Savéant, 1971; Nelson *et al.*, 1973; Parker, 1981a,b) and in liquid ammonia (Amatore *et al.*, 1979a; Savéant and Thiebault, 1978b), from pulse radiolysis in water (Neta and Behar, 1981), from γ-ray irradiation and low-temperature esr detection in apolar or weakly polar solid matrixes (Symons, 1981) and also in the gas phase (Steelhammer and Wentworth, 1969; Wentworth *et al.*, 1967). The lifetime of the anion radicals that have been measured in the liquid phase varies from minutes to nanoseconds according to the nature of the halogen and of the aryl moiety.

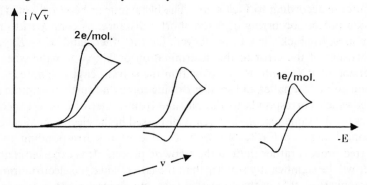

Fig. 9 Electrochemical reductive cleavage of aryl halides in a poor H-atom donor solvent. Cyclic voltammetry as a function of the scan rate, v. E, Electrode potential; i, current. Reduction (cathodic) currents are represented as being upwards.

The simplest way of generating and observing aryl halide anion radicals is to use an electrochemical technique such as cyclic voltammetry. With conventional microelectrodes (diameter in the millimetre range), the anion radical can be observed by means of its reoxidation wave down to lifetimes of 10^{-4} s. Under these conditions, it is possible to convert, upon raising the scan rate, the irreversible wave observed at low scan rates into a one-electron chemically reversible wave as shown schematically in Fig. 9. Although this does not provide any structural information about $RX\cdot^-$, besides the standard potential at which it is formed, it does constitute an unambiguous proof of its existence. Under these conditions, the standard potential of the $RX/RX\cdot^-$ couple as well as the kinetics of the decay of $RX\cdot^-$ can be derived from the electrochemical data. Peak potential shifts (Fig. 9) can also be used for this purpose as well as many other electrochemical techniques with about the same limitations concerning the anion radical lifetime when the same type of microelectrode is used. These limitations mainly derive from the fact that the ohmic drop becomes exceedingly large at high scan rates. They can be partly overcome by using the recently developed ultramicroelectrode (electrode diameter in the micrometer range) techniques (Amatore *et al.*, 1987; Andrieux *et al.*, 1988a,b,c, 1990d; Fitch and Evans, 1986; Garreau *et al.*, 1989, 1990; Hapiot *et al.*, 1990; Howell and Wightman, 1984; Montenegro and Pletcher, 1986; Wightman and Wipf, 1989). Lifetimes in the submicrosecond range can be reached in this way. Examples of the application of these techniques to aryl halides are the reductions of 3-bromoacetophenone and 9-bromoanthracene in *N,N*-dimethylformamide (Andrieux *et al.*, 1988a).

As represented in Fig. 9, the irreversible reduction of aryl halides at low scan rate is a two electron per molecule process, at least in poor H-atom donor solvents such as liquid ammonia (Amatore *et al.*, 1979a; Savéant and Thiebault, 1978). This is due to the fact that aryl radicals, produced upon cleavage of $RX\cdot^-$, are very easy to reduce, around -0.3 V *vs* SCE (Jaun *et al.*, 1980), much more than the starting aryl halides (from about -1 to -2.8 V *vs* SCE). It follows that $R\cdot$, as soon as produced in (47), is immediately reduced into the corresponding carbanion, R^- (71 and/or 72), which is eventually protonated (73) by the strongest acid present in the

$$RX + e^- \rightleftharpoons RX\cdot^- \tag{46}$$

$$RX\cdot^- \xrightarrow{k} R\cdot + X^- \tag{47}$$

"ECE" $$R\cdot + e^- \rightleftharpoons R^- \tag{71}$$

"DISP" $$R\cdot + RX\cdot^- \xrightarrow{k_D} R^- + RX \tag{72}$$

$$R^- + H^+ \rightarrow RH \tag{73}$$

reaction medium (e.g. residual water, ammonia) yielding the final hydrogenolysis product, RH. The second electron transfer may occur at the electrode surface or in the solution (from $RX\cdot^-$), giving rise, respectively, to an "ECE" (71) or "DISP" (72) process (Amatore *et al.*, 1984a; Andrieux and Savéant, 1986a). There is, in fact, a competition between these two pathways which is governed by the parameter p_E, defined in (74), where θ is the measurement time in potential-step experiments; $\theta = RT/Fv$ in cyclic voltammetry and $\theta = \delta^2/D$ in rotating disc electrode voltammetry. Of particular importance is the effect of the rate constant of the cleavage reaction, k, on the outcome of the competition. When k is large $R\cdot$ is formed

$$p_E = \frac{k_D[RX]}{k^{3/2}} \theta^{-1/2} \quad \begin{array}{l} \nearrow \quad 0 \qquad \text{ECE} \\ \\ \searrow \quad \infty \qquad \text{DISP} \end{array} \tag{74}$$

close to the electrode surface. It thus diffuses back to the electrode surface rapidly where it is immediately reduced before having time to react with incoming $RX\cdot^-$ species. The ECE pathway then tends to predominate. Conversely, when k is small, $R\cdot$ is formed far from the electrode surface. It will thus be reduced by incoming $RX\cdot^-$ molecules on its way to the electrode surface. The DISP pathway will then predominate. It may appear that this distinction between the two pathways is rather casuistic, since, in both cases, the final product, RH, is the same. However, it becomes quite important as soon as the intermediate $R\cdot$ is involved in a third competing reaction. The outcome of the resulting three-cornered competition then heavily depends upon whether an ECE or a DISP regime is established as concerns the electron-transfer reduction of $R\cdot$. Important differences ensue, not only in microelectrolytic regimes, such as those of the usual electrochemical kinetic techniques, but also on the preparative scale.

$$RX + e^- \rightleftharpoons RX\cdot^- \tag{46}$$

$$RX\cdot^- \xrightarrow{k} R\cdot + X^- \tag{47}$$

Electron transfer pathway: *H-atom transfer pathway:*

$$R\cdot + e^- \rightleftharpoons R^- \tag{71}$$

$$R\cdot + SH \xrightarrow{k_H} RH + S\cdot \tag{75}$$

$$S\cdot + e^- \rightleftharpoons S^- \tag{76}$$

and/or: and/or

$$R\cdot + RX\cdot^- \xrightarrow{k_D} R^- + RX \tag{72}$$

$$S\cdot + RX\cdot^- \xrightarrow{k_D} S^- + RX \tag{77}$$

$$R^- + H^+ \longrightarrow RH \tag{73}$$

$$S^- + H^+ \longrightarrow SH \tag{78}$$

Scheme 1

An example of such a third competing step is the reaction of the intermediate aryl radical with H-atom donors present in the reaction medium, possibly the solvent itself. Indeed, aryl radicals are good H-atom scavengers and reduction of aryl halides is often carried out in organic solvents such as acetonitrile (ACN), N,N-dimethylformamide (DMF), dimethyl sulphoxide (DMSO), and ethers, that are good H-atom donors

(Bridger and Russell, 1963). The final product is again the hydrogenolysis hydrocarbon and, in ACN and DMSO, the overall electron stoichiometry remains two electrons per molecule, indicating that, if formed upon H-atom abstraction, the solvent radical, $S\cdot$, is easier to reduce than the starting aryl halide (M'Halla *et al.*, 1978, 1980) (Scheme 1). The use of the standard microelectrolytic techniques is thus of little help for unravelling the reaction mechanism. This could be established by preperative-scale deuterium incorporation experiments carried out in heavy water–light organic solvent mixtures and vice versa. Such experiments are based on the fact that water is a much poorer H-atom donor than the organic solvent but a much better acid (M'Halla *et al.*, 1980). The yield of RD obtained in heavy water–light organic solvent mixtures is thus a measure of the electron-transfer pathway, whereas the yield of RH is a measure of the H-atom transfer pathway. The opposite applies to the experiments carried out in light water–heavy organic solvent mixtures The three-cornered competition is a function of two parameters (79). Typical results are displayed in Fig. 10 in the form of a

$$\frac{k_D[RX]\delta}{k^{1/2}k_H D^{1/2}} \quad \text{and} \quad \frac{k}{k_H} \tag{79}$$

diagram showing the zones where one of the competing pathways predominates over the other two (the boundary lines are drawn on the basis of a 50–50 participation of each pair of competing pathways). Among the various experimental parameters that affect the competition, as summarized in the top right-hand corner of Fig. 10, the effect of k is again of particular interest. Upon increasing k, the system passes from DISP to "HAT" (H-atom transfer) and to ECE. With 1-substituted naphthalene and 4-substituted benzonitrile derivatives, k is so large that, for all halogens, the competition involves practically only HAT and heterogeneous electron transfer, whereas, with 9-substituted anthracene derivatives, the reaction passes progressively from a DISP–HAT competition to an ECE–HAT competition. It was found, accordingly, that, with the latter compounds, deuterium incorporation is a function of the rate of agitation of the solution (the thickness of the diffusion layer is a decreasing function of the rate of agitation) and of the concentration of RX, whereas this is not the case in the naphthalene and benzonitrile series.

A somewhat more complex mechanism takes place with other H-atom donors, such as primary and secondary alcohols, either added to the liquid ammonia solution or used as the solvent (Andrieux *et al.*, 1987). Instead of being totally reduced, the hydroxyalkyl radical, resulting from the H-atom abstraction from the alcohol, partly deprotonates, generating the anion radical of the parent carbonyl compound. The latter is then generated by

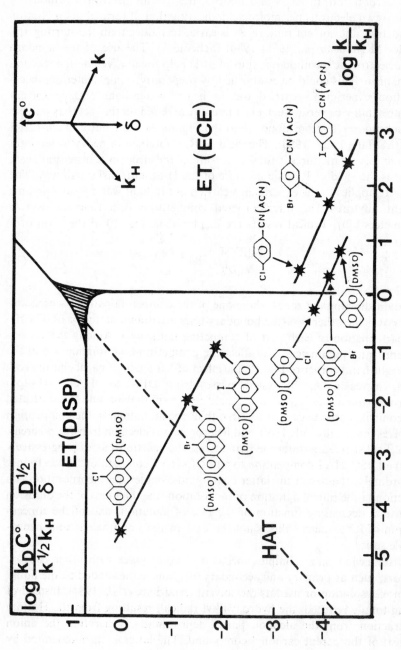

Fig. 10 Competition between H-atom transfer and electron transfer in the reduction of some aryl halides in DMSO and ACN. The symbols are defined in the text. (Adapted from M'Halla et al., 1980.)

$$R\cdot + H\!\!-\!\!\!|\!\!-\!\!OH \longrightarrow RH + \text{\textbackslash}\!\!\!\rangle\!\!-\!\!OH \qquad (75)$$

Reduction: *Deprotonation*:

$$\rangle\!\!-\!\!OH + e^- \rightleftharpoons H\!\!-\!\!\!|\!\!-\!\!O^- \quad (71) \qquad \rangle\!\!-\!\!OH + Z^- \rightleftharpoons \rangle\!\!-\!\!O^- + ZH \quad (80)$$

and/or:

$$\rangle\!\!-\!\!OH + RX\cdot^- \longrightarrow H\!\!-\!\!\!|\!\!-\!\!O^- + RX \quad (72) \qquad \rangle\!\!-\!\!O^- - e^- \rightleftharpoons \rangle\!\!=\!\!O \quad (81)$$

and/or:

$$H\!\!-\!\!\!|\!\!-\!\!O^- + ZH \rightarrow H\!\!-\!\!\!|\!\!-\!\!OH + Z^- \quad (73) \qquad \rangle\!\!-\!\!O^- + RX \rightarrow \rangle\!\!=\!\!O + RX\cdot^-$$

[ZH: strongest acid present in the medium]   (82)

Scheme 2

electron-transfer reoxidation. The reaction scheme of the overall process thus starts with the same reactions (46) and (47) as before. There is again competition between the electron-transfer pathway (71–73) and the H-atom-transfer reaction, but the latter is now followed by a second competition, namely that between electron-transfer reduction and deprotonation of the hydroxyalkyl radical (Scheme 2). Thus, when the H-atom transfer (75) to R· prevails over its electron-transfer reduction (71, 72), and when the reaction medium is sufficiently basic for reaction (80) to predominate over the reduction of the hydroxyalkyl radical (71, 72), the electron stoichiometry tends toward zero electrons per molecule. The whole process is then electrocatalytic (Andrieux and Savéant, 1986a) in the sense that the overall reaction (83), which does not consume electrons, is catalysed by electron

$$H\!\!-\!\!\!|\!\!-\!\!OH + RX + Z^- \longrightarrow \rangle\!\!=\!\!O + RH + X^- + ZH \qquad (83)$$

injection. The electron "borrowed" by the system for the generation of the intermediate R· is returned when the carbonyl anion radical is oxidized. On the whole, the alcohol is oxidized into the corresponding carbonyl compound by the aryl halide which is itself reduced into the corresponding hydrocarbon. Such behaviour has been observed in liquid ammonia (Andrieux *et al.*, 1987), whereas, if the alcohol is taken as solvent with no base added, a one electron per molecule stoichiometry is observed when the H-atom pathway prevails over the electron pathway; this corresponds to the overall reaction (84). The reduction of the hydroxyalkyl radicals (71, 72)

$$\tfrac{1}{2}H\!\!-\!\!\!|\!\!-\!\!OH + RX + e^- \longrightarrow \tfrac{1}{2}\rangle\!\!=\!\!O + RH + X^- \qquad (84)$$

then yields the base Z^- (82) required to deprotonate the same radicals, leading ultimately to the carbonyl compound. The resulting one-electron stoichiometry indicates that the deprotonation step (80) by the electrogenerated base (OH^- and/or the alcoholate) is more rapid than the reduction steps (71, 72), but that deprotonation requires these strong bases to be effective. Overall, there is thus a balanced combination of two-electron and zero-electron stoichiometries. This is confirmed by the observation that the electron stoichiometry falls below one as soon as a strong base, such as OH^-, is added to the alcoholic solution.

In DMF, a commonly used solvent in reductive organic electrochemistry, the electron stoichiometry varies as a function of the starting RX compound (Andrieux et al., 1984a, 1987). With 4-bromobenzophenone and 9-bromoanthracene, the situation is similar to that which occurs in alcohols, as just described (electron stoichiometry equal to one when no base or acid is added to the solution, falling below one upon addition of a strong base, and increasing up to two upon addition of an acid). Although the product to which DMF is converted has not been identified so far, these findings suggest a similar type of reaction as in alcohols, involving H-atom transfer, followed by reduction, on one hand, and deprotonation and reoxidation on the other. More generally, it has been found in the investigation of a large series of aryl halides in DMF (Andrieux et al., 1984a, 1987), that the electron stoichiometry is equal to two for compounds having a standard potential, $E^0_{RX/RX\cdot^-}$, below -2 V vs SCE and an $RX\cdot^-$ cleavage rate constant, k, larger than $10^7 \, s^{-1}$. Going to more positive values of $E^0_{RX/RX\cdot^-}$ and to smaller values of k, the electron stoichiometry decreases from two to one. This can be rationalized in the framework of the same reaction scheme as proposed above for alcohols. A large value of k leads to a fast reduction of $R\cdot$ at the electrode surface in the context of an ECE regime; the aryl radical that could survive this reduction could abstract an H-atom from DMF, but the resulting DMF radical would then be generated under very reducing conditions and would thus be reduced at the electrode surface rather than deprotonate into an oxidizable anion radical.

That reductive cleavage or aryl halides by an outer sphere electron donor does involve the anion radical $RX\cdot^-$ as an intermediate, rather than occurring along a concerted electron-transfer–bond-breaking process, clearly appears in the cases where the cleavage rate constant, k, is lower than $10^6 \, s^{-1}$. This is then proved by the observation that the ArX cyclic voltammetry wave becomes a one-electron reversible wave upon raising the scan rate (using conventional or ultramicroelectrodes). In the case of larger rate constants for cleavage, indirect evidence that $RX\cdot^-$ is an intermediate can be gained from the effect of H-atom donors as just described. For example, the characterization of the anion radicals of the 1-substituted naphthalene and

4-substituted benzonitrile derivatives as presumed intermediates in the reaction is out of reach of cyclic voltammetry or any other direct electrochemical technique in the present state of the art. However, the very fact that, in the deuterium incorporation experiments summarized in Fig. 10, these compounds obey the predictions based on Scheme 1, clearly points to the conclusion that $RX\cdot^-$ is indeed an intermediate. In the opposite case of a concerted process, $R\cdot$ would have been formed directly at the electrode surface and been immediately reduced there, since its reduction potential is much more positive relative to that of the starting RX. Whatever the halogen, no deuterium incorporation should then have been observed in the presence of the perdeuterated solvent and H_2O, and total incorporation should have been observed in the presence of D_2O and of the non-deuterated solvent. Similar indirect evidence can be obtained from the electron stoichiometry as a function of the addition of acids and bases, as well as from the identification of products deriving from solvent molecules in preparative-scale electrolyses in solvents such as primary and secondary alcohols. The reduction of aryl halides in the presence of alcoholates and the observation of the corresponding carbonyl compounds among the electrolysis products, as well as an electron stoichiometry tending toward zero, likewise provides indirect evidence that $RX\cdot^-$ is an intermediate (Amatore et al., 1982a) (Scheme 3). The overall reaction (88) is an electrocatalytic process, i.e. it does not consume electrons. A similar approach consists in carrying out the

$$RX + e^- \rightleftharpoons RX\cdot^- \tag{46}$$

$$RX\cdot^- \xrightarrow{k} R\cdot + X^- \tag{47}$$

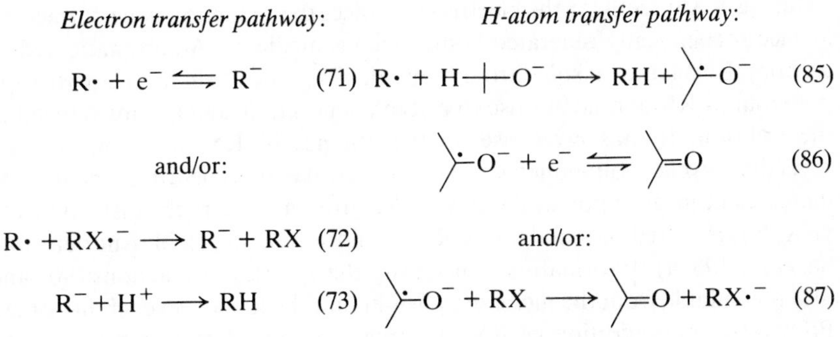

Electron transfer pathway: *H-atom transfer pathway:*

$$R\cdot + e^- \rightleftharpoons R^- \quad (71) \quad R\cdot + H\!\!-\!\!\!|\!\!-\!\!O^- \longrightarrow RH + {>}\!\!-\!O^- \tag{85}$$

and/or: $${>}\!\!\overset{\cdot}{-}\!O^- + e^- \rightleftharpoons {>}\!\!=\!O \tag{86}$$

$$R\cdot + RX\cdot^- \longrightarrow R^- + RX \quad (72) \qquad \text{and/or:}$$

$$R^- + H^+ \longrightarrow RH \quad (73) \quad {>}\!\!-\!O^- + RX \longrightarrow {>}\!\!=\!O + RX\cdot^- \tag{87}$$

Scheme 3

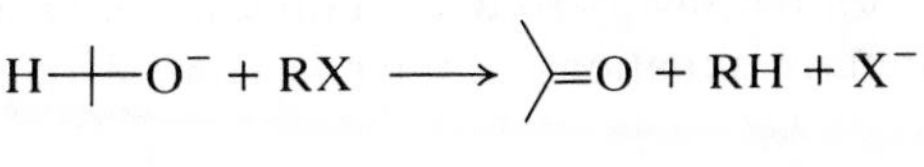

$$H\!\!-\!\!\!|\!\!-\!\!O^- + RX \longrightarrow {>}\!\!=\!O + RH + X^- \tag{88}$$

reduction of aryl halides in the presence of a nucleophile. This again is an electrocatalytic process leading to an electrochemically induced $S_{RN}1$ aromatic substitution. These reactions are discussed in detail in Section 3.

In the case of rapidly cleaved anion radicals, there are, however, less indirect ways of proving that they are intermediates, using as outer sphere electron donors electrochemically generated aromatic anion radicals. At the same time, this allows the determination of the standard potential, $E^0_{RX/RX\cdot^-}$, and of the cleavage rate constant, k, in the most favourable cases and provides an estimate of $E^0_{RX/RX\cdot^-}$ for very fast cleavages. It has also the advantage of providing examples of the reactivity of homogeneous outer sphere electron donors toward aryl halides in addition to that of heterogeneous outer sphere electron donors (inert electrodes). The principle of this indirect electrochemical approach is illustrated in Scheme 4. Instead of

Direct electrochemistry: *Indirect electrochemistry:*

$$P + e^- \rightleftharpoons Q \tag{89}$$

$$RX + e^- \rightleftharpoons RX\cdot^- \tag{46} \qquad RX + Q \underset{k_+}{\overset{k_-}{\rightleftharpoons}} RX\cdot^- + P \tag{90}$$

$$RX\cdot^- \xrightarrow{k} R\cdot + X^- \tag{47}$$

Scheme 4

reducing RX directly at the electrode surface, the reduction is performed by an electrochemically generated homogeneous mediator. An aromatic hydrocarbon, P, having a standard potential, $E^0_{P/Q}$, positive to the reduction potential of RX and giving rise to a stable anion radical, Q, is introduced to the solution. It thus gives rise, in the absence of RX, to a one-electron reversible cyclic voltammetry wave. The wave becomes irreversible and increases in height upon addition of RX. From an electrochemist's point of view, this is a "redox" catalytic process (Andrieux *et al.*, 1978; Andrieux and Savéant, 1986a). Information concerning the kinetics of reactions (90) and (47) is available from the increase in height and the loss of reversibility of the P/Q wave upon addition of RX. The regeneration of P, which causes the change of the wave, is indeed kinetically controlled by these two reactions. Since we deal with fast cleavages (47) and with an endergonic preceding electron transfer (90), the steady-state assumption can be applied to $RX\cdot^-$. There are two limiting situations to kinetic control of the system. One is when $k_-[RX] \gg k$. Then the rate-determining step is the forward electron transfer (90). Conversely, if $k_-[RX] \ll k$, the rate-determining step is the cleavage reaction (47), with (90) acting as a pre-equilibrium. In the first case, analysis of the changes of the P/Q wave observed upon introduction of RX leads to the value of k, while, in the second, it leads to the value of kk_+k_-. Whether the system falls into one situation or the other is reflected by the effect of the concentration of the mediator. In the first case, the overall rate constant, and thus the catalytic effect, increases as the concentration of P increases, since the rate-controlling forward electron transfer (90) is a bimolecular reaction, whereas, in the second, it is independent of the concentration of P since the rate-controlling cleavage (47) in monomolecular. One way of making the system shift from the first situation to the second is to decrease the mediator concentration, since backward electron transfer (90) is a bimolecular reaction while (47) is a monomolecular one.

A good example of the possibilities offered by the "redox catalysis" approach is the reduction of 1-chloronaphthalene by the electrochemically generated anion radical of 4-methoxybenzophenone in DMSO (Andrieux *et al.*, 1980). The kinetic control passes from forward electron transfer (90) to a mixed control situation upon increasing the RX concentration. This indicates that $RX^{\cdot-}$ is an intermediate, since a concerted electron-transfer–bond-breaking pathway should only exhibit the first type of dependency toward the RX concentration with no change upon raising [RX]. The values of k_+ and of k/k_- can thus be derived from the redox catalysis data. On the other hand, the cyclic voltammetry data indicate that, in the scan rate range $0.1–1000 \text{ V s}^{-1}$, the overall electrochemical reaction is kinetically controlled by the cleavage reaction (47). This also shows that the reaction goes through $RX^{\cdot-}$ and, in addition, provides the value of $E^0_{RX/RX^{\cdot-}} + (RT/F) \ln(k)$. Since, in the relationship $(RT/F) \ln(k_+/k_-) = E^0_{RX/RX^{\cdot-}} - E^0_{P/Q}$, $E^0_{P/Q}$ is known, one can derive all four unknowns, $E^0_{RX/RX^{\cdot-}}$, k_+, k_- and k, from the combination of the above four relationships. Thus k_- is found to be equal to the diffusion limit. In most cases such a favourable situation is not met. However, as long as the redox catalysis data are such that the system passes from one of the two limiting controls to the mixed control, one can conclude that the reaction does go through $RX^{\cdot-}$ and the value of k/k_- can be obtained. Since, in most cases, k_- can be proved to be at the diffusion limit, k ensues. Cleavage rate constants as high as $5 \times 10^8 \text{ s}^{-1}$ have been determined in this way (Andrieux *et al.*, 1980). The cleavage rate constants that are accessible by this method range from about 10^5 to 10^9 s^{-1}. At the lower end of this range, the values of the rate constants have been found to be in quite satisfactory agreement with those obtained by means of the ultra-microelectrode techniques mentioned earlier, which have themselves an upper limit of 10^6 s^{-1} (Andrieux *et al.*, 1988a).

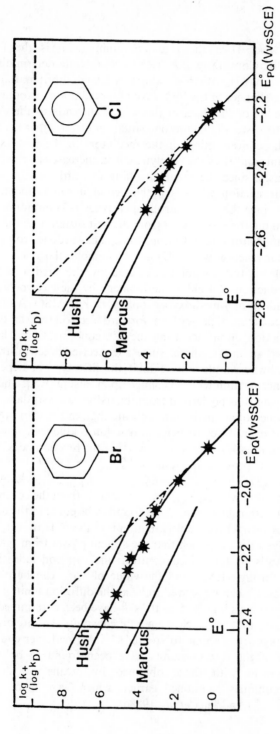

Fig. 11 Forward electron transfer (90) rate constant, k_+, versus the standard potential, $E^0_{P/Q}$, of a series of aromatic anion radicals for rapidly cleaved aryl halide anion radicals (DMF, 20°C). k_D is the bimolecular diffusion limit. (Adapted from Andrieux *et al.*, 1979.)

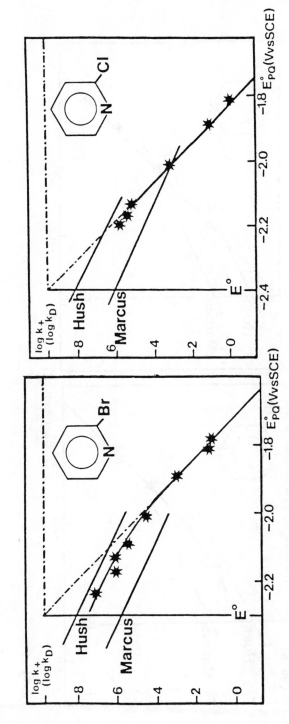

Fig. 11 Cont.

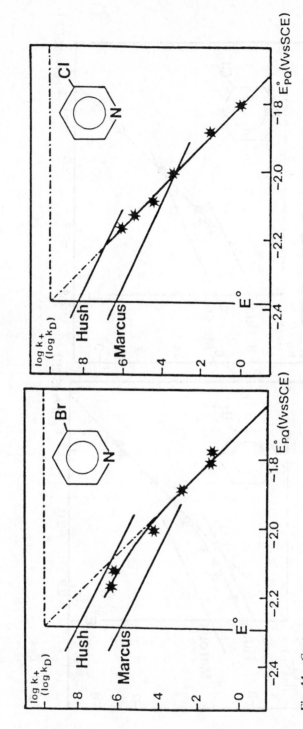

Fig. 11 Cont.

For even faster cleavages, the above procedures break down because the rate-determining step becomes forward electron transfer (90), thus hiding any manifestation of the follow-up cleavage reaction (47). The only available information is the value of the rate constant for the forward electron transfer (90), k_+. It is, however, still possible to know whether or not electron transfer and bond breaking are stepwise or concerted and to get an approximate estimate of the standard potential of the $RX/RX^{\cdot-}$ redox couple. Typical examples are shown in Fig. 11, under the form of a $\log(k_+)$ vs $E^0_{P/Q}$ plot (Andrieux et al., 1979). This can be regarded with a good approximation as a true Brønsted plot involving, for each halide, a single intrinsic barrier, since the aromatic anion radicals used as electron donors have a small (about 0.16 eV) and almost constant self-exchange intrinsic barrier (Kojima and Bard, 1975). Two distinct regions appear with slopes corresponding to activation and "counter-diffusion" control of the forward electron transfer (90), respectively, as discussed on pp. 34–37. The very existence of a region that is "counter-diffusion" controlled points to the intermediacy of the radical $RX^{\cdot-}$ and the standard potential $E^0_{RX/RX^{\cdot-}}$ can be estimated as indicated in Fig. 11. As discussed earlier, this method may overestimate $E^0_{RX/RX^{\cdot-}}$ because it assumes that the cleavage reaction takes place outside the solvent cage while, in fact, it may occur inside the solvent cage when it is very fast. The error that might ensue is, however, rather small, 80 mV at most. Once $E^0_{RX/RX^{\cdot-}}$ is known, the electrochemical and the homogeneous intrinsic barriers of RX can be obtained, using, in the latter case, the data points belonging to the activation region. The correlation between the heterogeneous and homogeneous intrinsic barriers thus derived falls, as expected, in between the predictions of the Marcus and Hush models, respectively. As suggested in a recent study of related compounds, viz., vinyl halides (Gatti et al., 1987), the complete $\log(k_+) - E^0_{P/Q}$ plots could be treated, assuming activation control all the way, according to the Marcus quadratic law. The values of $E^0_{RX/RX^{\cdot-}}$ and of the intrinsic barrier thus obtained would be close to those resulting from the previous treatment. This is an a posteriori confirmation that what was taken as the "counter-diffusion" region in the first treatment was actually so and therefore justifies its validity.

All the evidence gathered so far points to the conclusion that the $RX^{\cdot-}$ radicals are intermediates in the reductive cleavage of aryl halides by outer sphere electron donors in polar solvents. This was an already established conclusion for the reduction of several aryl halides in the gas phase (Steelhammer and Wentworth, 1969; Wentworth et al., 1967), and for iodobenzene in apolar or weakly polar matrixes from γ-irradiation studies with esr detection at low temperatures (Symons, 1981). It might, however, not have been true in the polar media used in direct and indirect electrochemistry investigations, requiring this conclusion to be established independently for polar media and for driving force conditions commonly used in direct and indirect electrochemistry. Insofar as the reduction of aryl radicals follows a stepwise pathway, electron transfer (46) is an example of an outer sphere reaction, a not too frequently established event in organic systems [see Section 2 (p. 15)].

It cannot be excluded, however, that reduction, under low driving force conditions, of the aryl halides that give rise to the fastest cleaving anion radicals, could occur along a concerted electron-transfer–bond-breaking pathway as discussed earlier and illustrated in Fig. 6. Attempts to observe such a transition from a stepwise to a concerted pathway using aromatic anion radicals as electron donors in DMF have so far failed because the reaction of interest is then so slow that side-reactions destroying the anion-radical interfere to too large an extent in the kinetics of its decay (Andrieux and Savéant, 1988).

Combining the direct electrochemical techniques with the redox catalytic method, it is possible to cover a vast range of cleavage rate constant values. Figure 12 shows such data gathered in the same solvent, DMF (Andrieux et al., 1984a). It appears that, given the aryl group, the anion radicals of the chlorides decompose more slowly than those of the bromides, as expected from the better leaving group ability of Br^- as compared to Cl^-. The largest variations are observed upon changing the Ar moiety while keeping the same halogen. It is interesting to note that, in Fig. 12, where the logarithms of the cleavage rate constants have been plotted against the standard potential of the $RX/RX^{\cdot-}$ couple, a rough linear correlation with a slope close to 0.5, in terms of free energies (0.48 for the chlorides and 0.53 for the bromides) is obtained. This surprising Brønsted relationship between the kinetics of the cleavage of $RX^{\cdot-}$ and the thermodynamics of another reaction, viz., the outer sphere electron transfer to RX, can be explained as follows (Savéant, 1988). The driving force of the cleavage reaction, $-\Delta G^0_{RX^{\cdot-} \rightleftharpoons R^\cdot + X^-}$ can be decomposed in terms of the free energy of bond dissociation in the parent aryl halide, $-\Delta G^0_{RX \rightleftharpoons R^\cdot + X^\cdot}$, and of the standard potentials of the $RX/RX^{\cdot-}$ and $X^\cdot/X^-$ redox couples (91). Given the halogen, the free energy of

$$-\Delta G^0_{RX^{\cdot-} \rightleftharpoons R^\cdot + X^-} = -\Delta G^0_{RX \rightleftharpoons R^\cdot + X^\cdot} - E^0_{RX/RX^{\cdot-}} + E^0_{X^\cdot/X^-} \tag{91}$$

dissociation does not vary very much upon changing R, and thus the driving force of the cleavage of the anion radical varies approximately as the standard potential of the $RX/RX^{\cdot-}$ couple. The plots in Fig. 12 can thus be regarded as true Brønsted relationships between the kinetics and the thermodynamics of the same reaction. On the other hand, the cleavage reaction can be viewed as an intramolecular, concerted electron-transfer–bond-breaking

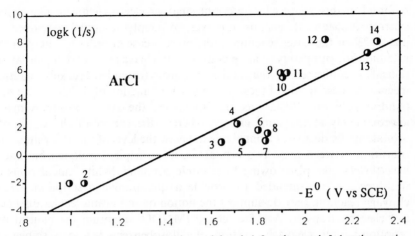

1, 4-Nitrophenyl; 2, 2-nitrophenyl; 3, 4-benzoylphenyl; 4, 9-anthracenyl; 5, 1-anthracenyl; 6, 2-anthracenyl; 7, 4-(2-(4-pyridyl)vinyl)phenyl; 8, 3-acetylphenyl; 9, 4-quinolyl; 10, 4-acetylphenyl; 11, 2-quinolyl; 12, 4-cyanophenyl; 13, 1-naphthyl; 14, 2-naphthyl.

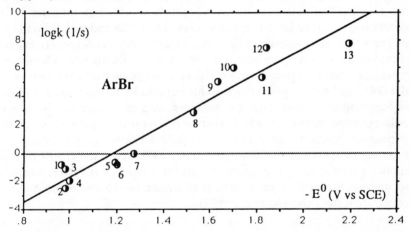

1, 2-Isopropyl-4-nitrophenyl; 2, 4-nitrophenyl; 3, 2-methyl-3-nitrophenyl; 4, 2-methyl-4-nitrophenyl; 5, 3-fluorenyl; 6, 1-fluorenyl; 7, 2,6-dimethyl-4-nitrophenyl; 8, 3-benzoylphenyl; 9, 4-benzoylphenyl; 10, 9-anthracenyl; 11, 3-acetylphenyl; 12, 4-acetylphenyl; 13, 1-naphthyl.

Fig. 12 Cleavage rate constant, k, of aryl chlorides and bromides as a function of their standard potential, E^0. (Adapted from Andrieux *et al.*, 1984a.)

process, in which the unpaired electron, accommodated at first in the π^*-orbital of the Ar moiety, would be transferred into the C—X σ^*-orbital in concert with the cleavage of the bond. It is thus not surprising that an activation–driving force relationship with an average symmetry factor of 0.5

is found in this case as for intermolecular or electrochemical dissociative electron transfers. It remains, however, to adapt the model developed on pp. 21–23 for the latter reaction to the present case in order to interpret the rate data quantitatively. This picture of the cleavage of aryl halide anion radicals is confirmed by the results of a gross (extended Hückel) quantum mechanical description of these molecules as a function of the C—X distance (Andrieux *et al.*, 1984a): as this increases, the Ar π^*-orbital remains approximately at the same energy, whereas the energy of the C—X σ^*-orbital rapidly decreases from above to below the level of the π^*-orbital. The transition between these two orbitals is, in principle, forbidden but could nevertheless take place owing to vibronic coupling. Whilst calculations of this type are not intended to provide a quantitative description of the cleavage reaction, they do support the notion of an intramolecular dissociative electron transfer. A more refined (MNDO) quantum mechanical description of the cleavage reaction of chlorobenzene led to a somewhat different picture (Casado *et al.*, 1987). The unpaired electron still initially in the aryl π^*-orbital would, as the C—X bond is stretched, be transferred into the C—X σ^*-orbital, the energy of which would first increase before decreasing at rather large C—X distances. The anion radical, at first of the π-type would thus acquire σ-character before cleavage. Strangely enough, these more elaborate calculations are in poorer qualitative and quantitative agreement than the crude extended Hückel description with the experimental data obtained not only in polar solvents but also in the gas phase (see Fig. 3 in Steelhammer, 1969). They do, however, lead to linear correlations with polarographic potentials which themselves represent undissected combinations of thermodynamic and kinetic factors characterizing the electron-transfer step (46) and the cleavage reaction (47). So do the results of other similar attempts using a different calculation technique (CNDO) which led to the conclusion of a substantial σ character of the anion radical in its equilibrium state (Beland *et al.*, 1977).

ALKYL HALIDES

Since the pioneering polarographic work of Von Stackelberg and Stracke (1949), the electrochemical reduction of alkyl halides has been widely investigated both on the analytical and preparative scales (for reviews see Becker, 1983; Hawley, 1980). However, most of the work was carried out using mercury electrodes, which cannot be considered as inert towards the alkyl halides and the radicals that may be produced during their reduction. Although interactions with gold and platinum are certainly less important than with mercury, glassy carbon appears as the best approximation to an inert electrode material. The reduction of alkyl halides on glassy carbon has been investigated only recently both at the microelectrode (Andrieux et al., 1986a) and at the preparative level (Cleary et al., 1986).

The cyclic voltammetry of n-, s- and t-BuI and BuBr, for example, at a glassy carbon electrode in DMF with a tetraalkylammonium supporting electrolyte, exhibits one or two irreversible cathodic waves according to the relative reducibility of the alkyl halide RX and of the radical R· (Andrieux et al., 1986a). The corresponding chlorides show waves that strongly overlap with the discharge current of the supporting electrolyte and cannot therefore be investigated accurately. The effective reduction potential of the butyl radicals varies in the order $n > s > t$ (Andrieux et al., 1989) and the butyl iodides are easier to reduce than the bromides. Thus, a one-electron irreversible wave followed by a small irreversible wave typical of the reduction of a transient intermediate (Andrieux et al., 1989; Andrieux and Savéant, 1989), here the butyl radical, is observed with s- and t-BuI whereas n-BuI, and n-, s- and t-BuBr show a single two-electron irreversible wave.

The apparent transfer coefficient, as derived from the peak width and the variation of the peak potential with the scan rate, is small (between 0.2 and 0.3) in all cases. This rules out the occurrence of a stepwise mechanism (46, 47), in which the follow-up, bond-breaking step would have been

$$RX + e^- \rightleftharpoons RX·^- \qquad (46)$$

$$RX·^- \xrightarrow{k} R· + X^- \qquad (47)$$

rate determining. The apparent transfer coefficient is thus a true transfer coefficient, i.e. the rate-determining step is an electron-transfer step. This either is of the outer sphere type (46) occurring prior to the bond-breaking step (47) or is a concerted electron-transfer–bond-breaking elementary step (48). These two possibilities can be distinguished from each other on the

$$RX + e^- \longrightarrow R· + X^- \qquad (48)$$

basis of the observations that the transfer coefficient is small, much smaller than 0.5, and appears to vary with the electrode potential (Andrieux et al., 1986a,b; Savéant, 1987). In the framework of a quadratic activation–driving force relationship, the standard potential should consequently be much positive of the reduction potential. By itself, this does tell whether electron transfer and bond breaking are concerted or stepwise. We know that the standard potential of the RX/R· + X$^-$ couple, as estimated from thermo-chemical data (Andrieux et al., 1986a; Eberson, 1982; Hush, 1957; Savéant, 1987), fulfills this requirement, but what about the standard potential of the RX/RX·$^-$ couple? The latter is certainly negative with respect to the standard potential for the formation of the corresponding halobenzene π-anion radical. Since this is negative with respect to the effective reduction potential of the corresponding butyl halide, it follows that the standard potential of the rate-determining step cannot be that of the outer sphere electron transfer (46). One is thus led to conclude that the reaction does not follow the stepwise pathway but rather the concerted pathway (see Fig. 6).

This conclusion falls in line with the fact that the anion radical could neither be detected after collision of the parent halide with alkali metal atoms in the gas phase (Compton *et al.*, 1978) nor upon γ-irradiation in apolar or weakly polar solid matrixes at 77 K by esr spectroscopy (Symons, 1981). However, these observations are not absolute proofs that the anion radicals do not exist: they might exist and be too short lived to be detectable. On the other hand, the reaction medium and the driving force conditions are quite different from those in the electrochemical experiments, which rendered necessary an independent investigation of the problem in the latter.

Chemically prepared alkali salts of aromatic anion radicals have been used to reduce alkyl halides and several rate constants have been measured (Bank and Juckett, 1975; Garst *et al.*, 1968; Garst, 1971; Garst and Barton, 1974; Garst and Abel, 1975; Malissard *et al.*, 1977). The largest body of kinetic data, however, comes from the reaction of electrochemically generated aromatic anion radicals (Andrieux *et al.*, 1986a; Lund and Lund, 1986) after more qualitative earlier investigations of the reaction (Britton and Fry, 1975; Margel and Levy, 1974; Sease and Reed, 1975; Simonet *et al.*, 1975). They have been used in the same way as for aryl halides. The reaction is, however, not exactly of the same type in the sense that alkyl radicals formed after electron transfer and bond breaking react, unlike aryl radicals, with the mediator anion radical giving rise to coupling products competitively with their reduction yielding the hydrogenolysis hydrocarbon. Another difference is that the reduction of the alkyl radical by means of H-atom abstraction from the solvent instead of electron transfer is much less important than in the case of aryl radicals. On the other hand, the initial electron-transfer and bond-breaking steps are now concerted instead of stepwise. This results from the fact that the driving force offered by the mediator aromatic anion radicals is smaller than that of the electrochemical reduction, such that the latter was shown to follow the concerted pathway, and that a decrease of the driving force tends to shift the reacting system from the stepwise to the concerted pathway (see pp. 30, 31 and particularly Fig. 6). The overall reaction mechanism can thus be summarized as in Scheme 5, where the aromatic hydrocarbon mediator is schematically represented by a benzene molecule. The rate-determining step is, in all cases, the dissociative electron transfer (92). If the electron-transfer reduction of the alkyl radical (93) prevails over its coupling with the aromatic anion radical (94), a catalytic current is observed as in the case of aryl halides, with the difference,

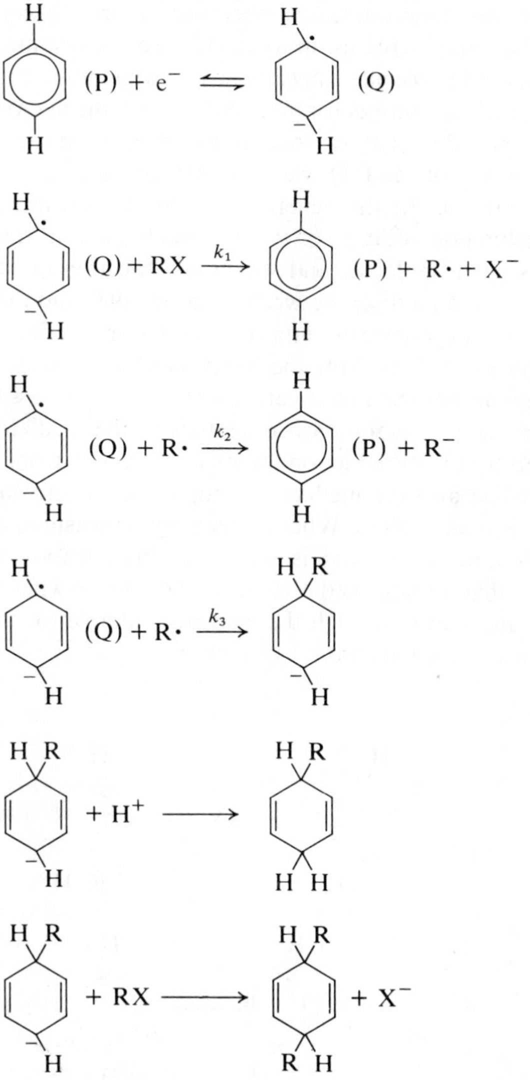

$$(P) + e^- \rightleftharpoons (Q) \qquad (89)$$

$$(Q) + RX \xrightarrow{k_1} (P) + R\cdot + X^- \qquad (92)$$

$$(Q) + R\cdot \xrightarrow{k_2} (P) + R^- \qquad (93)$$

$$(Q) + R\cdot \xrightarrow{k_3} \qquad (94)$$

$$+ H^+ \longrightarrow \qquad (95)$$

$$+ RX \longrightarrow + X^- \qquad (96)$$

Scheme 5

however, that the rate-determining step is here always the first electron-transfer step (92) since this is purely dissociative. Then, for example in cyclic voltammetry, k_1 can be derived from the increase and the loss of reversibility of the P/Q wave using a working curve appropriate for a 2Q/RX pure catalytic situation (Andrieux and Savéant, 1986a). In the converse case where the coupling reaction (94) is faster than the reduction step (93), the

P/Q wave no longer increases indefinitely upon addition of RX. It loses its reversibility and increases up to a height corresponding at maximum to two electrons per P molecule. The kinetics corresponding to this reaction scheme are formally the same as for a "DISP1" mechanism (Andrieux and Savéant, 1986a). The value of k_1 can again be derived from the increase and the loss of reversibility of the P/Q wave using a working curve appropriate for this reaction scheme. In the general case, the peak current does not increase indefinitely upon addition of RX but reaches a limit that is higher than two electrons per P molecule and which is a function of the ratio k_2/k_3 which features the competition between reactions (93) and (94). A two-parameter fitting of the experimental data then allows one to derive both k_1 and k_2/k_3 (Nadjo et al., 1985). With the butyl iodides, bromides and chlorides (the chlorides can now be investigated, which was not possible in direct electrochemistry, since the standard potentials of the mediators are positive with respect to the direct reduction wave), k_2/k_3 ranges from zero to 20 according to the halide and the mediator, being close to zero in a number of cases (Andrieux et al., 1986a). When a meaningful measurement of k_2/k_3 could be carried out, i.e. when k_2/k_3 is neither too big nor too small, it was found, as expected, that its value is the same for the same R whatever the nature of the halogen atom, even though the mediator suitable for the measurement was not the same in each case.

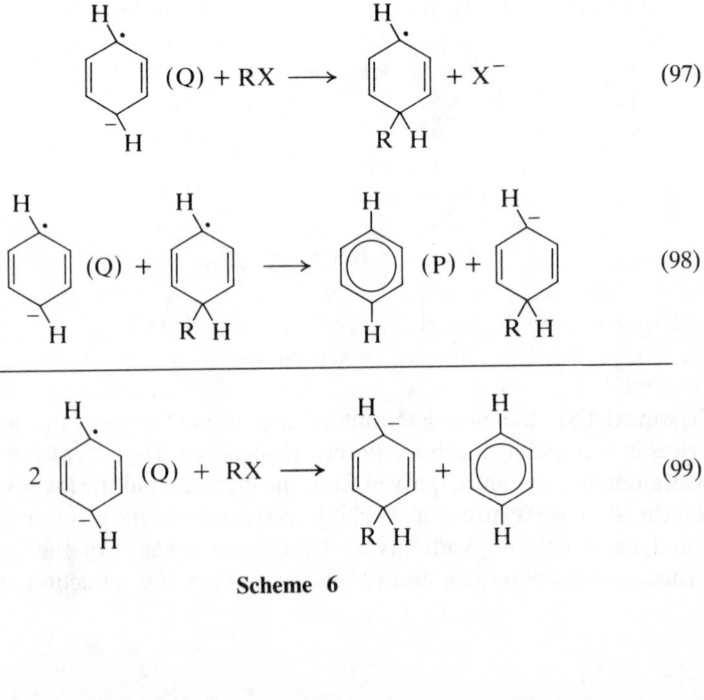

$$(Q) + RX \longrightarrow \quad + X^- \tag{97}$$

$$(Q) + \quad \longrightarrow \quad (P) + \tag{98}$$

$$2 \quad (Q) + RX \longrightarrow \quad + \tag{99}$$

Scheme 6

In the numerous cases where $k_2/k_3 = 0$, the overall reaction amounts to a substitution of the halogen in the halide by the aromatic anion radical followed by the electron-transfer reduction of the ensuing radical by another aromatic anion radical, i.e. the sum of steps (92) and (94) might rather occur as in Scheme 6.

There are several reasons to believe that pathway (92, 94) is followed rather than pathway (97, 98), in other words, that aromatic anion radicals behave as outer sphere electron donors in their reaction with aliphatic alkyl halides rather than as nucleophiles in an S_N2 reaction. One of these derives from the observation, noted earlier, that members of this family give rise, at least partly, to catalytic currents besides the coupling reaction. In these cases, the alkyl radical is formed as an intermediate and its coupling with the aromatic anion radical yielding the substitution product is likely to be a fast reaction. Another one derives from stereochemical experiments. When optically active 2-octyl iodide, bromide and chloride react with electrochemically generated anthracene anion radicals in DMF, almost total racemization occurs. The amount of inversion is of the order of 10% and increases slightly upon going from the iodide to the bromide to the chloride (Hebert *et al.*, 1985). This indicates that the outer sphere, dissociative electron-transfer pathway (92 + 94) prevails over the S_N2 pathway (97 + 98), at least in the proportion 9:1, if the electron transfer pathway leads to complete racemization, as seems likely if the coupling reaction (92) takes place exclusively outside the solvent cage in which the preceding electron-transfer step (94) occurred. In the opposite case, it might be that the aromatic anion radical, even though functioning as an outer sphere electron donor, would attack the carbon side of the RX molecule in a somewhat privileged manner, leading to some inversion when the coupling reaction occurs inside the solvent cage. At any rate, we can conclude that if the reaction follows an S_N2 pathway, this occurs to a quite small extent, 10% at maximum, and can therefore be neglected to a first approximation as far as kinetics are concerned.

The experimental kinetic data obtained with the butyl halides in DMF are shown in Fig. 13 in the form of a plot of the activation free energy, $\Delta G^{\neq}$, against the standard potential of the aromatic anion radicals, $E^0_{P/Q}$. The electrochemical data are displayed in the same diagrams in the form of values of the free energies of activation at the cyclic voltammetry peak potential, E_p, for a 0.1 V s^{-1} scan rate. Additional data have been recently obtained by pulse radiolysis for n-butyl iodide in the same solvent (Grimshaw *et al.*, 1988) that complete nicely the data obtained by indirect electrochemistry. In the latter case, indeed, the upper limit of obtainable rate constants was 10^6 M^{-1} s^{-1}, beyond which the overlap between the mediator wave and the direct reduction wave of n-BuI is too strong for a meaningful measurement to be carried out. This is about the lower limit of measurable

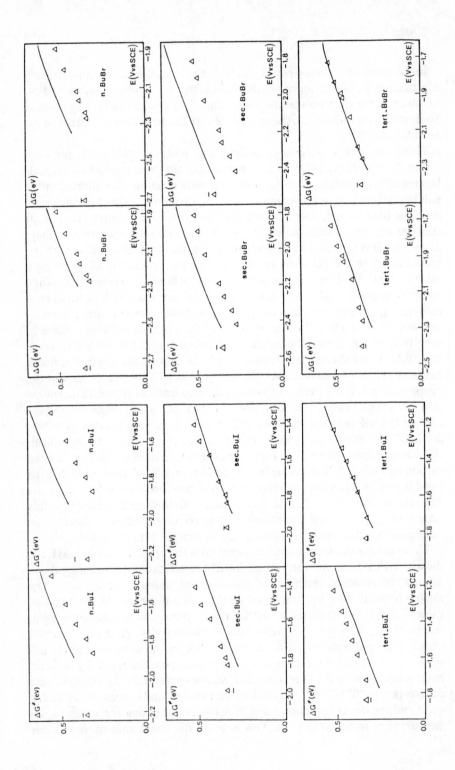

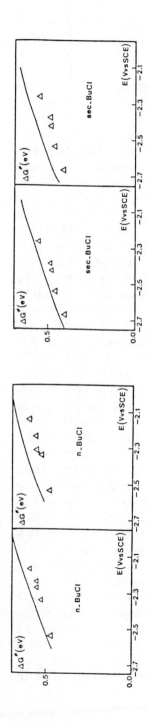

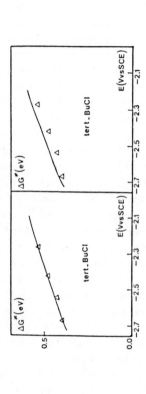

Fig. 13 Electrochemical (▲) and homogeneous (△) reduction of butyl halides in DMF. Variation of the activation free energy, $\Delta G^{\neq}$, with the standard potential of the aromatic anion radicals, $E^0_{P/Q}$, in the homogeneous case and the cyclic voltammetry peak potential, E_p, at 0.1 V s^{-1} in the electrochemical case. The solid lines represent the predictions of the model for dissociative electron transfer (pp. 21–23) in both the homogeneous case and the electrochemical case. In the estimation of $E^0_{RX/R\cdot + X^-}$, the free energies of solvation of RX and R· are either assumed to be the same (left-hand side of diagrams) or that of R· is assumed to be the same as that of RH (right-hand side of diagrams). (Adapted from Andrieux et al., 1986a; Savéant, 1987.)

rate constants in pulse radiolysis which, by contrast, permits the determination of much higher rate constants, allowing one to observe the attainment of the diffusion limit upon going to strongly reducing aromatic anion radicals.

Since alkyl halides undergo, as discussed earlier, dissociative rather than stepwise electron transfer upon reduction by homogeneous and heterogeneous outer sphere electron donors, they provide a good opportunity to test the theoretical model of dissociative electron transfer described on pp. 21–23. The comparison between experimental data and theoretical predictions is shown in Fig. 13 for the reduction of butyl iodides, bromides and chlorides by aromatic anion radicals and for the electrochemical reduction of butyl iodides and bromides at a glassy carbon electrode in DMF as plots of $\Delta G^{\neq}$ vs $E^0_{P/Q}$ and E_p, respectively. The construction of the theoretical curves requires a knowledge of the standard potential of the $RX/R\cdot + X^-$ couple, which is not directly derivable from electrochemical experiments it can, however, be approximated from thermochemical data (Andrieux et al., 1986a; Eberson, 1982; Hush, 1957; Savéant, 1987). In these estimations, the free energies of solvation of the butyl radicals are not available, but can be bracketted by the free energies of solvation of the parent halides and hydrocarbons, respectively. The results obtained with these two assumptions are displayed in Fig. 13, the left-hand curves corresponding to the first assumption and the right-hand curves to the second. On average, the fit between theory and experiment is within ± 90 meV, which is quite satisfactory in view of the crudeness of the model and of the approximations made in the estimation of the standard potential of the $RX/R\cdot + X^-$ couples. It is of the same order as for the experimental testing of the Marcus–Hush theory of outer sphere electron transfer with organic systems. The model of dissociative electron transfer discussed above has recently been criticized on the basis of the data obtained for the reduction of n-BuI by aromatic anion radicals in DMF (see Fig. 13) by indirect electrochemistry and by pulse radiolysis (Grimshaw et al., 1988). The experimental $\log k_1$ vs $E^0_{P/Q}$ plot was considered to be linear rather than parabolic as predicted by the theory, and this was explained by introducing into the model a resonance interaction energy between the σ- and σ^*-orbitals at the transition state that would stabilize it more and more as the driving force increases. It should be noted, however, that a parabola with a small aperture, as predicted by the theory, would fit the experimental data as well as a straight line, taking into account the experimental uncertainty and the fact that the self-exchange free energy of activation of the electron donors is not exactly the same when passing from one aromatic anion radical to the other in a series (see the discussion about obtaining Brønsted plots on pp. 13, 14 and 23). On the other hand, a significant interaction between the σ HOMO and σ^* LUMO is unlikely for symmetry reasons (Albright *et al.*, 1985). There are other, more convincing, reasons for the approximate character of the adherence between the experimental data and the theoretical predictions. These are (i) the crudeness of the representation of the $R \cdot + X^-$ potential energy curve by the dissociative part of the RX Morse curve, (ii) the crudeness of the Morse curve representation of RX itself, (iii) the neglect of other internal changes such as that of the angles between the functional carbon to α-carbon bonds or their integration in the Morse curve description, and (iv) the assumed additivity of the RX bond-stretching and solvent-reorganization contributions and the crude Born hard sphere estimation of the latter (Savéant, 1987). What is actually surprising is that, in spite of these rather drastic approximations, such a satisfactory prediction of the experimental activation–driving force relationships could be reached.

OTHER EXAMPLES

In other classes of organic halides, for example perfluoroalkyl and vinyl halides, the distinction between stepwise and concerted electron-transfer–bond-breaking upon reduction by outer sphere heterogeneous and/or homogeneous electron donors is less unambiguous than in the case of aryl and alkyl halides. As discussed in Section 3, they also present the interest of being active substrates in $S_{RN}1$ reactions.

The electrochemical reduction of several perfluoroalkyl halides at glassy carbon electrodes as well as their reduction by aromatic anion radicals have been investigated in non-aqueous solvents (Andrieux *et al.*, 1990b). The most significant results concern trifluoromethyl bromide and iodide. We first note that, unlike alkyl halides, they both give rise to a $CF_3X \cdot^-$ anion radical in the gas phase upon collision with alkali metal atoms (Compton *et al.*, 1978) and in apolar or weakly polar solid matrixes at $77\,K$ upon γ-irradiation as revealed by esr detection (Hasegawa and Williams, 1977; Hasegawa *et al.*, 1977). The anion radical thus exists in both cases, but the question is to know if the reduction by outer sphere electron donors in polar solvents proceeds through the anion radical or if the concerted pathway is energetically more advantageous. Polar solvents are expected to accelerate the cleavage of the anion radical as compared to the gas phase and apolar or weakly polar matrixes. In addition, the driving forces offered by the reduction at inert electrodes or by aromatic anion radicals are smaller than in the two preceding cases, a factor that should favour the concerted pathway *vis-à-vis* the stepwise pathway (see Fig. 6 and the discussion on pp. 30, 31). The values of the transfer coefficient derived from the cyclic voltammetry data are much smaller than 0.5, indicating that the standard

$$CF_3X + e^- \rightleftharpoons CF_3X^{\cdot-} \qquad (100)$$

$$CF_3X + e^- \rightleftharpoons CF_3{}^\cdot + X^- \qquad (101)$$

potential of the rate-determining step (100) or (101) is much positive of the reduction potential. Reaction (101) certainly fulfills this requirement, but it is not *a priori* certain, unlike the case of alkyl halides (pp. 55–56), that reaction (101) may not do so as well. However, the magnitudes of the standard potential and of the intrinsic barrier that can be derived from the experimental data according to a quadratic activation–driving force relationship indicate that the concerted pathway is more likely than the stepwise pathway. The difficulty in estimating the standard potential of the $CF_3X/CF_3{}^\cdot + X^-$ couple from thermochemical data derives from the uncertainty about the difference in the free energies of solvation of CF_3X and $CF_3{}^\cdot$. It is certainly more in favour of a stabilization of the radical than in the case of simple alkyl halides discussed above, in view particularly of its permanent dipolar character (Chase *et al.*, 1985). The standard potential of reaction (101) is thus clearly more consistent with the experimental value than that of reaction (100). On the other hand there is a good agreement between the experimental value of the intrinsic barrier and that derived from the model of dissociative electron transfer described on pp. 21–23, whereas it is much too high to be compatible with the outer sphere electron transfer (100). The reduction by a series of aromatic anion radicals, having standard potentials positive of the electrochemical reduction potential, has also been investigated (Andrieux *et al.*, 1990b). Since the driving force offered by these homogeneous outer sphere electron donors is less than in the electrochemical case, the reaction is more likely, even than in the electrochemical case, to be of the concerted rather than of the stepwise type. Again, a good agreement between the experimental value of the intrinsic barrier and the value derived from the dissociative electron-transfer model (pp. 21–23) was found.

Similar questions arise for other polyhaloalkanes such as CCl_4, CBr_4 and CCl_3Br. Their reduction by two bulky coordination compounds (a polyoxometallate and a "sepulchrate") reputed to function as outer sphere electron donors have been recently investigated (Eberson and Ekström 1988a,b) as well as the reduction of CCl_4 by aromatic anion radicals (Eberson *et al.*, 1989). From the rate data thus obtained it is difficult to conclude whether the reaction goes through the intermediacy of the polyhalogenoalkane anion radical or rather follows a concerted pathway. The fact that the anion radicals have been detected under quite different medium and driving force conditions (Brickenstein and Khairutdinov, 1985; Mishra and Symons, 1973) does not provide unequivocal evidence for the stepwise pathway, as discussed earlier. Some evidence in favour of the latter mechanism was, however, found in the observation that the reduction, by the two complexes, of CBr_4 is somewhat faster than that of CCl_3Br (the difference in the free energies of activation is about 100 meV). Since the C—Br bond dissociation energies are practically the same in the two compounds, the standard potentials of the $CBr_4/CBr_3\cdot + Br^-$ and $CCl_3Br/CCl_3\cdot + Br^-$ couples were also considered to be the same, assuming that the difference of the free energies of solvation between the halide and the radical are the same for the two compounds (Eberson and Ekström, 1988a). We might add that, according to the model of dissociative electron transfer developed earlier, the contribution of bond breaking to the intrinsic barrier should also be the same in both cases. If it is further assumed that the contribution of solvent reorganization to the intrinsic barrier is the same for the two compounds as well, the intrinsic barriers would also be the same. Thus, in the context of the dissociative electron-transfer pathway, the rate constants should be the same. The fact that they are not was thus taken as an indication that the reaction follows the stepwise pathway instead. In fact, the estimated uncertainties in the determination of the bond dissociation energies (King et al., 1971; Mendenhall et al., 1973)$(-0.100 < E^0_{CCl_3Br/CCl_3\cdot + Br^-} - E^0_{CBr_4/CBr_3\cdot + Br^-} < 0.143 \text{ V})$ are rather large and it is not certain that the difference of the free energies of solvation between the halide and the radical are the same for the two compounds or that the contribution of solvent reorganization to the intrinsic barrier is the same in both cases. In addition, the adherence of the predictions of the model of dissociative electron transfer developed earlier with the experimental data is not expected to be better than about ± 0.1 eV. In other words, the observed difference between the free energies of activation for the reduction of the two compounds by the same complex is not sufficient to draw a definitive conclusion as to the intermediacy of the halide anion radical.

For the reduction of CCl_4 by electrochemically generated aromatic anion radicals in DMF (Eberson et al., 1989), the kinetics were treated according to a reaction scheme in which $CCl_4\cdot^-$ appears as an intermediate, but no strong evidence was offered that they would not fit a concerted pathway as well. As discussed on pp. 32, 33, it is indeed not possible to distinguish between the two pathways in such types of experiment when, in the stepwise pathway, the cleavage reaction is so fast that the electron-transfer step would be rate determining, a situation likely to be met in the present case.

The reduction of aryl-substituted vinyl halides by electrochemically generated aromatic anion radicals has also been investigated in DMF (Gatti et al., 1987). "Counter-diffusion" behaviour at low driving forces (pp. 34, 35) does not appear as clearly as in the case of aryl halides (Fig. 11). However, analysis of the log k vs E^0 plot according to a quadratic activation–driving force relationship gave standard potential and intrinsic barrier values that are more consistent with a stepwise than with a concerted pathway. These halides would thus behave like the aryl halides, as expected from the similarity of their electronic structure. It would be interesting to see if the same is also true with vinyl halides not substituted by aryl groups.

The electrochemical reduction of benzyl and other arylmethyl halides provides an interesting example of how the reaction can change from a stepwise to a concerted type upon a relatively small variation of the structure of the starting compound (Andrieux *et al.*, 1990e). Cyclic voltammetry, at a glassy carbon electrode in DMF, of both 4-nitrobenzyl and 4-cyanobenzyl bromides both exhibit a first irreversible wave corresponding to the reductive cleavage of the C—Br bond and a second reversible wave corresponding to the reduction of the hydrogenolysis hydrocarbon into its anion radical. There is, however, a striking difference between the two first reduction waves. As derived from its variations with the scan rate, the nitro-compound exhibits a mixed charge transfer–first order follow-up kinetic control behaviour. The two waves are located at potentials close to one another (0.35 V between the peak potentials at $0.5 \, \mathrm{V \, s^{-1}}$). This clearly shows the occurrence of a stepwise electron-transfer–bond-breaking pathway [reactions (46) and (47)]. With the cyano-derivative, charge-transfer control with a value of the transfer coefficient much smaller than 0.5 (0.3) is observed, while the peaks are much more distant ($1.02 \, \mathrm{V}$ at $0.5 \, \mathrm{V \, s^{-1}}$). This clearly indicates the occurrence of the concerted pathway (48) since the standard potential of the $RX/RX\cdot^-$ couple, being close to that of toluonitrile, is certainly much negative of the reduction potential. We are thus in the situation depicted in Fig. 6. Why is there a stepwise mechanism in the first case and a concerted mechanism in the second? This immediately results from the locations of the second wave that are approximate estimates of the standard potential of the $RX/RX\cdot^-$ couples. This is much more negative in the case of the cyano-derivative than in the case of the nitro-derivative (by 1.24 V), being a measure of the respective energies of the π^*-orbitals of the phenyl rings. In the nitro case the π^*-orbital is low enough for the electron to arrive there in a first stage before being transferred to the C—Br σ^*-orbital in the bond-breaking reaction. In the case of the cyano-derivative, the π^*-orbital is so high that the concerted pathway is energetically more advantageous. Other compounds in the series, such as unsubstituted benzyl bromide and chloride, exhibit the same behaviour as 4-cyanobenzyl bromide owing to the large energy required to locate an electron in the π^*-orbital of their aromatic portion.

Out of a series of careful studies by the Padua School of the reductive cleavage of diphenyl and alkylphenyl sulphides in DMF by heterogeneous and homogeneous electron donors (Arevalo *et al.*, 1987; Griggio, 1982; Griggio and Severin, 1987; Severin *et al.*, 1987, 1988), the investigation of

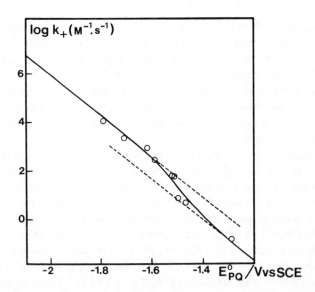

Fig. 14 Reduction of Ph_3CSPh by electrochemically generated aromatic anion radicals (in DMF at 25°C). Variation of the rate-determining step rate constant, k_1, with the standard potential of the aromatic anion radical, $E^0_{P/Q}$ (from left to right: azobenzene, benzo[c]cinnoline, 4-dimethylaminoazobenzene, terephthalonitrile, naphthacene, phthalonitrile, perylene, fluoranthene, 9,10-diphenylanthracene). The dotted lines are the theoretical limiting behaviours corresponding to the concerted (right) and stepwise (left) pathways. (Adapted from Severin *et al.*, 1988.)

the reduction of triphenylmethyl phenyl sulphide (Severin *et al.*, 1988) offers a quite interesting attempt to observe the passage from a stepwise to a concerted pathway upon changing the driving force. The electrochemical reduction shows charge-transfer control with a transfer coefficient somewhat smaller than 0.5 (0.4), consistent with a concerted pathway. The reduction by aromatic anion radicals, having standard potentials more positive than the electrochemical reduction potential, is also charge-transfer controlled. The variations of the rate constant with the driving force, as represented by the standard potential of the aromatic anion radical, $E^0_{P/Q}$, are shown in Fig. 14. There is a small but distinct change of the curve on which the data points are located upon decreasing the driving force of the reaction, as predicted (Andrieux and Savéant, 1986b; Andrieux *et al.*, 1985) for the passage from a stepwise to a concerted pathway (see Fig. 8(a) and the discussion on pp. 35–37). It is true that the variation is small (as expected for a fast follow-up cleavage) and that care should be exercised in deriving mechanistic conclusions from activation–driving force plots obtained with a series of homogeneous outer sphere electron donors for the reasons discussed at the beginning of this section. One of these is the possible variation of the electron donor self-exchange barrier in the series. Note in this connection, however, that if such a variation exists, the most positive electron donor couples should have the smallest barrier and, thus, that the correction for this factor would enhance the observed effect. It may thus be concluded that the effect seems real but that further confirmation involving other systems would be welcomed, possibly using a direct electrochemical approach whereby the variations in the driving force could be made more progressive as already discussed.

Coming back to the cleavage of carbon–halogen bonds, the reduction of vicinal dihaloalkanes offers an interesting opportunity to test the predictions of the model of dissociative electron transfer on stereochemical and not only on purely kinetic grounds (Lexa et al., 1990). The electrochemical reduction has been investigated at glassy carbon electrodes (Lexa et al., 1987, 1990) and at less inert electrodes, viz., mercury, gold and platinum (Bowyer and Evans, 1988; Casanova and Rogers, 1974; Evans and O'Connell, 1986; Inesi and Rampazzo, 1974; Klein and Evans, 1979; O'Connell and Evans, 1983; Von Stackelberg and Stracke, 1949; Zavada et al., 1963) in non-aqueous solvents. As in preceding cases, electrochemically generated (Lexa et al., 1987, 1990; Lund et al., 1987) or chemically prepared (Adam and Arce, 1972; Garst et al., 1975; Scouten et al., 1969) aromatic anion radicals have also been used as outer sphere homogeneous electron donors. The kinetic data obtained for the reduction of an extended series of cyclic and acyclic vic-dibromoalkanes at a glassy carbon electrode and by aromatic anion radicals (Lexa et al., 1990) indicate that the reaction proceeds through a concerted electron-transfer–bond-breaking rate-determining step leading to the β-bromoalkyl radical. In the electrochemical case this is then reduced rapidly, in a second step, most probably along another concerted electron-transfer–bond-breaking pathway which has a very positive standard potential and a very small barrier. In the homogeneous case, the β-bromoalkyl radical loses a bromine atom which is then reduced by the aromatic anion radicals. In the absence of steric constraints, the reduction goes entirely through the antiperiplanar conformer, since the resulting α-bromoalkyl radical is then stabilized by delocalization of the unpaired electron over the C—C—Br framework owing to favourable interaction between the p_z orbital of the radical carbon and the C—Br σ*-orbital. Application of the model of dissociative electron transfer to the analysis of the kinetic data indicates that the stabilization energy is of the order of 0.2–0.3 eV for the anti-conformers as compared to the syn-conformers or to what can be derived from thermochemical data according to additivity rules (Benson, 1976). This energy can also be taken as a measure of the barrier to rotation around the C—C bond in the α-bromoalkyl radical that makes it pass from an E-like to a Z-like conformation, and which is, in competition with

bromine expulsion in these two forms, responsible for the loss of stereospeci-
fity observed in the reduction of *threo-* and *erythro-*isomers by aromatic
anion radicals. There is a remarkably good agreement between the value of
the stabilization energy derived from the kinetic data using the model of
dissociative electron transfer, on the one hand, and that of the rotation
barrier derived from the stereoselectivity experiments, on the other. This
provides strong additional support for the model, independent of the direct
kinetic measurements concerning alkyl halides described on pp. 61, 62.

The reduction of arylalkyl halides of the triphenylmethyl type by electrochemically generated outer sphere single electron donors offers an example
of a sequencing of the bond-breaking and electron-transfer steps different
from what has been described before. The cleavage of the halide ion then
precedes electron transfer which thus involves the carbocation, a mechanism
reminiscent of S_N1 reactions (Andrieux *et al.*, 1984b) as shown in (102). The

$$P + e^- \xrightleftharpoons \quad Q, \quad RX \underset{k_{+c}}{\overset{k_{-c}}{\rightleftharpoons}} R^+ + X^-, \quad R^+ + Q \xrightarrow{k_E} R \cdot + P \qquad (102)$$

difference is that the S_N1 bond-forming step is here replaced by an outer
sphere electron-transfer step. This was found, for example, in the reduction,
in CH_3CN, of the two chlorides shown in Scheme 7 by electrochemically
generated substituted ferrocenes and anion radicals of tetracyanoethylene

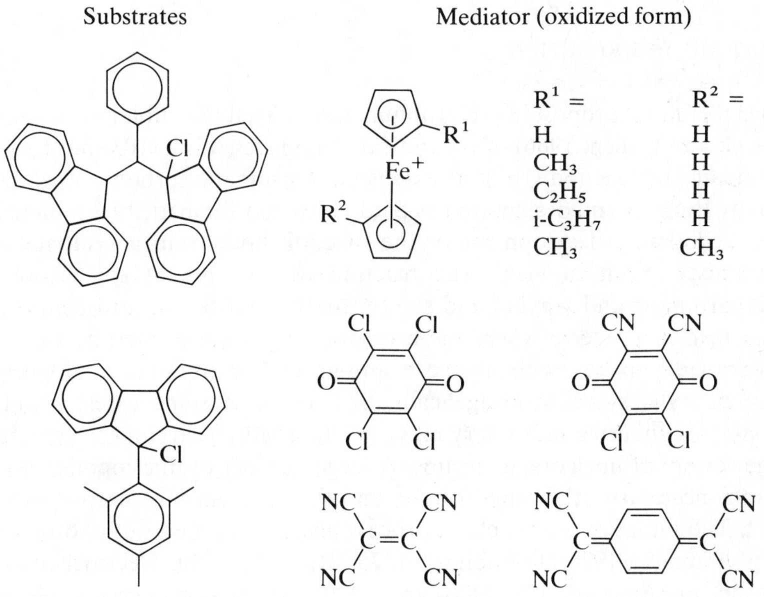

Substrates   Mediator (oxidized form)

Scheme 7

and of quinonoid compounds. In such systems the $R\cdot/R^-$ couple is reversible and so also is the $R^+/R\cdot$ couple in the absence of chloride ions (Andrieux et al., 1985; Merz and Tomahogh, 1979). In most cases, the rate-determining step was the dissociation of the chloride, as attested by the independence of the overall kinetics from both the standard potential and concentration of the mediator. Under these conditions, the follow-up electron-transfer step is at the diffusion limit, as can be inferred from the quite large difference in standard potentials between the reductants and the $R^+/R\cdot$ couple (Andrieux et al., 1985; Merz and Tomahogh, 1979). This is why the electron-transfer step is able to compete successfully with the coupling between R^+ and Cl^-, energetically a strongly downhill reaction. Mixed kinetic control by the cleavage reaction and the successive electron-transfer steps was observed only with the most positive mediator, viz., 2,3-dicyano-5,6-dichloroquinone, for which the electron-transfer rate constant falls below the diffusion limit. The rate and equilibrium constants of the cleavage reaction were derived from the kinetic data. A very similar mechanism has been recently found in the reduction of trityl halides by lithium dialkylamide bases in THF (Newcomb et al., 1990).

3 $S_{RN}1$ Substitutions

SCOPE OF THE REACTION

Since the initial proposal by Kornblum and Russell (Kornblum et al., 1966; Russell and Danen, 1966) of the radical chain mechanism (103) put forward to explain the reaction (104), its extension to aromatic carbon centres (105, 106) by Bunnett and designation as $S_{RN}1$ (Kim and Bunnett, 1970), there has been a booming development of the reaction both from a synthetic and mechanistic point of view. The reaction occurs both at sp^2 (aromatic, heteroaromatic and vinylic) and sp^3 (nitroalkyl, nitroallyl, nitrobenzyl and other benzylic systems where the aromatic ring is substituted by electron-withdrawing groups such as the cyano-group, heterocyclic analogues of these benzylic systems, bridgehead alkyl, even, in some cases, standard tertiary, secondary and primary alkyl, perfluoroalkyl) carbon centres with a large variety of nucleofugic groups. A large number of nucleophiles, which are not necessarily the same for sp^2 and sp^3 centres, have been shown to react. The initiation of the chain process has been carried out photochemically (Bunnett, 1978; Kornblum, 1975; Russell, 1970), electrochemically (Pinson and Savéant, 1974; Savéant, 1980a) or by homogeneous electron donors added in catalytic amounts. The nucleophile may itself belong to the

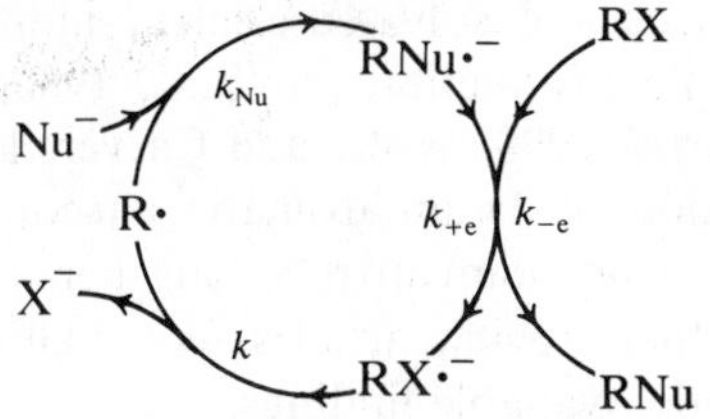

$$\text{(103)}$$

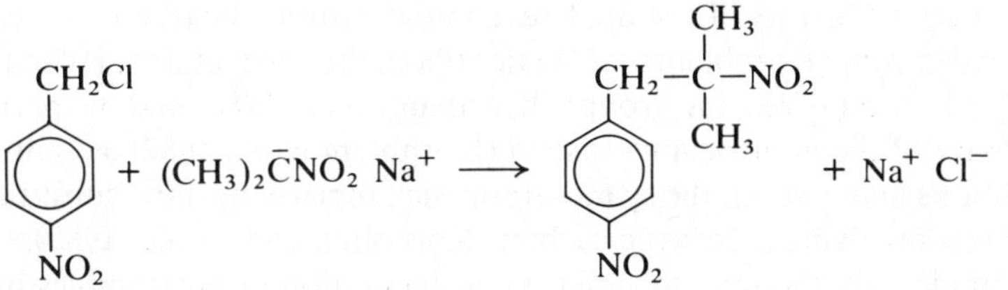

$$\text{(104)}$$

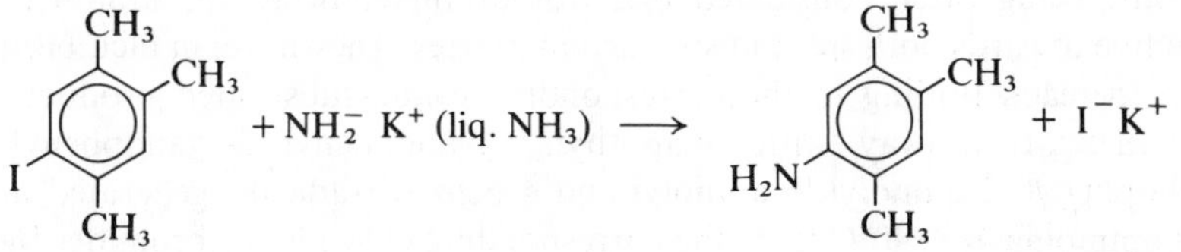

$$\text{(105)}$$

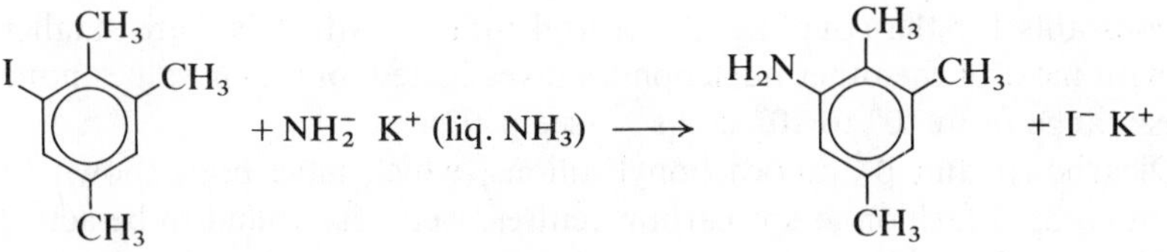

$$\text{(106)}$$

latter category if it is a sufficiently powerful reducing agent, able to reduce the substrate "thermally", thus leading to a so-called "thermal" initiation of the reaction. It may also come from another nucleophile than that participating in the substitution process, i.e. from an electron donor which is irreversibly oxidized in the reaction, thus leading to an "entrainment" process (Kornblum, 1975). A more controllable tuning of the electron-transfer reducing properties of the initiator is obtained when it is a member of a reversible redox couple, for example generated electrochemically (Amatore et al., 1984b; Swartz and Tenzel, 1984).

This very rich chemistry has been the subject of several comprehensive reviews, including recent ones: sp^3 carbon centres, general (Kornblum, 1975, 1982; Russell, 1970, 1987), photochemically induced reactions (Bowman, 1988a), substituted aliphatic nitro-compounds (Bowman, 1988b), alkyl mercurials (Russell, 1989); sp^2 carbon centres, general (Norris, 1983; Rossi and

Rossi, 1983), light-induced and solvated-electron-induced reactions (Bunnett, 1978), electrochemical induction (Savéant, 1980a), synthetic aspects (Beugelmans, 1984; Norris, 1990; Wolfe and Carver, 1978), photochemical induction of the reaction at heteroaromatic carbon centres (Lablache-Combier, 1988). There is no point in repeating here the extensive lists of reactions contained in these review articles. We shall just underline a few points and report recently available findings.

In a series of recent papers, Kornblum *et al.* have systematically investigated various factors of the $S_{RN}1$ reaction at saturated carbon, namely, the stereochemistry (Kornblum and Wade, 1987), the effect of light (Wade *et al.*, 1987), the effect of leaving groups (Kornblum *et al.*, 1988) and the particular reactivity of the *p*-nitrocumyl system (Kornblum *et al.*, 1987) as well as the conditions under which the cyano-group may replace the nitro-substituent in reactions involving a benzylic carbon (Kornblum and Fifold, 1989).

Cyanide ions continue to suffer a bad reputation as nucleophiles in $S_{RN}1$ reactions, being often considered (see, for example, Bowman, 1988a) as unreactive towards both sp^2 and sp^3 carbon centres. They have, in fact, been shown to react, leading to the corresponding cyano-substituted products, along an $S_{RN}1$ pathway, with 1-naphthyl, 2-cyanophenyl, 3-cyanophenyl, 4-cyanophenyl, 2-quinolyl, 4-quinolyl and 4-benzoyl radicals, generated in liquid ammonia at $-40°C$ from the corresponding chlorides or bromides by electrochemical induction (Amatore *et al.*, 1985a, 1986a). It is true that the rate constants for the coupling of the aryl radicals with CN^- are smaller than with most of the other nucleophiles investigated, but their values none the less range from 10^6 to $10^9 \, M^{-1} s^{-1}$ (see p. 92).

β-Dicarbonyl and β-cyanocarbonyl anions, which have been shown to react in an $S_{RN}1$ fashion at sp^3 carbon centres, were also found to be active toward aromatic and heteroaromatic carbons under photochemical (Beugelmans *et al.*, 1982) or electrochemical stimulation (Oturan *et al.*, 1989).

Alkoxide or aryloxide anions are also reputed to be inactive in $S_{RN}1$ reactions. There is, however, one example of such a reaction at an sp^3 carbon: the nitro-derivative of 4-nitrocumyl reacts with phenoxide and 1-methyl-2-naphthoxide ions yielding the corresponding ethers (Kornblum *et al.*, 1967). A similar reaction has been reported for halobenzenes in t-butyl alcohol upon stimulation by sodium amalgam (Rajan and Sridaran, 1977). This reaction could not, however, be reproduced (Rossi and Pierini, 1980) and other attempts to make phenoxide ions react at sp^2 carbons have been equally unsuccessful (Ciminale *et al.*, 1978; Rossi and Bunnett, 1973; Semmelhack and Bargar, 1980). It has been found, more recently, that phenoxide ions react with a series of aryl halides under electrochemical induction, but that the coupling occurs at the *p*- or *o*-phenolic carbon rather than at the phenolic oxygen (Alam *et al.*, 1988; Amatore *et al.*, 1988). This is an interesting way of synthesizing unsymmetrical biaryl compounds, which are themselves a source of valuable materials in the case where the biaryl compounds are substituted by one electron-withdrawing and one electron-donating group (Alam et al., 1987). A similar reaction has recently been shown to occur between aryl iodides and 2-naphthoxide ions in liquid ammonia under photochemical induction (Pierini et al., 1988). Tertiary alkyl alkoxide ions do not seem to react either at sp^2 (Bunnett, 1978) or sp^3 (Bowman, 1988a) carbons. Primary and secondary alkoxide ions react with aryl halides by an electron-transfer catalysed chain process which is, however, completely different from an $S_{RN}1$ reaction since it leads, through H-atom abstraction (Amatore et al., 1982a), to the overall oxidation of the alkoxide into the corresponding carbonyl compound (see Section 2, p. 45). Alcohols (Andrieux et al., 1987) react in a similar fashion.

Among the most striking differences between $S_{RN}1$ reactions at sp^2 and sp^3 carbon centres is the fact that the 2-nitropropane anion, a classic nucleophile (Bowman, 1988a,b; Kornblum et al., 1966; Kornblum, 1975; Russell and Danen, 1966; Russell 1970, 1989) for sp^3 carbons, seemed for a long time to be unreactive towards sp^2 carbons. In fact, it reacts with aryl radicals (for example phenyl and 4-benzoyl) generated electrochemically from the corresponding halides (Amatore et al., 1986b), leading to the adduct anion radical (Scheme 8). The start of the reaction is thus the same as in an $S_{RN}1$ reaction, but the benzylic radical anion so obtained (107) loses a nitrite anion (108) faster than being reoxidized to the substitution product at the electrode or in the solution (109). Overall, the reaction consumes electrons (two per aryl halide) and yields a substitution product in which NO_2 has been replaced by H (112).

$$ArX + e^- \rightleftharpoons ArX^{\cdot-} \qquad (46)$$

$$ArX^{\cdot-} \rightarrow Ar^{\cdot} + X^- \qquad (47)$$

$$Ar^{\cdot} + NO_2(CH_3)_2C^- \rightarrow ArC(CH_3)_2NO_2^{\cdot-} \qquad (107)$$

$$ArC(CH_3)_2NO_2^{\cdot-} \rightarrow ArC(CH_3)_2^{\cdot} + NO_2^- \qquad (108)$$

$$ArC(CH_3)_2NO_2^{\cdot-} - e^- \nrightarrow ArC(CH_3)_2NO_2 \qquad (109)$$

$$ArC(CH_3)_2^{\cdot} + e^- \rightarrow ArC(CH_3)_2^- \qquad (110)$$

$$\underline{ArC(CH_3)_2^- + H^+ \text{ (solvent, residual water)} \rightarrow ArC(CH_3)_2H \qquad (111)}$$

$$ArX + NO_2(CH_3)_2C^- + H^+ + 2e^- \rightarrow ArC(CH_3)_2H + NO_2^- + X^- \quad (112)$$

Scheme 8

There are few examples of $S_{RN}1$ reactions leading to organometallic compounds. This does not seem to result from lack of reactivity but rather from a limited number of investigations. The reaction of secondary and tertiary alkyl iodides with the $(\pi\text{-allylNiBr})_2$ complex has been suspected to follow an $S_{RN}1$ pathway (Hegedus and Miller, 1975) and so has the reaction of vinyl halides with $Co(CO)_4^-$ (Brunet et al., 1983). In both cases, however, the product is not an organometallic compound (allylation and carbonylation occur, respectively) and it is not clear whether or not the reaction goes through an organometallic intermediate resulting from substitution of the halogen by a metal-centred nucleophile. An example where an $S_{RN}1$ reaction has been suspected to take place leading to an organometallic compound as the main product is the reaction of primary alkyl bromides and iodides with $(CH_3)_3Sn^-$ (Ashby et al., 1985). In the proposed reaction scheme, there is, however, rather severe competition with other mechanisms, namely S_N2 and outer sphere dissociative electron transfer followed by radical coupling.

There is another example of a reaction leading to an organometallic complex where, this time, there is little doubt that the reaction follows exclusively an $S_{RN}1$ pathway. It deals with the substitution of the halogen by iron(I) porphyrins in their reaction with aryl and vinyl halides under electrochemical stimulation (Lexa and Savéant, 1982). The product of the reaction is the σ-aryl(vinyl)–iron complex which can be produced and observed in both its iron(III) and iron(II) oxidation states in cyclic voltammetry and thin-layer cell spectroelectrochemistry. That the reaction is of the $S_{RN}1$ type is shown by the fact that the substitution complex is obtained only if the electrode potential is held at a value corresponding to the reduction of the aryl(vinyl) halide, i.e. where the aryl(vinyl) radical is formed upon decomposition of the anion radical produced at the electrode surface. In this process, electrochemistry serves a second purpose, besides stimulation of the $S_{RN}1$ reaction, namely the in situ preparation of the nucleophile, the iron(I) porphyrin, from the starting iron(III) porphyrin by successive reversible addition of two electrons. The reaction is an attractive alternative to the other main route to σ-aryl(vinyl)–iron complexes, namely the reaction of the iron(III) complex with a Grignard or an alkyllithium reagent.

The electrochemical in situ generation of the nucleophile to be used in an $S_{RN}1$ reaction has not been much exploited so far in spite of the advantage that quite reactive species can be created in this way in the presence of the substrate, thus avoiding side-reactions that may occur during or after an ex situ generation. 2-, 3- and 4-Chlorobenzonitriles as well as 2-, 3- and 4-bromobenzophenones have been found to react successfully with electrochemically prepared phenylselenide and telluride (Degrand, 1986, 1987a,b; Degrand et al., 1987). The reaction was, however, carried out in two steps, electrochemical preparation of the nucleophile from the parent diselenide or telluride first, then addition of the aryl halide leading to the $S_{RN}1$ reaction. More similar to the procedure used in the above-cited arylations and vinylations of iron porphyrins is the $S_{RN}1$ synthesis of diaryldiselenides and tellurides from the corresponding halides by reduction at selenium and tellurium electrodes at which the primary nucleophile, $(Se—Se)^{2-}$, or $(Te—Te)^{2-}$, is generated (Degrand and Prest, 1990a).

$S_{RN}1$ reactions involving perfluoroalkyl halides as substrates have been mentioned in some of the review articles cited above (Bowman, 1988a; Russell, 1987). However, results of further investigations are now available. The $S_{RN}1$ mechanism was deemed to be operative in the reaction of perfluoroalkyl iodides with several anionic nucleophiles: thiolates (Boiko *et al.*, 1977; Feiring, 1984; Popov *et al.*, 1977, 1982), selenates (Voloschhuk *et al.*, 1977), sulphinates (Kondratenko *et al.*, 1977), 2-nitropropanate (Feiring, 1983), malonates (Chen and Qiu, 1986), ethyl acetoacetate (Chen and Qiu, 1987a), imidazolate (Chen and Qiu, 1987b) and the anion of 5-nitrotetra-hydro-1,3-oxazine (Archibald *et al.*, 1989). In most cases the reaction was carried out under photochemical stimulation and shown in several cases to be slowed by radical traps, viz., dinitrobenzenes and unsaturated compounds. Although *a priori* less reactive, perfluoroalkyl bromides have been shown to react with thiolates (Wakselman and Tordeux, 1984, 1985) and imidazolates (Chen and Qiu, 1987b). Trapping experiments again provided support for the $S_{RN}1$ mechanism. Indirect electrochemical induction of the reaction, by means of aromatic anion-radical mediators, of perfluoroalkyl iodides and bromides with imidazolate and substituted imidazolate anions has recently been reported, unambiguously showing by means of a cyclic voltammetric and preparative-scale analysis, the occurrence of an $S_{RN}1$ mechanism (Médebielle *et al.*, 1990).

QUALITATIVE ASSESSMENT OF THE MECHANISM

The first clue to the existence of the $S_{RN}1$ mechanism came from product studies both in aliphatic and aromatic cases. It was noticed that in the reaction of benzyl and substituted benzyl chlorides with the 2-nitropropane anion, oxygen alkylation, yielding the oxime and then the aldehyde, occurs exclusively in the case of benzyl chloride and 3-nitrobenzyl chloride, whereas, with 4-nitrobenzyl chloride, the yield of aldehyde is only 6% and the carbon-alkylated (104) product is obtained in 92% yield (Kornblum, 1975). This was interpreted as the result of a competition between S_N2 (*O*-alkylation) and $S_{RN}1$ (*C*-alkylation) reactions. In the aromatic case, it was observed that the reaction of 5- and 6-halopseudocumenes with KNH_2 in liquid ammonia (Kim and Bunnett, 1970) forms the 5- and 6-pseudocumidines in a ratio which is the same whether the starting compound is the 5- or 6-isomer in the case of the chloro- and bromo-derivatives, as expected from an aryne mechanism (Scheme 9), whereas much more non-rearranged products were found in the case of the iodo-derivatives. Since, in view of the electron-donating properties of the substituent, an S_NAr mechanism is unlikely, the mechanism shown in Scheme 9 was proposed (and designated by the $S_{RN}1$ initials). Its radical-chain character was confirmed when it was found that addition of a radical trap, tetraphenylhydrazine, largely depresses the non-rearranging pathway, whereas, by contrast, the addition of potassium metal yielding solvated electrons that are obviously able to produce the anion radical $ArX\cdot^-$ from ArX, steers the reaction entirely to unrearranged products.

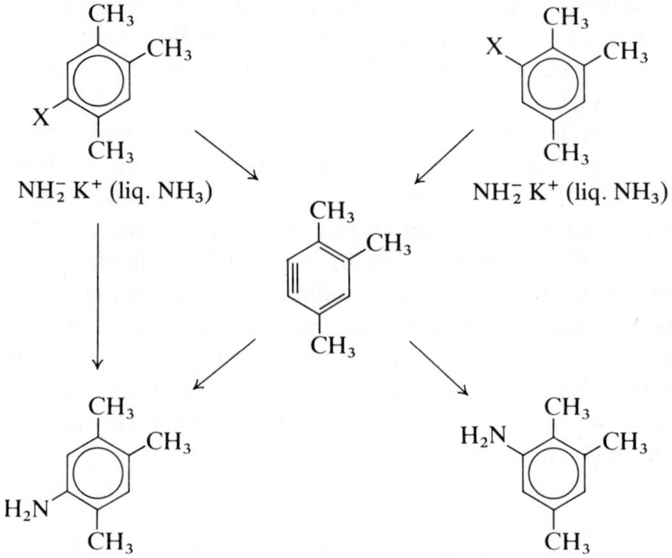

$$\text{X--C}_6\text{H}_2(\text{CH}_3)_3 \quad \xrightarrow{\text{NH}_2^- \text{K}^+ \text{(liq. NH}_3\text{)}} \quad (113)$$

Scheme 9

Inhibition by radical traps has been extensively used as providing evidence for the $S_{RN}1$ mechanism both with aliphatic (Bowman, 1988a,b; Kornblum, 1975) and aromatic (Bunnett, 1978; Rossi and Rossi, 1983) substrates. Inhibitors are basically of two types, viz., compounds adding irreversibly the intermediate radical R· (di-t-butyl nitroxide is one of the most popular traps in this connection) and good reversible electron acceptors such as dinitrobenzenes. In the latter case, the inhibitor may intercept the anion radical $RNu\cdot^-$, thus interrupting the propagation of the chain, but other possibilities exist such as one-electron transfer from the anion radical $ArX\cdot^-$ or destruction of the single electron donor species that participate in the initiation process. Dioxygen, a commonly used inhibitor, may act in one way or the other according to circumstances, leading to the superoxide anion radical in the second case. A lack of inhibition does not necessarily disprove the existence of an $S_{RN}1$ pathway: R· may react faster with the nucleophile than with the trap, or $RNu\cdot^-$ may transfer its odd electron faster to RX (one of the propagation steps) than to the inhibitor. With putative inhibitors of the latter type that are members of reversible redox couples, as are for example nitro- and dinitro-benzenes, effective inhibition mostly depends upon the difference in standard potentials of the inhibitor couple and of the $RNu/RNu\cdot^-$ couple being large enough for the reaction (114) to overcome

$$\text{inhibitor} + RNu\cdot^- \;\rightleftharpoons\; \text{inhibitor}\cdot^- + RNu \tag{114}$$

$$RX + RNu\cdot^- \;\rightleftharpoons\; RX\cdot^- + RNu \tag{115}$$

the propagation step (115). Note that, if the inhibitor/inhibitor$\cdot^-$ standard potential is not positive enough, not only is reaction (114) not fast enough to compete efficiently with (115) but the inhibitor anion radical may serve as an initiator for the $S_{RN}1$ reaction by triggering the generation of $RX\cdot^-$ as in (116). An ambident behaviour of the inhibitor/inhibitor$\cdot^-$ couple might be

$$\text{inhibitor}\cdot^- + RX \;\rightleftharpoons\; \text{inhibitor} + RX\cdot^- \tag{116}$$

the cause of the unusual effect of nitrobenzene on the dark reaction of iodobenzene with pinacolone enolate in DMSO: the reaction is inhibited initially but then becomes faster than in the absence of nitrobenzene (Scamehorn and Bunnett, 1977).

Another piece of evidence in favour of the $S_{RN}1$ mechanism is the stimulation of the reaction by various means that all have in common that they involve the catalytic injection of electrons into the system. The addition of alkali metals in the aromatic substitutions carried out in liquid ammonia has proved to be an efficient way to trigger or stimulate the reaction in a number of cases (Bunnett, 1978; Rossi and Rossi, 1983). The solvated electron thus generated, being a strong reductant, is indeed able rapidly to generate $RX\cdot^-$ from most of the RX substrates. Although much less commonly used in substitutions at sp^3 carbons, addition of an alkali metal (in hexamethylphosphoramide) has been shown to produce the same effect (Kornblum, 1971). Stable anion radicals, for example sodium naphthalenide, have been used similarly to stimulate substitutions at sp^3 carbons (Kornblum, 1971; Russell et al., 1971). It is interesting to note, as discussed earlier, that the anion radical of nitrobenzene and the superoxide ion have been used as initiators (Kornblum, 1971; Russell et al., 1971), whereas their oxidized forms work as inhibitors for other systems. As discussed in more detail in the next section, in situ electrochemically generated anion radicals have been used as initiators in many cases in substitution of aromatic and heteroaromatic halides (Amatore et al., 1984b; Swartz and Tenzel, 1984) as well as perfluoroalkyl halides (Médebielle et al., 1990).

Triggering or stimulation of the reaction by shining light on to the reaction mélange has been observed in a number of cases both with sp^3 (Bowman, 1988a; Kornblum, 1975; Russell, 1970, 1987) and sp^2 (Bunnett, 1978; Rossi and Rossi, 1983; Lablache-Combier, 1988; Norris, 1983, 1990; Wolfe and Carver, 1978) carbon centres. Very often, quantum yields much larger than one have been measured, pointing to the chain character of the reaction and hence providing support for the proposed mechanism. Although photochemical induction has become one of the most popular modes of performing an $S_{RN}1$ reaction on a preparative scale, the mechanism of the photochemical initiation is not fully understood. Three types of possible initiation steps [(117), (118) and (119)] have been proposed (Hoz,

$$RX \xrightarrow{h\nu} R\cdot + X^- \tag{117}$$

$$RX \xrightarrow{h\nu} RX^* \xrightarrow{Nu^-} RX\cdot^- + Nu\cdot \tag{118}$$

$$[RX \cdots Nu^-] \xrightarrow{h\nu} [RX\cdot^- \cdots Nu\cdot] \longrightarrow RX\cdot^- + Nu\cdot \tag{119}$$

and Bunnett, 1977), and it has been shown, by wavelength-dependent quantum-yield measurements, that the pathway involving the photoexcitation of the nucleophile–substrate charge-transfer complex (119) appears to prevail in the reaction of acetone enolate with iodo- and bromo-benzenes in DMSO, whereas this is not the case with potassium diethylphosphite (Fox et al., 1983).

Electrochemical induction of the reaction (Pinson and Savéant, 1974), as shown to occur in a number of cases with aromatic and heteroaromatic substrates (Savéant, 1980a, 1986, 1988), also provides evidence for the $S_{RN}1$ mechanism. The electrochemical approach allows, in addition, a quantitative analysis of the mechanism and reactivity problems and will be described in this connection in the next subsection.

Stereochemistry has also been used as a diagnostic tool. If an $S_{RN}1$ reaction occurs at an optically active sp^3 carbon centre, racemization should take place. This has indeed been observed in the reaction of 2-(4-nitrophenyl)-2-nitrobutane with benzenethiolate, benzenesulphinate, 2-nitropropanate and also nitrite ions (Kornblum, 1975; Kornblum and Wade, 1987) as well as in the reactions of 2-(4-nitrophenyl)-2-chloroethane with the N,N-diethyl-α-aminopropionitrile and phenlyacetonitrile anions (Cabaret et al., 1985).

QUANTITATIVE APPROACH TO MECHANISMS AND REACTIVITY IN THE
CASE OF AROMATIC SUBSTITUTION. TERMINATION STEPS

The first experiments (Pinson and Savéant, 1974) demonstrating the electrochemical induction of $S_{RN}1$ reactions concerned the reaction of 4-bromo-
benzophenone with benzenethiolate ions. It was shown, by cyclic voltammetry and preparative-scale electrolysis, that the reaction is triggered by
setting the electrode potential at a value where the substrate is reduced (this
value was derived from a separate investigation of the electrochemical
reduction of the halide in the absence of the nucleophile, see Section 2,
pp. 37–42) and that the consumption of electrical charge is small, much
smaller than for the reduction in the absence of the nucleophile. Evidence
was thus provided that the $RX \cdot^{-}$ anion radicals are intermediates and that
the electrons supplied by the electrode play a catalytic role. Similar observations were then made for a series of other substrate–nucleophile couples
(Pinson and Savéant, 1978) and the mechanism of the reaction analysed on
quantitative bases (Savéant, 1980a). The typical example of the reaction of
2-chloroquinoline with benzenethiolate ions in liquid ammonia (Amatore et
al., 1979a) is represented in Fig. 15. In the absence of benzenethiolate ions,
the voltammogram exhibits a first irreversible two-electron reduction wave
corresponding, as discussed in detail in Section 2, pp. 37–42, to the one-
electron formation of $RX \cdot^{-}$ (46). Its decomposition leads to R· (47), and its
reduction at the electrode (71) or in solution (72) ultimately yields quinoline
(73), the reversible reduction of which to its anion radical is responsible for
the second, one-electron, reversible wave. Upon addition of the nucleophile,
these two waves decrease while a third, one-electron, reversible wave appears
and increases accordingly, corresponding to the reversible reduction of the
substituted product, 2-phenylthioquinoline, into its own anion radical as
checked with an authentic sample. The phenomena taking place at the first
wave are summarized in Scheme 10. After its initial formation at the
electrode surface, $RX \cdot^{-}$ diffuses towards the solution while decomposing
into the R·, which then reacts with the nucleophile yielding $RNu \cdot^{-}$. This is
eventually oxidized into the final substitution product, RNu, at the electrode
surface or in the solution. The latter process is identical to the chain process
put forward for explaining homogeneous $S_{RN}1$ reactions. Whether one or
other possibility predominates depends primarily upon the rates at which
$RX \cdot^{-}$ decomposes and the ensuing R· reacts with the nucleophile. If these
reactions are very rapid, $RNu \cdot^{-}$ is formed close to the electrode surface and
will thus be predominantly oxidized there to RNu. Conversely, if they are
less rapid, the oxidation of $RNu \cdot^{-}$ will predominantly take place in the
solution in the chain process. The competition between these two pathways
is similar to the ECE–DISP competition already discussed when describing

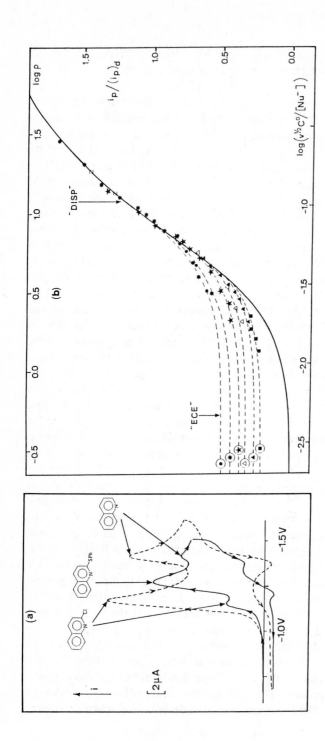

Fig. 15 (a) Cyclic voltammetry of 2-chloroquinoline (2.5 mM) in liq. NH$_3$ + 0.1 M KBr at $-40°$C in the absence (- -) and presence (—) of benzenethiolate (56 mM). Scan rate, 0.2 V s^{-1}. (b) Peak current of the first 2-chloroquinoline wave, i_p, normalized to the one-electron reversible peak height, $(i_p)_d$, as a function of the concentrations of RX and Nu$^-$. Key: [RX] = 1.5 mM, [Nu$^-$] = 13 mM ($\square$); [RX] = 1.5 mM, [Nu$^-$] = 15.4 mM ($\bullet$); [RX] = 2.65 mM, [Nu$^-$] = 21.2 mM ($*$); [RX] = 2.65 mM, [Nu$^-$] = 31.8 mM ($\star$); [RX] = 2.65 mM, [Nu$^-$] = 40.8 mM ($\triangle$); [RX] = 2.65 mM, [Nu$^-$] = 66.2 mM ($\blacktriangle$); [RX] = 2.65 mM, [Nu$^-$] = 103 mM ($\blacksquare$). Scan rate, 0.2 V s^{-1}. The solid line represents the predictions for an S$_{RN}$1–DISP competition and the dashed lines represent the influence of the ECE pathway for each value of [Nu$^-$]. (Adapted from Amatore *et al.*, 1979a.)

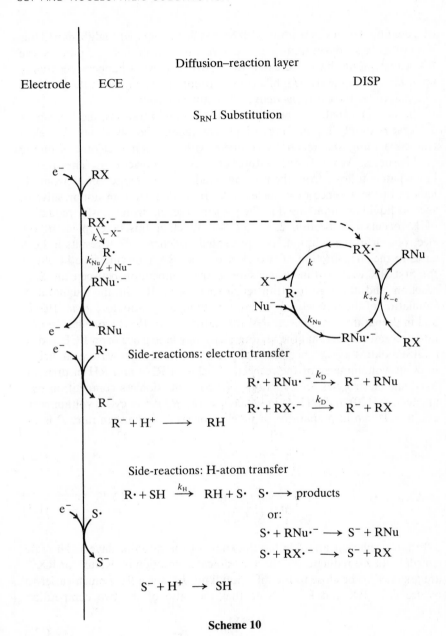

Scheme 10

the reduction of aryl halides in the absence of nucleophiles. It is known, from previous studies of the substrate under these conditions that, in low scan rate voltammetry and on the preparative scale, the overall process, triggered by the decomposition of $RX\cdot^-$, is fast compared to diffusion. Thus if no other reactions were to occur competitively with the $S_{RN}1$ process, the RX waves should have decreased to zero. This is what happens eventually upon increasing the nucleophile concentration, stressing again the catalytic role played by the electrons delivered by the electrode.

The very fact that, for smaller nucleophile concentrations, the RX waves decrease without disappearing completely unambiguously shows that electron-consuming side-reactions compete with the non-electron-consuming $S_{RN}1$ process. What is the nature of these side-reactions? The answer immediately follows from the previous studies of the reduction of aromatic halides in the absence of a nucleophile. In a poor H-atom donor solvent, such as liquid ammonia used in the present case, electron transfer reduction of $R\cdot$ occurs (see Section 2, pp. 37–54). Whether this takes place at the electrode or in the solution, as represented in Scheme 10, it results in both cases in the consumption of two electrons per RX molecule. The height of the first RX wave is thus a measure of the competition between the $S_{RN}1$ reaction and the electron-transfer reduction of $R\cdot$. In the quantitative simulation of the competition, in cyclic voltammetry (Amatore et al., 1979b) and in preparative-scale electrolysis (Amatore et al., 1981a), one has to take into account the fact that the various steps may take place in an ECE and/or a DISP context as sketched in the above scheme. The first wave peak height in cyclic voltammetry or the respective yields of RNu and RH in preparative-scale electrolyses are functions of two dimensionless competition parameters, σ (120) and ρ (121) in which $\theta = (Fv/RT)^{1/2}$ in cyclic voltammetry and $\theta = D^{1/2}/\delta$ in preparative-scale electrolysis (v is the scan rate, D is the

$$\sigma = \frac{k}{k_{Nu}[Nu^-]} \tag{120}$$

$$\rho = \frac{k_D}{k^{1/2}k_{Nu}} \frac{[RX]}{[Nu^-]} \theta^{-1/2} \tag{121}$$

diffusion coefficient and δ is the thickness of the diffusion layer). The rate constants of the reduction of the aryl radical in solution by $RNu\cdot^-$ or $RX\cdot^-$ are assumed to be close to the diffusion limit, k_D, since $R\cdot$ is much easier to reduce than RX and RNu. Note that the ratio of the two competition

$$\frac{\rho}{\sigma} = \frac{k_D[RX]}{k^{3/2}} \theta^{-1/2} = p_E \tag{122}$$

parameters (121) is the same as the ECE–DISP competition parameter, p_E [see (74)], in the reduction of aryl halides in the absence of nucleophile. Figure 15(b) shows that there is an excellent agreement between the predictions of the kinetic model and the experimental data, thus providing quantitative support for the $S_{RN}1$ mechanism. Most of the data points fall in the DISP rather than in the ECE framework, in keeping with the fact that the cleavage rate constant of the 2-chloroquinoline anion radical is rather small ($2 \times 10^4 \, s^{-1}$). In contrast, an $S_{RN}1$–ECE competition is found in the case of 2-iodoquinoline with the same nucleophile (Amatore et al., 1979a), as expected from the cleavage of the halide in the $RX^{\cdot -}$ anion radical being faster ($k = 3 \times 10^6 \, s^{-1}$) than with the chloro-derivative.

The reaction of 2-chloroquinoline with benzenethiolate ions in liquid ammonia (Amatore et al., 1979a) is a good example for testing the possible occurrence of an $S_{RN}2$, instead of $S_{RN}1$, mechanism, in which the formation of $RNu^{\cdot -}$ from $RX^{\cdot -}$ and Nu^- would involve concerted, rather than stepwise, bond breaking and bond formation (Scheme 11). This mechanism

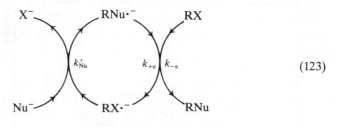

(123)

Scheme 11

has been suggested for rationalizing product distributions in the reaction of 2-substituted 2-nitropropanes with enolate ions (Russell et al., 1981, 1982a,b). The $i_p/(i_p)_0$ ratio at the first wave is predicted to be a function of the parameter $(k_D[RX]/k^{1/2}k_{Nu}[Nu^-]) \, (Fv/RT)^{1/2}$ in the $S_{RN}1$ case and of the parameter $k/k'_{Nu}[Nu^-]$ in the $S_{RN}2$ case $\{i_p/(i_p)_0 = k/(k + k'_{Nu}[Nu^-])\}$. The experimental variations of the $i_p/(i_p)_0$ ratio with the scan rate and the RX concentration [Fig. 15(b)] are obviously in accord with the $S_{RN}1$ mechanism and not with the $S_{RN}2$ mechanism.

Many other substrate–nucleophile couples have been similarly investigated in liquid ammonia and found to conform to the $S_{RN}1$ mechanism quantitatively. Among them, several systems give rise to unusual cyclic voltammetry patterns, exhibiting dips between the two waves of the starting halide (Pinson and Savéant, 1978) or trace crossing upon scan reversal (Amatore et al., 1980a). Examples of such behaviour are shown in Fig. 16.

In the first case, the substrate gives rise, as discussed earlier, to a rapidly cleaved anion radical $RX^{\cdot -}$ and thus to a $S_{RN}1$–ECE competition. As can be seen from the height of the first wave, the $S_{RN}1$ reaction is not very efficient under these conditions owing to the fact that the aryl radical is formed very close to the electrode surface where it is rapidly reduced. As the electrode

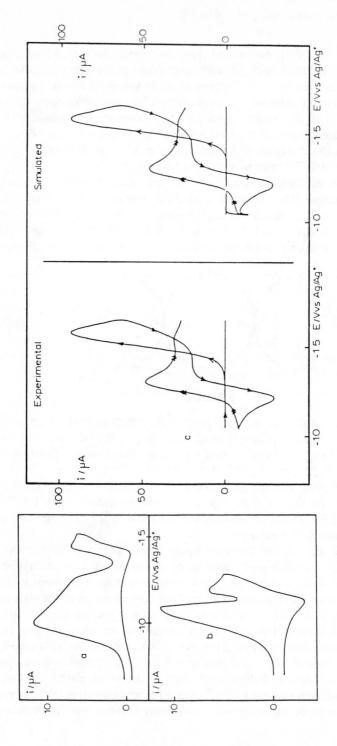

Fig. 16 Examples of current dips: (a) 2-iodoquinoline (1.6 mM) with benzenethiolate ions (44 mM); (b) 2-iodoquinoline (5.2 mM) with diethylphosphite ions (100 mM). Scan rate, 0.2 V s⁻¹. Example of trace crossing: (c) 4-chlorobenzonitrile (1.55 mM) with diethylphosphite ions (663 mM). Scan rate, 0.215 V s⁻¹. In liq. NH₃/0.1 M KBr at −40°C. (Adapted from Amatore *et al.*, 1980a,b.)

potential comes closer to the second wave, which corresponds to the $RNu/$ $RNu^{\cdot-}$ couple, the equilibrium is more and more in favour of $RNu^{\cdot-}$, which then diffuses in the solution and triggers the chain process (DISP conditions) against which the electron-transfer side-reactions (termination steps) are less efficient than in the previous ECE context. The $S_{RN}1$ reaction is thus accelerated and the current falls accordingly.

In the trace-crossing example, the cleavage of $RX^{\cdot-}$ is again fast $(k = 9 \times 10^8\,s^{-1})$ and thus the reaction takes place within a $S_{RN}1$–ECE context, along the first cathodic scan. Unlike the preceding case, the standard potential of the $RNu/RNu^{\cdot-}$ couple is located in front of the first RX wave. The small amount of $RNu^{\cdot-}$ formed during the first scan (small because the $S_{RN}1$ reaction competes unfavourably with the ECE electron-transfer side-reaction) diffuses away from the electrode surface and starts the chain process in the solution, thus producing RNu. This process goes on after scan reversal. RNu thus formed in the solution diffuses back to the electrode where it is reduced, hence giving rise to a cathodic current during the anodic scan, and, therefore, to trace crossing. These unusual forms of cyclic voltammetric behaviour are thus in perfect agreement with the $S_{RN}1$ mechanism sketched in the above scheme, not only on qualitative grounds, as just explained, but also quantitatively (Amatore et al., 1980a,b). The preceding examples stressed the importance, besides other factors such as the rapidity of $RX^{\cdot-}$ cleavage, of the relative location of the $RNu/RNu^{\cdot-}$ standard potential and of the RX reduction potential. If the $RNu/RNu^{\cdot-}$ couple is much positive of the RX first wave, the propagation step (124) in the chain

$$RNu^{\cdot-} + RX \rightleftharpoons RNu + RX^{\cdot-} \qquad (124)$$

process becomes slow because of its unfavourable thermodynamics. Substitution may well occur but may stop at the level of the anion radical $RNu^{\cdot-}$. Under such conditions, the substitution process is non-catalytic in the sense that the formation of the substituted product, viz., $RNu^{\cdot-}$, consumes one electron per molecule of substrate. An example of such a situation is provided by the reaction of 4-bromobenzophenone (which gives rise to a relatively slowly cleaved anion radical) with cyanide ions. There is again an excellent quantitative agreement between the experimental data and the theoretical predictions based on this modified mechanism in which the essential step of $S_{RN}1$ reactions, viz., the coupling of the aryl radical with the nucleophile is still the key step of the reaction (Amatore et al., 1985a).

The experiments in liquid ammonia described above and the corresponding theoretical analyses have allowed a quantitative confirmation of the $S_{RN}1$ mechanism and the discovery of the nature of the side-reactions, i.e. the termination steps in the chain process. As discussed in Section 2, pp. 40–45, another means of destroying aryl radicals in organic solvents, besides electron transfer reduction, is H-atom abstraction from the solvent, which also leads to the hydrogenolysis hydrocarbon. This is indeed what was found in the investigation of the reaction of a series of substrate–nucleophile couples in ACN and DMSO (Amatore et al., 1982b). The complete analysis of the whole system is rather complex since it deals with a four-cornered competition at the level of the aryl radical involving its attack by the nucleophile, its electron-transfer reduction at the electrode, and in the solution, and the abstraction of an H-atom from the solvent as sketched in Scheme 10. Results quantitatively consistent with this reaction scheme and with the H-atom abstraction characteristics derived from previous experiments carried out in the absence of a nucleophile (Section 2, pp. 40–45) were obtained with the following systems: 9-chloro-, 9-bromo- and 9-iodo-anthracenes with cyanide and benzenethiolate ions in DMSO, 4-bromobenzophenone with the same nucleophiles in DMSO and CH_3CN, 1-bromo-naphthalene with benzenethiolate ions in the same solvents, 4-iodo- and 4-bromo-benzonitrile with the same nucleophile in CH_3CN (Amatore et al., 1982b).

As discussed earlier, given the radical and the nucleophile, the competition between the $S_{RN}1$ reaction and the electron-transfer reduction of the aryl radical is more in favour of the $S_{RN}1$ reaction if the aryl radical is produced in the solution, under DISP conditions, rather than close to the electrode surface, under ECE conditions. It was found, along the same lines, that indirect electrochemical induction, by means of an electrochemically generated mediator, is often more advantageous than direct electrochemical induction both in microelectrolytic techniques, such as cyclic voltammetry, and on the preparative scale (Alam et al., 1988; Amatore et al., 1984b, 1985b; Swartz and Tenzel, 1984). The same aromatic anion radicals as in the indirect electrochemical reduction of aryl halides in the absence of nucleophile were also used in the present case. The same approach has also been recently employed with perfluoroalkyl iodides and bromides (Médebielle et al., 1990). A typical example is shown in Fig. 17. One starts with the reversible wave of the mediator alone (P and Q designate the oxidized and reduced forms of the mediator, respectively). Upon addition of the aryl halide, the wave of the mediator loses its reversibility and increases in height as the result of a redox catalytic process, as described in Section 2. Upon addition of the nucleophile, the P/Q wave decreases and its reversibility is restored, pointing to the catalytic character of this indirect electrochemical induction of the reaction. The mechanism of the whole reaction, whereby an "electrocatalytic process" is "redox catalysed", is depicted in Scheme 12. The reduced form of the mediator, Q, reduces the aryl halide which then cleaves off the halide ion, giving rise to the aryl radical which reacts with the nucleophile thus providing $RNu^{\cdot-}$. This can enter the chain propagation loop or be reoxidized to the final substitution product by the oxidized form of the mediator. Termination steps in poor H-atom donor solvents, such as liquid ammonia, involve the reduction of the aryl radical by electron transfer from $RNu\cdot^-$ and $RX\cdot^-$ and also from the reduced form of the mediator, Q. In usual organic solvents, H-atom abstraction (not represented in Scheme 12 for clarity) should also be taken into account, as discussed in the case of direct electrochemical induction. One thus ends up with a fairly complicated overall kinetic scheme which is difficult to master in the general case. Quantitative support for the above reaction mechanism has been reached under simplifying conditions: no H-atom-transfer side-reaction, prevalent reoxidation of $RNu\cdot^-$ by the oxidized form of the mediator rather than by RX (chain process), electron-transfer reduction of the aryl radical by the reduced form of the mediator, rather than by $RNu\cdot^-$ and $RX\cdot^-$ (Amatore *et al.*, 1984b). Here again, an $S_{RN}2$ type of mechanism would not fit the experimental data since the competition between substitution and electron-transfer reduction of the aryl halide would hinge on the parameter $k_2'[Nu^-]/k$ not depending on the substrate concentration, unlike the $S_{RN}1$ case where the competition parameter is $k_2[Nu^-]/k_D[RX]$.

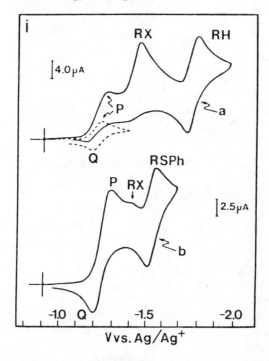

Fig. 17 Indirect electrochemical induction of $S_{RN}1$ reactions. Cyclic voltammetry of (a) 4-cyanopyridine (2.2 mM) in the absence (- - -) and presence (—) of 8.9 mM 2-chlorobenzonitrile; (b) 4-cyanopyridine (6.6 mM) in the presence of 8.9 mM 2-chlorobenzonitrile and 35 mM PhS⁻. (Adapted from Amatore *et al.*, 1984b.)

Electrode | Diffusion–reaction layer

S$_{RN}$1 Substitution

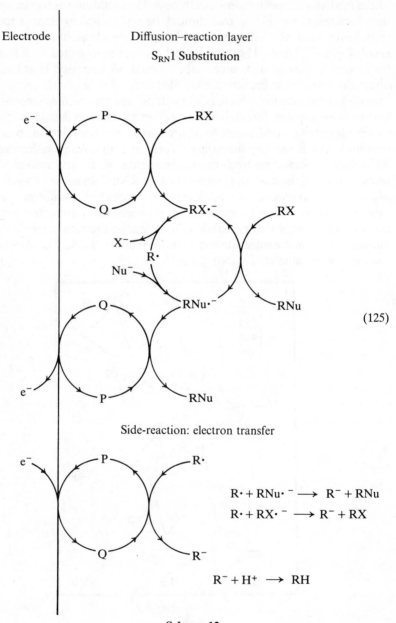

(125)

Side-reaction: electron transfer

$R\cdot + RNu\cdot^- \longrightarrow R^- + RNu$

$R\cdot + RX\cdot^- \longrightarrow R^- + RX$

$R^- + H^+ \longrightarrow RH$

Scheme 12

At this point, it can be concluded that the direct and indirect electro-chemical approach of the reaction in the case of aryl halides has provided a quantitative kinetic demonstration of the $S_{RN}1$ mechanism and the establishment of the nature of the side-reactions (termination steps in the chain process). In poor H-atom donor solvents, the latter involve electron-transfer reduction of the aryl radical.

Since these radicals are very easy to reduce, any reductant taking part in the inititation step or in the chain process is able to convert them rapidly into the corresponding carbanions. In other words, the very species that are essential to the triggering of the reaction are potentially able to destroy its propagation. This is the reason why the reaction achieves success only when the nucleophile possesses a large affinity towards the aryl radical, as will be seen on quantitative bases in the next section. This side-reaction also takes place in H-atom donor solvents, but, H-atom transfer from the solvent to the aryl radical is an additional major pathway competing with the $S_{RN}1$ process. Dimerization of two aryl radicals yielding the corresponding biaryl, a conceivable additional termination step, has not been observed so far in $S_{RN}1$ reactions. This is explicable by the fact that, with the stationary concentration of the aryl radicals being very low, this bimolecular reaction is likely to be outrun by their electron transfer and H-atom-transfer reduction (Amatore et al., 1981b). It might, however, come into play in poor H-atom donor solvents, if the electrolysis could be carried out at not too negative a potential. Some evidence of this possibility is provided by the formation of a small amount (6%) of 4,4′-dicyanobiphenyl in the indirect electrochemical reduction of 4-chlorobenzonitrile by means of 4,4′-bipyridine anion radicals in the presence of styrene in liquid ammonia (Chami et al., 1990). In this type of reaction, which aims at adding aryl radicals on to double bonds, a major side-reaction is the electron-transfer reduction of the aryl radical as in $S_{RN}1$ reactions. A firm conclusion concerning this possible termination step should, however, await confirmation that biaryls can be obtained in other systems and elucidation of the mechanism by which they might be formed.

There have been few quantitative investigations and models of the kinetics of $S_{RN}1$ reactions triggered by other means than electrochemical ones. This is presumably because the system involving the propagation loop, the initiation process and the various possible termination steps is rather complex from a kinetic-modelling point of view. On the other hand, most of the available experimental data concern product distribution and its variation upon changing factors such as the nature of the leaving group and of the nucleophile and, in a few cases, the measurement of an overall half-reaction time. Such a quantitative study has been performed for reactions initiated homogeneously by the anion radical of the substituted product itself rather than by an external electron donor (Amatore et al., 1981b). The speed at which the propagation loop rotates has been expressed as a function of the pertinent rate constants, k, k_{Nu}, k_{+e} and k_{-e}, and of the starting RX and RNu·⁻ concentrations, with the effect of the termination steps estimated in an approximate fashion. This analysis allowed the interpretation of several previous experimental observations on semi-quantitative grounds. For example, in the reaction of substituted benzenes with acetone enolate ions in liquid ammonia, the overall rate of the reaction under photostimulation decreases in the following order (Galli and Bunnett, 1981): PhI > PhBr > PhN⁺(CH₃)₃ > PhSPh > PhCl > PhF > PhOPh.   When solvated electrons are used as the initiator in the same series, the amount of dissolved potassium metal required to reach a given conversion of the starting halide increases from left to right, together with the production of the side-product RH (Bard *et al.*, 1980; Tremelling and Bunnett, 1980). This can be explained by a progressively slower turnover of the propagation loop versus the electron-transfer termination steps caused by the decrease of k and k_{+e} in the series. The latter factor is likely to prevail over the former since k is large with substituted benzenes. It was also noted that, in the reaction initiated by solvated electrons, the ketone resulting from the substitution, $PhCH_2COCH_3$, is partially reduced into the corresponding alcohol, $PhCH_2CHOHCH_3$, and the alcohol/ketone ratio roughly parallels the yield of RH and the required amount of dissolved potassium metal. The alcohol is likely to result from one or other following electron–proton transfer reductions. Product selection takes place at the level of RNu·⁻ as a result of the competition between the electron transfer from RNu·⁻ to RX, on the one hand, and reactions (126 + 127) or (128 + 129), on the other.

$$(RNu\cdot^-)\ \ PhCH_2\overset{\textstyle\cdot}{\underset{\textstyle O}{C}}CH_3 + NH_3 \ \rightleftharpoons\ PhCH_2\overset{\textstyle\cdot}{\underset{\textstyle OH}{C}}CH_3 + NH_2^- \qquad (126)$$

$$PhCH_2\overset{\textstyle\cdot}{\underset{\textstyle OH}{C}}CH_3 + e_{solv}^- \ \longrightarrow\ PhCH_2\underset{\textstyle O}{C}HCH_3 \qquad (127)$$

or

$$(RNu\cdot^-)\ \ PhCH_2\overset{\textstyle\cdot}{\underset{\textstyle O}{C}}CH_3 + e_{solv}^- \ \longrightarrow\ PhCH_2\overset{\textstyle\cdot\cdot}{\underset{\textstyle O}{C}}CH_3 \qquad (128)$$

$$PhCH_2\overset{\textstyle\cdot\cdot}{\underset{\textstyle O}{C}}CH_3 + NH_3 \ \rightleftharpoons\ PhCH_2\underset{\textstyle O}{C}HCH_3 + NH_2^- \qquad (129)$$

The increase of the alcohol/ketone ratio from the left to the right of the series therefore derives from the fact that the RNu·⁻ + RX electron transfer becomes slower and slower in the same direction, resulting in a less and less efficient formation of the ketone as compared to the alcohol.

The consequences of fragmentation of the anion radical $RNu\cdot^-$ in the course of the substitution process have been actively investigated and shown to be qualitatively consistent with the $S_{RN}1$ mechanism. (A detailed discussion of this can be found in Rossi, 1982; Rossi and Rossi, 1983.) The bond that breaks in $RNu\cdot^-$ may be one belonging originally to the substrate or to the nucleophile. In the first case, the question that arises is that of mono- versus di-substitution. This has been shown, on semi-quantitative grounds, to depend primarily on the competition between the rates of fragmentation and of electron transfer shown in Scheme 13. A recent quantitative analysis of the problem (Amatore et al., 1989) has shown that electron transfer should be considered not only to YRX, but also to YRNu, leading to two imbricated propagation chains.

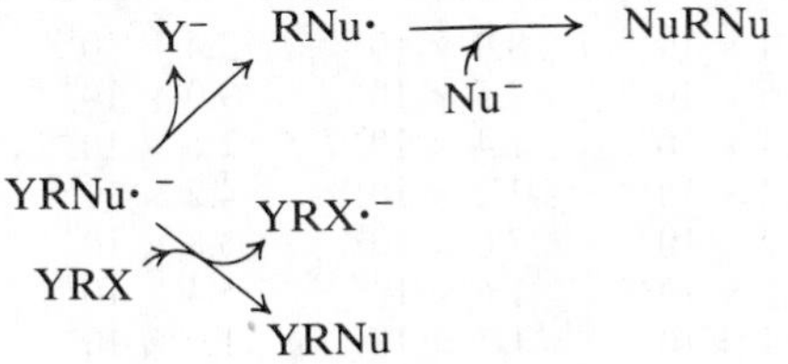

$$\begin{array}{ccc}
Y^- \quad RNu\cdot \quad \xrightarrow{\qquad} \quad NuRNu \\
\nearrow \diagup \qquad \uparrow \\
\diagup \qquad Nu^- \\
YRNu\cdot^- \quad YRX\cdot^- \\
\nearrow \diagdown \\
YRX \qquad \diagdown \\
\qquad YRNu
\end{array} \tag{130}$$

Scheme 13

COUPLING OF ARYL RADICALS WITH NUCLEOPHILES AS A CONCERTED ELECTRON-TRANSFER–BOND-FORMING REACTION

Several methods have been employed to extract the rate constant of the addition of nucleophiles to the aryl radicals from the kinetics of $S_{RN}1$ reactions. Relative reactivities of two nucleophiles towards the same aryl radical have been obtained from the ratio of the two substitution products after preparative-scale reaction of the substrate with a mixture of the two nucleophiles under photochemical or solvated-electron induction (Galli and Bunnett, 1981).

A quick and non-destructive method for measuring relative reactivities of two nucleophiles towards the same aryl radical derives from repetitive cyclic voltammetry of a solution containing the substrate and the two nucleophiles (Amatore et al., 1985b). The ratio of the peak heights corresponding to the two substitution products provides straightforwardly the ratio of the two rate constants.

On the other hand, analysis of the cyclic voltammetric data pertaining to the direct and indirect electrochemical induction of $S_{RN}1$ reactions by means of the appropriate working curves, as discussed in the last subsection, allows the determination of k_{Nu} (Amatore *et al.*, 1985b, 1986a, and references cited therein), leading to absolute values. The currently available data (in $M^{-1} s^{-1}$), obtained by one or the other of these electrochemical methods, and pertaining to the same conditions, i.e. liquid ammonia + 0.1 M KBr at $-40°C$, are listed in Table 1. A few additional data pertaining to the nucleophiles PhSLi, PhSeLi and PhTeLi in CH_3CN have appeared very recently (Degrand and Prest, 1990b).

Table 1   Rate constants ($M^{-1} s^{-1}$) for the reaction of nucleophiles with aryl radicals in liquid ammonia at $-40°C$.

Aryl radical	Nucleophile			
	PhS$^-$	(EtO)$_2$PO$^-$	CH$_2$COCH$_2$$^-$	CN^{-1}
2-Cyanophenyl	1.4×10^{10}	8.0×10^{10}	2.4×10^{10}	9.6×10^8
3-Cyanophenyl	1.5×10^{10}	7.5×10^9	9.0×10^9	5.0×10^7
4-Cyanophenyl	3.4×10^9	1.4×10^9	2.6×10^9	3.3×10^7
1-Naphthyl	2.0×10^{10}	3.2×10^{10}	4.2×10^{10}	1.0×10^6
3-Quinolyl	1.9×10^9	7.6×10^8	3.8×10^9	—
4-Quinolyl	3.2×10^9	1.6×10^9	5.4×10^9	6.0×10^7
2-Quinolyl	1.4×10^7	1.7×10^7	1.0×10^8	2.0×10^6
3-Pyridyl	1.0×10^{10}	7.0×10^9	1.6×10^{10}	—
2-Pyridyl	1.0×10^8	$>10^8$	—	—
Phenyl	2.6×10^7	3.8×10^8	2.7×10^8	$>10^8$

A large number of these values are close to the diffusion limit. This is not actually very surprising since the coupling of the aryl radical with the nucleophile has to compete with quite rapid side-reactions, if only its electron-transfer reduction, for the substitution to be effective. When taking place homogeneously, the latter reaction itself at the diffusion limit and the parameter that governs the competition is $k_{Nu}[Nu^-]/k_D[RX]$. This is the reason why a discussion of structure–reactivity relationships is necessarily restricted to a rather narrow experimental basis.

The addition of the nucleophile to the aryl radical is the reverse of the cleavage of substituted aromatic anion radicals that we have discussed in Section 2 in terms of an intramolecular concerted electron-transfer–bond-breaking process and illustrated with the example of aryl halides. The present reaction may thus be viewed conversely as an intramolecular concerted electron-transfer–bond-forming process. The driving force of the reaction can be divided into three terms as in (131). The first of these, the

$$-\Delta G^0_{R\cdot + Nu^- \rightleftharpoons RNu^{\cdot-}} = \Delta G^0_{RNu \rightleftharpoons R\cdot + Nu\cdot} + E^0_{RNu/RNu^{\cdot-}} - E^0_{Nu\cdot/Nu^-} \qquad (131)$$

free energy of dissociation of RNu, does not vary much with the nature of R as far as aryls are concerned. Thus, for a series of aryl radicals opposed by the same nucleophile, the driving force of the addition reaction increases with the standard potential of the $RNu/RNu\cdot^-$ couple: the easier the introduction of an electron in RNu is, the stronger the driving force. The rate constant of the addition reaction is expected to increase accordingly with a symmetry factor of the order of 0.5, as discussed earlier for the reverse reaction (Section 2, pp. 52–54). It is noteworthy that among the data listed in Table 1, the phenyl radical exhibits the lowest reactivity as compared to the other aryls (with the exception of the 2-quinolyl and 2-pyridyl radicals), whatever the nucleophile. Phenyl derivatives have standard potentials more negative than those of the other aryl derivatives listed in the table (Amatore *et al.*, 1985b, 1986a), either because of the presence of electron-withdrawing groups or of fused aromatic rings both of which decrease the energy of the π^*-orbital. The low reactivity of the phenyl radical is thus an illustration of the above rule of thumb: the easier it is to inject an electron into the substituted product, the faster is the addition of the nucleophile to the aryl radical. The deviation from this rule exhibited by the 2-quinolyl and 2-pyridyl radicals has been rationalized in terms of electronic *ortho*-effects (Amatore *et al.*, 1985b).

There are no large differences between the reactivities of PhS^-, $(EtO)_2PO^-$ and $CH_3COCH_2^-$ with the same aryl radical, but CN^- appears to be significantly less reactive. It is not easy to evaluate the respective role of the bond dissociation free energy and of the $Nu\cdot/Nu^-$ standard potential in equation (13) in this connection because of the paucity of available data concerning these two quantities. An explanation of the low reactivity of CN^- should thus await the availability of such data as well as that of a precise expression of the intrinsic barrier in a model of these intramolecular concerted electron-transfer–bond-breaking (or forming) reactions.

ADDITIONAL REMARKS

It is remarkable that the initial description of the $S_{RN}1$ mechanism, although based on qualitative evidence, has so successfully passed the test of time and of quantitative investigations. The latter have allowed a more precise description of the kinetics and the assignment of the termination steps, but the core of the mechanism, i.e. the propagation loop (103), has essentially remained untouched (see, however, the discussion at the end of this section).

A different mechanism for the $S_{RN}1$ reaction has been proposed in an attempt to unify conceptually the S_N2, $S_{RN}1$ and S_N1 mechanisms (Shaik, 1985). As sketched in Scheme 14, one electron is transferred, thermally or

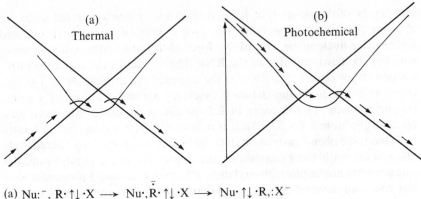

(a) Nu:⁻, R·↑↓·X ⟶ Nu·,R̄·↑↓·X ⟶ Nu·↑↓·R,:X⁻

(b) Nu:⁻, R·↑↓·X ⟶ Nu·,R̄·↑↓·X ⟶ Nu·↑↓·R,:X⁻

Scheme 14

under photochemical induction, from the nucleophile to the halide, yielding, in an intermediate stage, the radical Nu· and the anion radical RX·⁻. The latter would then cleave off the halide ion and the resulting radical R· would simultaneously couple with the radical Nu·, yielding the final substitution product. This mechanism seems difficult to reconcile with essential experimental features of the $S_{RN}1$ reaction, not least the chain character of the reaction. This implies the formation of RNu·⁻ as a necessary intermediate for the propagation of the chain. The essential role of the nucleophile in $S_{RN}1$ reactions is not to provide electrons to the system but rather to couple with R· and thus produce RNu·⁻. This is attested by the results of entrainment experiments, where the nucleophile by which the leaving group is substituted is different from the nucleophile that provides one electron to initiate the reaction, and also by the very existence of electrochemical induction of the reactions as well as its induction or stimulation by reversible one-electron donors. Furthermore, as discussed in the following, there are cases where RX·⁻ is not an intermediate along the reaction pathway, whereas RNu·⁻ is still formed and carries on the propagation loop through its reaction with the starting halide. In other words, although it would seem conceptually attractive to unify the S_N2 and $S_{RN}1$ mechanisms in the common scheme (132), the above-sketched mechanism, although not inconceivable, does not

$$RX + Nu^- \xrightarrow{\ a\ } RX^{·-} + Nu^· \xrightarrow{\ b\ } R^· + Nu^· + X^- \xrightarrow{\ c\ } RNu + X^- \qquad (132)$$

[S_N2: a, b, c concerted; "$S_{RN}1$": a, b, c successive or possibly (a, b concerted), c successive]

correspond to the chain reactions commonly designated as $S_{RN}1$ reactions and has not so far been experimentally documented. Indeed, each time the anion radical of the substrate has been proven to be an intermediate, as in the case of aromatic and nitrobenzylic substrates, it has been shown that the substitution process involves its decomposition into the R· which then reacts with the nucleophile yielding RNu·⁻, which starts the $S_{RN}1$ propagation loop upon reacting with RX. Conversely, each time step c in the above reaction scheme has been shown to take place, it happened to follow a dissociative electron-transfer step, not going through RX·⁻, or else to occur in concert with it (see Section 4). The closest experimental example is found in the reaction of 4-nitrobenzyl halides with cobalt(II) Schiff's base complexes [Section 4, reactions (160–162)], a non-chain process. However, even then, the reaction does not quite follow the above mechanism since the coupling step does not involve R· and Nu· but rather R· and Nu⁻.

As mentioned earlier, the "$S_{RN}2$" mechanism sketched in Scheme 11 has been suggested to explain the effect of a change of leaving group on product distribution in the reaction of 2-substituted 2-nitropropanes with enolate ions (Russell et al., 1981, 1982a,b). It has been proposed that the bimolecular substitution step (133) would involve, rather than S_N2 substitution at the

$$RX \cdot^- + Nu^- \longrightarrow (Nu \cdot, R^-, X^-) \longrightarrow RNu \cdot^- + X^- \qquad (133)$$

functional carbon of RX·⁻, its dissociative electron-transfer reduction by the nucleophile, followed by cage recombination of the Nu· and R⁻. This would require that RNu·⁻ be very easy to reduce. Examples of such a possibility have been reported in the reaction of 1,2- and 1,4-dinitrobenzenes with hydroxide ions, in which OH⁻ appears to react with the dinitrobenzene anion radical before the cleavage of the nitrite ion (Abe, 1973; Abe and Ikegamo, 1976, 1978). Additional investigations seem desirable, however, to establish firmly the occurrence of this mechanism and its scope.

Based on photostimulation and radical trapping experiments and on the observation of R–Ph scrambling in several cases, an $S_{RN}1$ mechanism has been proposed for the reaction of 1-haloadamantanes with PhS⁻, PhSe⁻, PhTe⁻, Ph_2P^-, Ph_2As^-, Se_2^{2-} and Se_2^{2-} (Palacios et al., 1985; Rossi, 1982, Rossi et al., 1982, 1984a), of neopentyl bromide with PhS⁻, Ph_2P^- and Ph_2As^- (Pierini et al., 1985), of 7-bromonoracane with Ph_2P^- and Ph_2As^- (Rossi et al., 1984b), and of 9-bromo- and 9,10-dibromo-triptycene with Ph_2P^- (Rossi et al., 1984a). The usual $S_{RN}1$ mechanism in which RX·⁻ is an intermediate in the propagation loop (103) and in the initiation step, was invoked for these reactions. However, there is little doubt, by reference to the behaviour of other aliphatic halides towards single electron donors (Section 2, pp. 54–56), that the carbon–halogen bond undergoes dissociative electron transfer in these compounds under the same reaction conditions, and thus that the reaction does not go through $RX\cdot^-$. It is thus much more likely to follow the mechanism in Scheme 15.

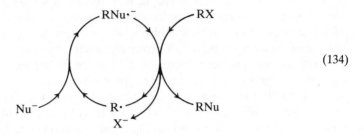

$$(134)$$

Scheme 15

Similarly, several examples of reactions of perfluoroalkyl halides have been demonstrated to follow an $S_{RN}1$-type mechanism (Section 3, p. 75). Since, here too, dissociative electron transfer is likely under the conditions of the reactions [Section 2, pp. 54–56 (Andrieux *et al.*, 1990b)], the substitution process most probably also follows mechanism (134) rather than (103). The same is also likely to be true with alkyl mercurials (Russell, 1989).

This nuance of the original $S_{RN}1$ mechanism may thus occur in quite a number of cases. Nomenclature purists may consider it necessary to find other symbols to name this mechanism and, presumably, to question the adequacy of the "1" in this case. Beyond symbols, if the $S_{RN}1$ mechanism is viewed as an "outer sphere electron-transfer-induced nucleophilic substitution", a possible designation of the mechanism under discussion might be "dissociative electron-transfer-induced nucleophilic substitution". The original designation of these reactions as "nucleophilic reactions proceeding via anion radical intermediates" (Kornblum, 1975) would still apply to both nuances of the mechanism since, in the present case, $RNu\cdot^-$ is an essential intermediate in the reaction, even if $RX\cdot^-$ is not.

4 S_N2 Substitution versus single electron transfer

WHAT IS THE PROBLEM?

The question we address now is that of the possible role of single electron transfer in substitution reactions that, unlike $S_{RN}1$ reactions, are not catalysed by electron injection. The problem is twofold. One side of it consists in answering the questions: do bond breaking and bond formation belong to two different and successive processes, i.e. (135) *followed* by (136), or, more

$$RX + D^-(Nu^-) \longrightarrow [RX^{\cdot-} + D{\cdot}(Nu{\cdot})] \longrightarrow R{\cdot} + X^- + D{\cdot}(Nu{\cdot}) \qquad (135)$$

$$R{\cdot} + D{\cdot}(Nu{\cdot}) \longrightarrow RD(RNu) \qquad (136)$$

likely, in the case where the electron donor (nucleophile) is an ion radical (137), *followed* by (138); or do bond breaking and bond formation occur concertedly as in (139)? In reactions (135) and (137), electron transfer and

$$RX + D^{\cdot-}(Nu^{\cdot-}) \longrightarrow [RX^{\cdot-} + D(Nu)] \longrightarrow R{\cdot} + X^- + D(Nu) \qquad (137)$$

$$R{\cdot} + D^{\cdot-}(Nu^{\cdot-}) \longrightarrow RD^-(RNu^-) \qquad (138)$$

$$RX + D^-(Nu^-) \longrightarrow RD(RNu) + X^- \qquad (139)$$

bond breaking may be concerted or stepwise (see Section 2). In both cases the nucleophile functions as an outer sphere electron donor. In short, and according to a commonly accepted terminology, the problem can be posed as: does the reaction follow an ET or an S_N2 mechanism?

Another side of the problem, different from the preceding one, is as follows. In the case of a concerted bond-breaking–bond-formation reaction, as represented by (139), i.e. a classical S_N2 reaction, where $R{\cdot}$ is not an intermediate along the reaction pathway, the reaction may be thought as an inner sphere single electron transfer (Lexa *et al.*, 1981). The process is inner sphere in the sense that bond breaking and bond formation are concerted with the transfer of one electron. This formal equivalence between S_N2 and inner sphere electron-transfer mechanisms has more recently been discussed in greater detail, based on valence bond configuration mixing, using "electron shift" rather than inner sphere electron transfer as a terminology (Pross and Shaik, 1982a,b, 1983; Pross and McLennan, 1984; Pross, 1985). If the two resonant forms (valence bond configurations) represented in (140)

$$^-Nu : R{\cdot} \cdot X \longleftrightarrow Nu{\cdot} \cdot R : X^- \qquad (140)$$
$$\text{(a)} \qquad\qquad\qquad \text{(b)}$$

predominate not only in the reactant and product states [(a) for the reactant and (b) for the product] but also in the transition state, which would thus be a resonance combination of the two, then the reaction is better described as a single "electron shift" from Nu^- to $X{\cdot}$ than by the classical electron-pair shift notion. With methyl halides, this appears to be a quite reasonable qualitative picture of the reaction. Consideration of additional resonant forms, such as those in (141), to be mixed with (a) and (b), allows the qualitative interpretation of various substituent effects.

$$^-\mathrm{Nu:R}^+ :\mathrm{X}^- \quad \text{and} \quad \mathrm{Nu}\cdot \, ^-\mathrm{R:}\cdot \mathrm{X} \qquad\qquad (141)$$
$$\quad\text{(c)} \qquad\qquad\qquad \text{(d)}$$

This conception of an S_N2 reaction as an electron-shift process is obviously equivalent to its conception as an inner sphere electron transfer, i.e. a single electron transfer concerted with the breaking of the R—X bond and the formation of the R—Nu bond. Faced with an experimental system, however, the first question—ET or S_N2?—still remains, whatever intimate description of the S_N2 reaction one may consider most appropriate. If this is thought of in terms of inner sphere electron transfer, the question thus raised is part of the more general problem of distinguishing outer sphere from inner sphere electron-transfer processes (Lexa et al., 1981), an actively investigated question in other areas of chemistry, particularly that of coordination complex chemistry (Taube, 1970; Espenson, 1986).

S_N2 SUBSTITUTION VERSUS SINGLE OUTER SPHERE
DISSOCIATIVE ELECTRON TRANSFER

Three main sources of information are available for solving the ET versus S_N2 problem, namely, comparative kinetic studies, stereochemistry and cyclizable radical-probe experiments.

Among comparative kinetic studies, the "kinetic advantage" method has been used systematically in several cases. It has been developed for the first time for investigating the ET versus S_N2 problem in the reaction of iron(I) and cobalt(II) porphyrins with primary butyl halides (Lexa et al., 1981), yielding the corresponding σ-butyl-iron(III) and cobalt(III) complexes according to the overall reaction (142).

$$\mathrm{Fe,Co(I)Porph}^- + \mathrm{RX} \xrightarrow{\ k\ } \mathrm{RFe,CO(III)Porph} + \mathrm{X}^- \qquad (142)$$

[In the case of iron, the Fe(III) alkyl complex is reduced, in a successive step, into the Fe(II) alkyl complex by the starting iron(I) porphyrin, whereas this reaction does not occur with cobalt. This difference is related to the fact that $E^0_{\mathrm{Fe(II)Porph/Fe(I)Porph}^-} < E^0_{\mathrm{RFe(III)Porph/RFe(II)Porph}^-}$ whereas $E^0_{\mathrm{Co(II)Porph/Co(I)Porph}^-} > E^0_{\mathrm{RCo(III)Porph/RCo(II)Porph}^-}$.] The principle of the method is illustrated in Fig. 18 with the example of reaction (142). The rate constants obtained with the investigated nucleophiles (or with single electron donors—that is the question!) are compared to those of the reaction of a series of anion radicals with the same alkyl halide in the same medium. As discussed on p. 59, aromatic anion radicals behave in this reaction as outer sphere electron donors and the alkyl halide undergoes a dissociative electron transfer. For each halide, the $\log k$–$E^0_{P/Q}$ plot thus provides a reference activation–driving force relationship characterizing dissociative electron transfer to the alkyl halide from outer sphere electron donors that possess the property that their intrinsic contribution to the activation free energy is small and practically constant in the series. The driving force of the reaction is $E^0_{RX/RX\cdot^-} - E^0_{P/Q}$, but the method does not require a knowledge of $E^0_{RX/RX\cdot^-}$ since the comparison is made for each alkyl halide individually and consists in seeing how much the rate constant of the investigated nucleophile, here an iron(I) or a cobalt(I) porphyrin, differs from that of the aromatic anion radical that would have the same standard potential. In the present case, the standard potential of the nucleophile investigated is readily obtained from the cyclic voltammetry of the iron(III) or cobalt(II) porphyrin from which the iron(I) or cobalt(I) nucleophiles are generated electrochemically and which exhibits a

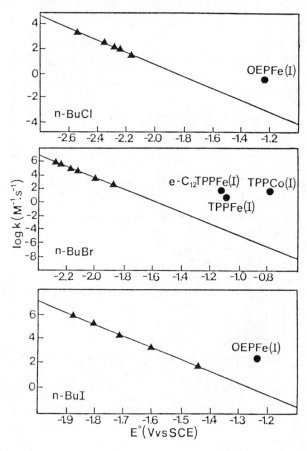

Fig. 18 *Cont.*

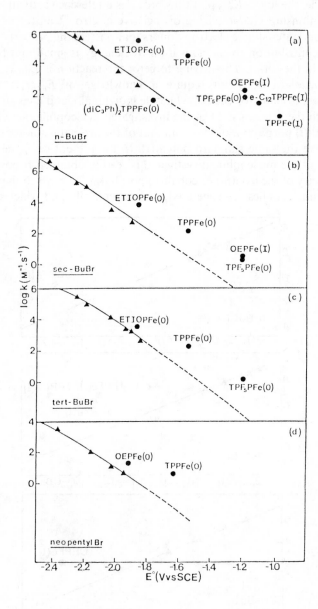

Fig. 18 *Cont.*

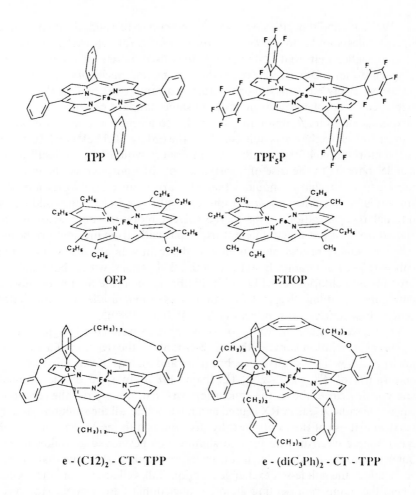

Fig. 18 Rate constants for the reaction of electrochemically generated iron(0), iron(I) and Co(I) porphyrins (●) and aromatic anion radicals (▲) with aliphatic halides as a function of their standard potentials. (Adapted from Lexa *et al.*, 1981, 1988.)

perfectly reversible Fe,Co(II)/Fe,CO(I)⁻ wave. As seen in Fig. 18(a), the rate constants for the iron(I) and cobalt(I) porphyrins fall well above the outer sphere reference line, thus pointing to the occurrence of an inner sphere mechanism in which the transition state is stabilized by bonding interactions as compared to that of the concerted outer sphere electron-transfer–bond-breaking process taking place with the aromatic anion radicals. It was concluded that this inner sphere mechanism is an S_N2 mechanism (Lexa *et al.*, 1981). It should be noted that for the method to be valid, the investigated nucleophiles should not have an intrinsic barrier for self-exchange significantly smaller than that of the compounds used as reference outer sphere electron donors. Otherwise, even if the investigated nucleophile behaved as an outer sphere electron donor, its rate constant could be larger than the reference outer sphere electron donor of same standard potential. A distinct advantage of aromatic anion radicals in this connection is that their intrinsic barrier for self-exchange is quite small, of the order of 0.16 eV (Fig. 3), and it is therefore unlikely that other electron donors would have a significantly smaller barrier. In the case of iron(I) and cobalt(I) porphyrins, the intrinsic barrier to self-exchange should actually be larger, since axial ligation by the solvent is broken when passing from Fe,Co(II) to Fe,Co(I)⁻, thus adding an internal reorganization factor to the solvent reorganization factor which should not be itself much different from that with aromatic anion radicals.

Quite similar results have been found recently in the reaction of the cobalt(I) form of vitamin B_{12} ($B_{12}s$) with alkyl halides: with n-butyl iodide, bromide and chloride, ethyl bromide and benzyl chloride the representative data point of vitamin $B_{12}s$ falls several orders of magnitude above the outer sphere dissociative electron-transfer line (Walder, 1989).

The above results concerning iron(I) porphyrins were later integrated into a general correlation between the rate constants of the reaction of a series of nucleophiles with alkyl chlorides, bromides and iodides of various structures and values of $E^0_{RX/RX^{\cdot-}} - E^0_{P/Q}$ (Eberson, 1982). In spite of the original analysis described above, the conclusion was that, apart from the reactions where the same X^- as in RX is used as a nucleophile, all the reactions, and in particular those of the iron(I) porphyrins, conform to the Marcus model for outer sphere electron transfer and should thus be viewed as following an outer sphere dissociative electron-transfer mechanism. It is true that in this correlation, the halide-self exchange reactions fall well apart from the region where the points representing the other nucleophiles are located. However, the correlation between the latter data and the $E^0_{RX/RX^{\cdot-}} - E^0_{P/Q}$ values is very rough [± four orders of magnitude scatter in most of the investigated domain (see Fig. 6 in Eberson, 1982)], much too rough for a sound conclusion to be reached. These views have now been modified (Eberson, 1987), reaching a tacit agreement with the earlier analysis of the reaction. It should also be noted that the Marcus theory of outer sphere electron transfer is not suited to these cases, since, even for those of the investigated nucleophiles that would actually behave as outer sphere electron donors, the acceptors react in an inner sphere manner.

More data were gathered later concerning the reaction of iron(I) and also iron(0) porphyrins with various aliphatic bromides (Lexa *et al.*, 1988). In the case of iron(0) porphyrins, the σ-alkyl-iron(II) complex is obtained directly

$$Fe(0)Porph^{2-} + RX \xrightarrow{k} R(\text{II})Porph^{-} + X^{-} \tag{143}$$

as in (143). As seen in Fig. 18, with n-butyl bromide, the porphyrin data points fall well above the outer sphere dissociative electron-transfer reference line, pointing to the occurrence of an inner sphere mechanism. Increasing steric hindrance at the carbon reacting centre in the n-, s-, t-isomer series and the blocked structure of neopentyl bromide decrease the kinetic advantage of the porphyrins over the aromatic anion radicals, practically cancelling it in the case of ETIOP and OEPFe(0). Steric hindrance at the iron centre also has a dramatic effect: the data point for the e-(diC$_3$Ph)$_2$TPPFe(0) falls on the outer sphere reference line in accord with the fact that the rotation of the two phenyl groups in the basket-handle chains severely inhibits bonded interactions in the transition state, although the σ-alkyl-iron complex is nevertheless eventually obtained (Gueutin et al., 1989).

All these observations point to the occurrence of a S_N2 rather than an outer sphere, dissociative electron-transfer mechanism in cases where steric constraints at the carbon or metal reacting centres are not too severe. It is, however, worth examining two other mechanistic possibilities. One of these is an electrocatalytic process of the S_{RN}-type that would involve the following reaction sequence. If, in the reaction of the electron donor (nucleophile), D_{red}, the bonded interactions in the transition state are vanishingly small, the alkyl radical is formed together with the oxidized form of the electron donor, D_{ox}. Cage coupling (144) may then occur, if their mutual affinity is

$$D_{red} + RX \xrightarrow{k_c} (D_{ox} + R\cdot + X^-) \longrightarrow RD_{ox} + X^- \tag{144}$$

very large, yielding the oxidized form of the substitution product and thus giving rise to the ET mechanism considered so far. If not, R· diffuses out of the solvent cage and may then react with other D_{ox} molecules (145), yielding

$$D_{ox} + R\cdot \xrightarrow{k_c} RD_{ox} \tag{145}$$

the oxidized form of the substitution product, RD_{ox}, again following an ET mechanism. RD_{ox} may or may not be further reduced (146) according to the

$$RD_{ox} + D_{red} \rightleftharpoons RD_{red} + D_{ox} \tag{146}$$

value of the difference $E^0_{RD_{ox}/RD_{red}} - E^0_{D_{ox}/D_{red}}$, between the standard potentials of the two couples. But, out of the solvent cage, R· may also react with the reduced form of the electron donor (147) rather than with its oxidized form, thus yielding the reduced form of the substitution product. If this is able to

$$D_{red} + R \cdot \xrightarrow{k_c'} RD_{red} \tag{147}$$

reduce RX as in (148) faster than the reduced form of the starting electron donor (the nucleophile), a chain process of the S_{RN} type, quite similar to that

$$RD_{red} + RX \xrightarrow{k_e'} RD_{ox} + R \cdot + X^- \tag{148}$$

shown in Scheme 15, is opened up. Thus if k_c is below the diffusion limit and is smaller than k_c', and, if at the same time k_e' is larger than k_e, the overall rate constant would be larger than expected for a simple ET mechanism due to the amplification effect of the chain process.

In the case of the iron and cobalt porphyrins discussed above, $E^0_{RD_{ox}/RD_{red}}$ $-E^0_{D_{ox}/D_{red}} \ll 0$ with cobalt(I) and iron(0) whereas the opposite is true for iron(I). $E^0_{RD_{ox}/RD_{red}} - E^0_{D_{ox}/D_{red}}$ is the difference in driving forces between reactions (147) and (145) and also that between reactions (144) and (148). Thus, $k_c > k'_c$ for the isoelectronic Co(I) and Fe(0), which matches the radical character of Co(II) and Fe(I) (Lexa et al., 1981), and $k_e > k'_e$ for Fe(I); hence the occurrence of the chain mechanism is unlikely in all cases. In the case of iron(0) and iron(I), this conclusion falls in line with the observed effect of steric constraints which should not appear in this outer sphere, dissociative electron-transfer-induced chain mechanism.

Another possibility is halogen atom transfer (149), again generating the

$$D_{red} + RX \longrightarrow XD_{ox} + R \cdot \tag{149}$$

alkyl radical that could then couple with D_{ox} or D_{red}. Bonded interactions in the transition state would again tend to favour this mechanism over the outer sphere dissociative electron-transfer pathway. The observation that the rate constant for each iron(0) and iron(I) porphyrin decreases when passing from n- to s- to t-alkyl halides, however, does not support this possibility since the opposite would be expected. It is not, however, excluded that the remaining extra reactivity over outer sphere, dissociative electron transfer observed for s- and t-butyl halides, especially with the weakest reducing porphyrins, is caused by some participation of this mechanism. Note, however, that there is no driving force for this reaction in the case of iron(0) since the resulting iron(I) porphyrin is a four-coordinated complex, and that the driving force in the case of iron(I) is rather weak (see Lexa et al., 1990, and references therein).

The kinetic advantage method, again using aromatic anion radicals as reference ET reagents, combined with the investigation of steric constraints at the electrophilic carbon centre, has also been used for examining the reaction of alkyl halides with several purely organic nucleophiles. The reactions in Scheme 16 have been investigated (Lund and Lund, 1986, 1987, 1988). The same trends as with the iron porphyrins were found: with primary halides, the data points fall well above the outer sphere reference

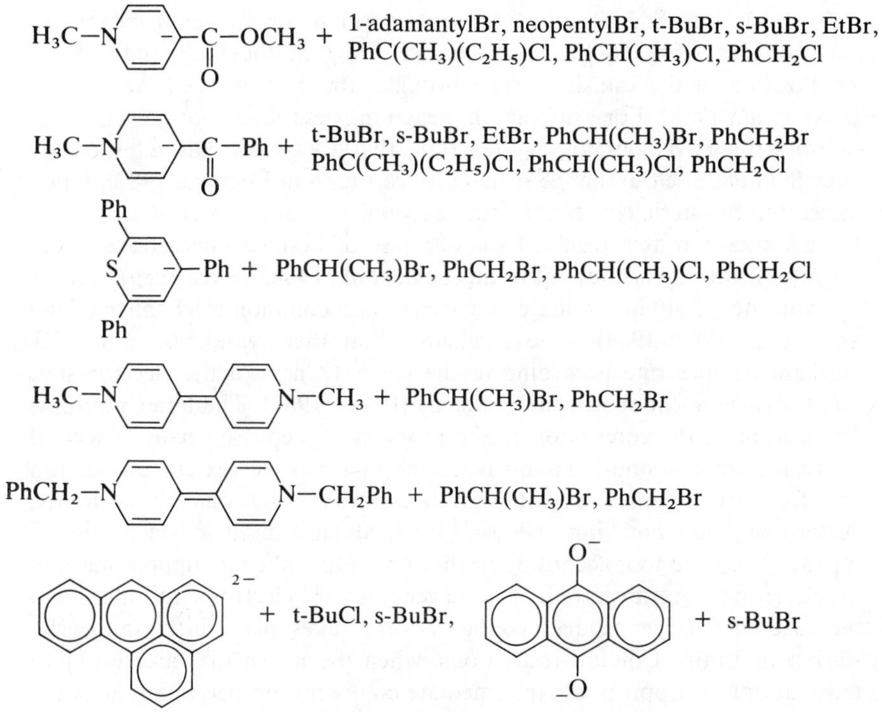

H_3C-N⟨ ⟩$-C-OCH_3$ + 1-adamantylBr, neopentylBr, t-BuBr, s-BuBr, EtBr,
‖
O PhC(CH_3)(C_2H_5)Cl, PhCH(CH_3)Cl, PhCH_2Cl

H_3C-N⟨ ⟩$-C-Ph$ + t-BuBr, s-BuBr, EtBr, PhCH(CH_3)Br, PhCH_2Br
‖
O PhC(CH_3)(C_2H_5)Cl, PhCH(CH_3)Cl, PhCH_2Cl

Ph
⟩
S ⟨ ⟩$-Ph$ + PhCH(CH_3)Br, PhCH_2Br, PhCH(CH_3)Cl, PhCH_2Cl
⟩
Ph

H_3C-N⟨ ⟩=⟨ ⟩$N-CH_3$ + PhCH(CH_3)Br, PhCH_2Br

$PhCH_2-N$⟨ ⟩=⟨ ⟩$N-CH_2Ph$ + PhCH(CH_3)Br, PhCH_2Br

[pyrene dianion]$^{2-}$ + t-BuCl, s-BuBr, [anthraquinone dianion, O⁻ ... O⁻] + s-BuBr

Scheme 16

line, while steric constraints at the electrophilic carbon bring them closer and closer to the line. Very similar observations were also made for the reaction of benzyl and diphenylmethyl halides with 9-substituted fluorene anions, where the effect of steric hindrance at both the electrophilic and nucleophilic carbon reacting centres was investigated (Bordwell and Hughes, 1983; Bordwell et al., 1987; Bordwell and Wilson, 1987; Bordwell and Harrelson, 1987, 1989).

Restricting ourselves for the moment to energy factors, all the observations described above converge towards the following picture. As soon as bonded interactions exist in the transition state, the reaction is of the S_N2 type since the activation energy is less than for the ET mechanism. Borderline cases may exist insofar as the corresponding gain in activation energy is too small to be detected experimentally. In this connection, it is remarkable that nucleophiles such as the above low-valent metalloporphyrins and

Lund's organic nucleophiles, which are all members of perfectly reversible redox couples and might therefore have been deemed to behave as outer sphere single electron donors, in fact react in an S_N2 fashion in the absence of too severe steric constraints.

It is true that S_N2 situations may be closer to or farther from the ET situation according to the case, i.e. according to the magnitude of the stabilization of the transition state brought about by bonded interactions (Lexa et al., 1988). For example, the reactions described above, even in the case of primary halides, are S_N2 reactions that are closer to an ET situation than the halide self-exchange reactions (see Fig. 6 in Eberson, 1982). Upon increasing the steric constraints, the reaction will come closer and closer to the ET situation as a result of the decrease of bonded interactions in the transition state. This does not mean (Lexa et al., 1988; Lewis, 1989) that the S_N2 and the ET situations are two extremes of a common mechanism (Lund and Lund, 1986, 1988) or, equivalently, that there would be a S_N2–ET mechanistic spectrum according to the concertedness of the electron-shift (i.e. transfer) and bond-formation steps (Pross, 1985). The latter notion is reminiscent of the conception of the problem of stepwise versus concerted electron-transfer–bond-breaking reactions in terms of "extent of electron transfer" accompanied by, but not necessarily synchronously with, the "extent of bond breaking" (Perrin, 1984) already discussed in Section 2 (pp. 31, 32). Here too, according to the Born–Oppenheimer approximation, the electron to be transferred, or more generally the electronic reshuffle from the reactant to the product configurations, takes place instantaneously, subject to Frank–Condon restrictions, when the nuclei that are slower to move adopt the appropriate intermediate configuration between that of the reactants and that of the products. The description of the nuclear reorganization leads to the introduction of a set of reaction coordinates representing bond breaking, bond formation, more minor changes in bond lengths and angles and solvent reorganization (Marcus, 1968). It is thus difficult to conceive what could be the parameter, the continuous variation of which would cause the passage from a "pure" ET to a "pure" S_N2 situation through a general mechanism combining the characteristics of the two and towards which the ET and S_N2 situations would appear as limiting asymptotic cases. Thus, on energy grounds only, a more realistic description appears to be that the S_N2 mechanism tends towards the ET mechanism as the bonded interactions in the transition state vanish. The extent of the borderline zone, i.e the zone in which the two mechanisms become indistinguishable in terms of energy is of course difficult to estimate with precision. In terms of experimental accuracy of rate determinations, it is of the order of a fraction of a kcal mol^{-1}. One has also to consider that for the ET reaction to proceed in an adiabatic manner, an interaction of the order of 1 kcal mol^{-1} is required (Eberson, 1987).

One essential factor among those governing the stabilization of the S_N2 transition state *vis-à-vis* the ET transition state is of an extrinsic nature, namely, the increase in driving force (DF) resulting from the fact that bond formation is concerted with bond breaking (Lexa *et al.*, 1988) and given by (150)–(152). There may well be changes in the intrinsic barriers upon passing

$$DF(\text{ET}) = E^0_{RX/R\cdot\,+\,X^-} - E^0_{Nu\cdot/Nu^-} = E^0_{X\cdot/X^-} - E^0_{Nu\cdot/Nu^-} - \Delta G^0_{RX\rightleftharpoons R\cdot\,+\,X\cdot} \qquad (150)$$

$$DF(S_N2) = E^0_{RX/R\cdot\,+\,X^-} - E^0_{RNu/R\cdot\,+\,Nu^-} = E^0_{X\cdot/X^-} - E^0_{Nu\cdot/Nu^-}$$

$$- \Delta G^0_{RX\rightleftharpoons R\cdot\,+\,X\cdot} + \Delta G^0_{RNu\rightleftharpoons R\cdot\,+\,Nu\cdot} \qquad (151)$$

$$DF(S_N2) - DF(\text{ET}) = \Delta G^0_{RNu\rightleftharpoons R\cdot\,+\,Nu\cdot} \qquad (152)$$

from the S_N2 to the ET case and, also, in the context of a S_N2 mechanism, when passing from one nucleophile to another. In addition, it is not completely certain, in spite of theoretical (Marcus, 1968) and experimental (Albery and Kreevoy, 1978; Lewis, 1986) justifications for it, that the form of cross-exchange relationship is exactly the same in all cases. One factor that obviously contributes to destabilizing the transition state is steric hindrance at the nucleophilic centre, thus favouring ET over S_N2 without affecting directly the value of $\Delta G^0_{RNu\rightleftharpoons R\cdot\,+\,Nu\cdot}$. In the absence of steric constraints, however, it is likely that changing the driving force scale from $DF(\text{ET})$ to $DF(S_N2)$, for the nucleophiles where $\Delta G^0_{RNu\rightleftharpoons R\cdot\,+\,Nu\cdot} \geq 0$, would bring the representative points of all nucleophiles and outer sphere electron donors much closer to a common line. The lack of $\Delta G^0_{RNu\rightleftharpoons R\cdot\,+\,Nu\cdot}$ data prevents this being done with a large number of nucleophiles for which rate data exist.

It is interesting to note in this connection, that, in the reaction of iron(I) and iron(0) porphyrins with n-butyl bromide, the kinetic data points fall approximately on a common line, roughly parallel to the outer sphere line. This seems to indicate that there is no large variation of the intrinsic barrier to self-exchange in the series. It also shows that the observation of a Marcus-type correlation between the free energies of activation and the standard potentials of the nucleophiles does not necessarily indicate the occurrence of an ET mechanism. Given the halide, making the nucleophile increasingly oxidizable in the thermodynamic sense, i.e. decreasing $E^0_{Nu\cdot/Nu^-}$, favours equally the ET and the S_N2 mechanisms in terms of driving forces unless the free energy of dissociation of the R—Nu bond varies simultaneously. The observation of a Marcus-type correlation may thus result from the near constancy of the bond dissociation free energy in the series [see (151)] on the one hand, and from the applicability of a Marcus-type law to the S_N2 reaction, at least within the same "family" of nucleophiles, on the other.

The same example also shows that the driving force factor is not the sole one to come into play. $\Delta G^0_{RNu \rightleftharpoons R \cdot + Nu \cdot}$ is not known for iron(I) and iron(0) porphyrins separately, but the difference can be estimated to be *ca.* 0.3 eV (from $E^0_{Fe(II)Porph/Fe(I)Porph^-} - E^0_{RFe(III)Porph/RFe(II)Porph^-}$). Using then the $DF(S_N2)$ scale instead of the $DF(ET)$ scale would bring the iron(I) line above the iron(0) line by *ca.* 0.15 eV, seeming to indicate that the weaker bond [Fe(III)—R] requires a smaller intrisic barrier than the stronger bond [Fe(II)—R] to be generated.

However, in the reaction of vitamin $B_{12}s$ with primary halides and with benzyl chloride, where the S_N2 driving force can be calculated from the standard potential of the Co(II)/Co(I) couple and from existing Co—C bond energy data, it appears that recasting the kinetic data point against $DF(S_N2)$ instead of $DF(ET)$ brings it close to the ET line (Walder, 1989).

The standard potential of the nucleophile can be taken as a measure of its "softness": the easier it is to oxidize, the softer the nucleophile. In terms of driving forces, this does not change, by itself, the tendency for the reaction to follow the ET rather than the S_N2 pathway if $\Delta G^0_{RNu \rightleftharpoons R \cdot + Nu \cdot}$ remains constant. What is true, however, when comparing a soft (easy to oxidize) with a hard (difficult to oxidize) nucleophile is that the radical of the latter needs a strong affinity toward R· ($\Delta G^0_{RNu \rightleftharpoons R \cdot + Nu \cdot} \gg 0$) for the two driving forces to be comparable.

In the dichotomy between the ET and the S_N2 pathways, it is often taken for granted that, given the nucelophile, the larger the electron-acceptor ability of the alkyl halide (I > Br > Cl), the larger the predominance of ET over S_N2 (Ashby, 1988; Eberson, 1987; Pross and Shaik, 1983). This might be true but does not rest on driving force effects since an increase of the electron-acceptor ability of the alkyl halide has the same favourable effect on the driving force of the two pathways (152).

Let us come back to the broad classification of nucleophiles into two categories—not too distant from ET and very far from ET—that we have invoked earlier. Even if it is now agreed that nucleophiles belonging to the first category do not necessarily react as ET reagents, the question nevertheless arises of the cause of this apparent dichotomy. In other words, is the lack of in-between nucleophiles a reflection of the lack of experimental data or is it due to more fundamental causes? It is noteworthy that the oxidized form of most nucleophiles belonging to the first category is stable, allowing the reduced form to be electrochemically generated *in situ* from them in a number of cases, whereas, in the second category, Nu· is, in a broad sense, unstable. It is therefore not too surprising that it exhibits a large affinity toward R·. In the latter case the weak reducing properties of the nucleophile are compensated, at least partly, by a large value of $\Delta G^0_{RNu \rightleftharpoons R \cdot + Nu \cdot}$. Along these lines it would not seem impossible to find nucleophiles falling within the gap.

Analysis of the transition state in terms of energy is certainly a key aspect of the S_N2–ET problem. Entropy considerations may, however, bring about additional information, possibly helping us to conceive better the transition between the two mechanisms. It was observed in this connection that, whereas the entropy of activation of both the anthracene anion radical and of the ETIOPFe(0) porphyrin (pp. 99, 100) (which have about the same standard potential) is close to zero in their reaction with s- and t-butyl bromides a definitely negative value, *ca.* -20 eu is obtained for the reaction of the porphyrin with n-butyl bromide (Lexa *et al.*, 1988). The same was found for the reaction of two other iron porphyrins, TPPFe(0) and OEP-Fe(I). These activation entropies were estimated from (153), where Z is

$$k = Z \exp\left(-\frac{\Delta H^{\neq}}{RT} + \frac{\Delta S^{\neq}}{R} \right) \qquad (153)$$

obtained, in the framework of the activated complex model, from the ratio of the partition function of the activated complex to the product of the partition functions of the reactants, all being approximated by equivalent hard spheres, which amounts to estimating Z as the bimolecular collision frequency. An approximate calculation showed that the negative value found in the t-BuBr/ETIOPFe(0) case is compatible with a geometry of the transition state in which the Fe, C and Br atoms are aligned. This suggests the reaction scheme sketched in Fig. 19. Bonded interactions in the transition state, leading to the S_N2 reaction with inversion of configuration, appear only when the three centres are aligned or close to being aligned, whereas the outer sphere, dissociative single electron transfer requires a much less precise relative orientation of the reactants to take place. The activation energy would then be a function of the angle between the three centres, having its minimum when they aligned. For clarity, this conception is pushed to the extreme in Fig. 19 where it is considered that the S_N2 pathway requires strict alignment of the three centres, whereas the outer sphere transfer does not suffer any geometrical requirement (i.e. is adiabatic in all directions) leading to a steep increase of the activation energy up to a plateau independent of the reaction angle. This is most probably an exaggerated picture and it is likely that the outer sphere pathway does possess some degree of orientational preference depending on the nature and symmetry of the nucleophile orbital where the electron to be transferred resides and which has to mix with the σ^*-orbital of the RX acceptor.

According to this picture, it is possible to restore the notion of a mechanistic spectrum of ET and S_N2 processes that the sole consideration of energetic factors did not permit. Within this framework, there is no real difference between the notions of comptetition between the two mechanisms

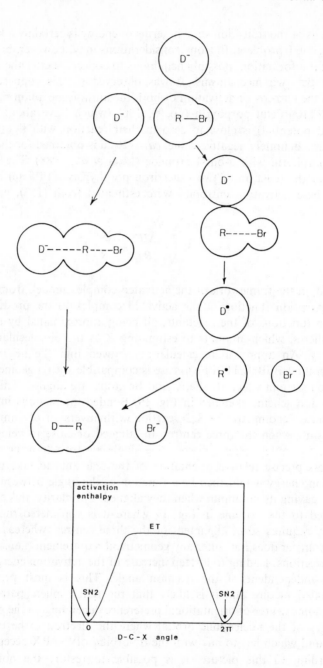

Fig. 19 S$_N$2 versus outer sphere, dissociative single electron transfer (in the case where this is adiabatic in all directions). (Adapted from Lexa *et al.*, 1988.)

and transition from one to the other. When bonded interactions in the transition state are large, the S_N2 pathway will predominate because the resulting small activation energy will overcompensate the unfavourable negative activation entropy. Conversely, its near-zero activation entropy will help the ET pathway to predominate when bonded interactions in the transition state are weak. When this is the case, the alkyl radical may couple, after loss of stereochemical integrity, inside the solvent cage or escape from the solvent cage and undergo the same reaction or other reactions. (In the case of aromatic anion radicals, coupling of the alkyl radical with the aromatic anion radical rather than with the aromatic hydrocarbon is likely, as discussed in Section 2.) According to this model, temperature appears as one determining factor for passing from one mechanism to the other, the general tendency being that the S_N2 pathway should be favoured over the ET pathway going to low temperatures and vice versa. Of course, most experimental systems will fall in one category or the other, but it is conceivable that particular systems could exist where this temperature-dependent transition might be observed. This was deemed to occur in the reaction of anthracene anion radical with n-BuBr where a biphasic Arrhenius plot was obtained upon determination of the rate constants by linear scan voltammetry (Lexa et al., 1988). However, it appears that this behaviour is not found when using cyclic voltammetry (Lund and Pedersen, 1989). Until the origin of the discrepancy is unravelled and while attempting to improve the statistical and systematic precision of the techniques, this example should not be regarded as fully established. Other examples ought to be sought, bearing in mind, however, that the transition is unlikely to be easy to observe within accessible temperature ranges since this requires the difference in activation energies to be small. It remains that the two limiting types of behaviour, corresponding to ET and S_N2, respectively, have been clearly observed, beyond experimental uncertainty, in the reaction of butyl bromides with low-valent iron porphyrins and the anthracene anion radical.

The above considerations should bear some relationship with the stereochemistry of the reaction. As indicated earlier (Section 2; Hebert et al., 1985), in the reaction of anthracene anion radicals with optically active 2-octyl bromide, racemization is mostly observed together with a small but distinct amount (ca. 10%) of inversion. In the context of the ET–S_N2 mixed mechanism sketched above, this can be rationalized in terms of a minor contribution of the latter pathway that would not detectably affect the overall rate constant of the reaction. The weakness of the bonded interactions in the transition state derives from the relatively poor affinity of the alkyl radical for the aromatic hydrocarbon. This is consistent with the fact that in those of the radical–anthracene pairs that were not favourably oriented for the S_N2 reaction to occur, the alkyl radical escapes from the solvent cage and couples with an anthracene anion radical rather than coupling with anthracene inside the cage. The latter is shown by a continuity argument, viz., with other more reducing, aromatic anion radicals, electron transfer reduction of the alkyl radical swamps coupling (Section 2; Andrieux *et al.*, 1986a). It is confirmed by cyclizable radical-probe experiments in the reaction of the anthracene anion radical with 6-bromo-1-hexene (Daasbjerg *et al.*, 1989); the ratio of uncyclized to cyclized coupling products in Scheme 17 was found to increase with the concentration of the anthracene anion radical as a result of the competition between the cyclization of the linear chain hexenyl radical and its coupling with the anthracene anion radical, whereas such a concentration dependence should not show up if cage coupling with anthracene occurred.

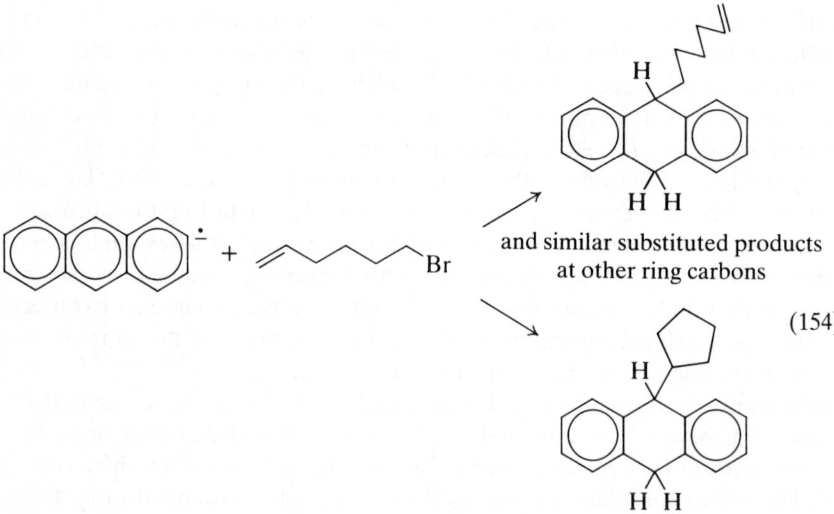

and similar substituted products
at other ring carbons

(154)

Scheme 17

The reaction of bornyl and isobornyl bromides with the nucleophile (Scheme 18) is another case where the amount of inversion is small and the rate constant close to that observed with an aromatic anion radical of the same standard potential (Daasbjerg *et al.*, 1989); it can therefore be rationalized along the same lines. Cyclizable radical-probe experiments carried out with the same nucleophile and 6-bromo-6-methyl-1-heptene, a radical clock presumably slower than the preceding one, showed no cyclized coupling product. It should be noted, on the other hand, that, unlike the case of the anthracene anion radical, the oxidized nucleophile is now a radical so its coupling with the alkyl radical within the solvent cage possesses a much more favourable driving force.

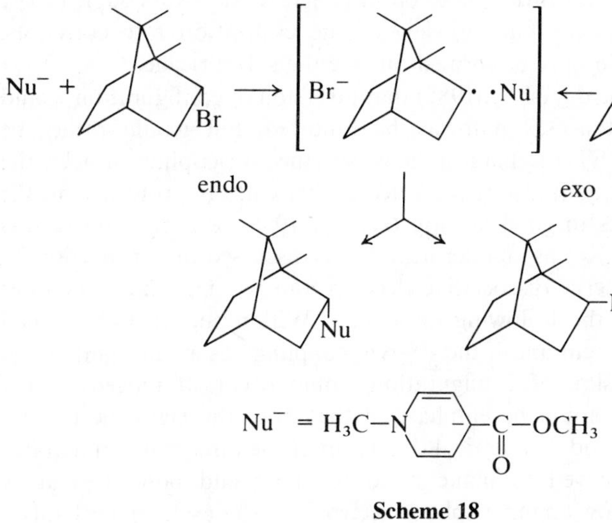

endo   exo

(155)

$$Nu^- = H_3C-N\langle\overline{}\rangle-\underset{\underset{O}{\|}}{C}-OCH_3$$

Scheme 18

Cyclizable radical-probe experiments have been extensively used in ET versus S_N2 investigations (see Ashby, 1988, and references cited therein). Attention has, however, been recently drawn to causes of possible misinterpretation, particularly in the case of iodides, where an iodine-atom-transfer chain mechanism is able to convert most of the starting linear iodide into the cyclized iodide, even if only a minute amount of linear-chain radical is present in ET–S_N2 reactions (Newcomb and Curran, 1988). Rather puzzling results were found in the reaction of $(CH_3)_3Sn^-$ ions with secondary bromides, which should not be involved in atom-exchange chain reactions

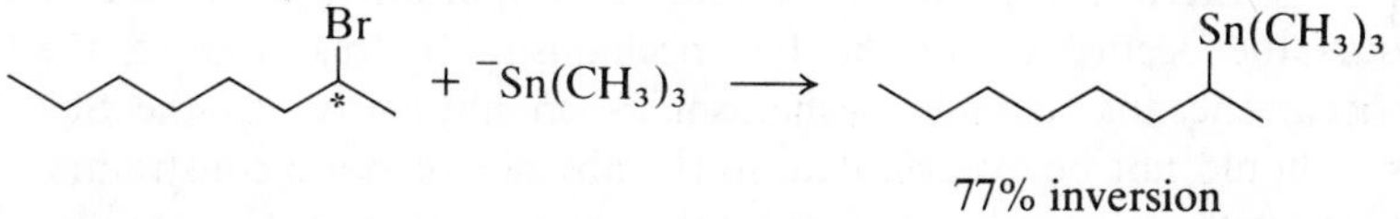

77% inversion

(156)

70%

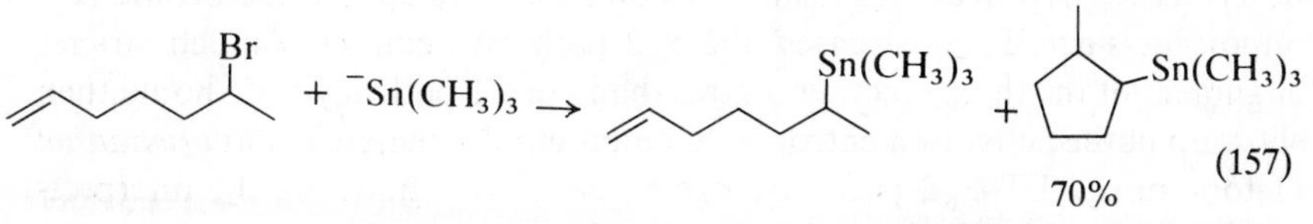

(157)

(Newcomb and Curran, 1988). It is indeed difficult to reconcile the two observations in (156) (Ashby, 1988; Kuivila and Alnajjar, 1982) and (157) (Ashby and DePriest, 1982; Ashby et al., 1984; Kitching et al., 1981, 1982); a minimal requirement for inversion to occur is that R· and Nu· couple before leaving the solvent cage but, in view of the cyclization rate constants, cyclized products can only be formed outside the solvent cage.

It has been suggested (Ashby, 1988) that inversion of configuration would not necessitate that a S_N2 pathway be followed, but would rather be consistent with an ET mechanism in which the nucleophile attacks the electrophilic carbon at its frontside since its backside is protected by the leaving group. In its original version (Ashby, 1988), the mechanism was thought to involve $RX·^-$ as an intermediate. As discussed earlier (Section 2), aliphatic halides undergo dissociative electron transfer. This, however, does not seriously affect the following discussion. Within such a well aligned geometrical arrangement and if the R·–Nu· coupling has a sufficiently large driving force, inversion of configuration would occur. If indeed such a situation is met, it should be emphasized that both the energetic (strong bonded interaction) and geometrical (lining up of the three centres) requirements are then achieved to make bond breaking and bond formation concerted, i.e. to follow an inner sphere electron transfer ($= S_N2$) mechanism instead of an outer sphere, dissociative electron-transfer mechanism. We are thus back to the model that we developed earlier. Consistent with this view is the observation, from Lund's bornyl–isobornyl bromide experiments, that substantial loss of stereochemical integrity occurs even within the solvent cage when the Nu–R–X centres do not have the right geometrical configuration!

In summary, a clear distinction should first be made between outer sphere dissociative electron transfer (ET) and S_N2, whatever the interest in the conception of the S_N2 reaction as an inner sphere dissociative electron transfer. In this framework, ET and S_N2 do not appear as extremes of a common mechanism purely on energy grounds, but rather ET occurs when the bonded interaction in the transition state vanishes. The ET–S_N2 dichotomy depends heavily, but not solely, upon the extrinsic factor $\Delta G^0_{RNu \rightleftharpoons R· + Nu·}$. Steric constraints constitute one important intrinsic factor that favours the occurrence of the ET mechanism. In this context, the possible occurrence of the ET mechanism as an alternative to the S_N2 mechanism should not be exaggerated: in the absence of steric constraints, nucleophiles that are members of reversible one-electron redox couples nevertheless react in an S_N2 fashion. Geometrical orientation factors are also important since, in most cases, the S_N2 pathway requires a much stricter alignment of the three reacting centres than the ET pathway and should then show up in the activation entropies. Combining the energetic and orientation factors, mixed ET–S_N2 pathways can be conceived, allowing the interpretation of reactions where partial racemization is observed.

ATOM-TRANSFER SUBSTITUTION

The possibility that substitution results from halogen-atom transfer to the nucleophile, thus generating an alkyl radical that could then couple with its reduced or oxidized form, has been mentioned earlier in the reaction of iron(I) and iron(0) porphyrins with aliphatic halides. This mechanism has been extensively investigated in two cases, namely the oxidative addition of various aliphatic and benzylic halides to cobalt(II) and chromium(II) complexes.

The reactions of several Co(II) complexes have been examined (Halpern, 1974), namely, pentacyanocobaltate(II) (Chock and Halpern, 1969; Halpern and Maher, 1964, 1965; Kwiatek and Seyler, 1965, 1968; Kwiatek, 1967), bis-(glyoximato)cobalt(II) (Schneider et al., 1969), cobalt(II) Schiff's base (Marzilli et al., 1970, 1971) and bis(dioximato)cobalt(II) (Halpern and Phelan, 1972) complexes. A halogen-atom-transfer mechanism has been proposed for most halides (158, 159), with the exception of the reaction of cobalt(II) Schiff's

$$L_5Co(II) + RX \longrightarrow L_5Co(III)X + R\cdot \text{ (rate determining)} \qquad (158)$$

$$L_5Co(II) + R\cdot \longrightarrow L_5Co(III)R \qquad (159)$$

base complexes with 4-nitrobenzyl halides, where the reaction proceeds by outer sphere electron transfer (160) leading to the anion radical. This decomposes into the nitrobenzyl radical (161) which then couples with the cobalt(II) complex, eventually yielding the organocobalt derivative (162).

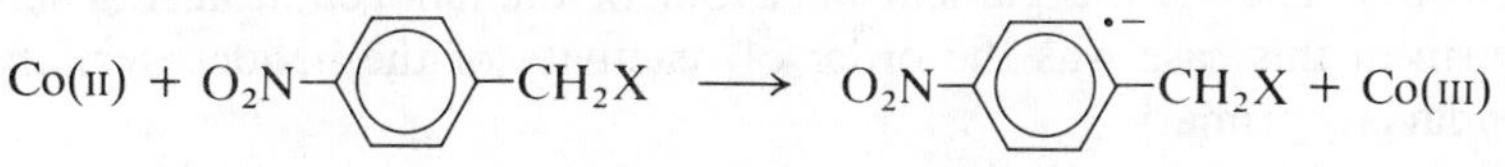

$$Co(II) + O_2N-\langle\bigcirc\rangle-CH_2X \longrightarrow O_2N-\langle\bigcirc\rangle-CH_2X^{\cdot-} + Co(III) \quad (160)$$

$$O_2N-\langle\bigcirc\rangle-CH_2X^{\cdot-} \longrightarrow O_2N-\langle\bigcirc\rangle-CH_2\cdot + X^- \qquad (161)$$

$$Co(II) + O_2N-\langle\bigcirc\rangle-CH_2\cdot \longrightarrow O_2N-\langle\bigcirc\rangle-CH_2-Co(III) \qquad (162)$$

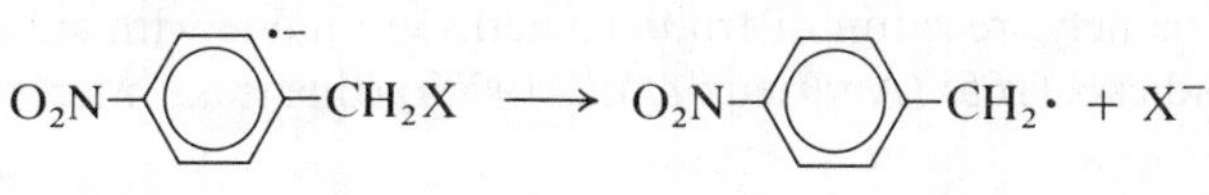

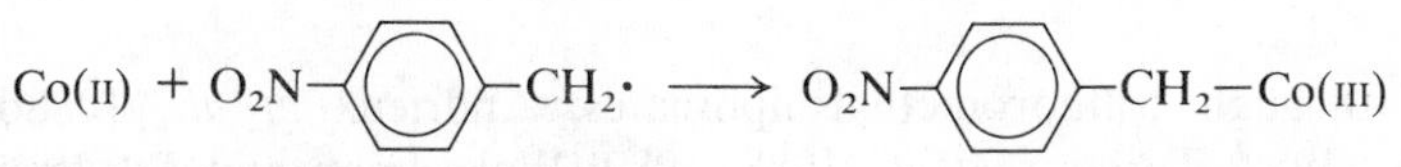

The fact that the anion radical is an intermediate in this case falls in line with the observation that it is also an intermediate in the reduction of the same substrates by homogeneous or heterogeneous outer sphere electron donors and also that nitrobenzyl halides are quite easy to reduce (see Section 2, p. 66). In the other cases, the generation of the R· radical has been assumed to proceed by halogen-atom transfer (158). It should, however, be noted that an outer sphere, dissociative electron-transfer reaction (163) would also

$$L_5Co(II) + RX \longrightarrow L_5Co(III) + R\cdot + X^- \text{ (rate determining)} \qquad (163)$$

$$L_5Co(II) + R\cdot \longrightarrow L_5Co(III)R \qquad (159)$$

$$L_5Co(III) + X^- \longrightarrow L_5Co(III)X \qquad (164)$$

generate $R\cdot$ according to the same second order kinetics. Several arguments can nevertheless be invoked in favour of the halogen-atom mechanism, i.e. in favour of the concertedness of reactions (163) and (164). One is that the latter reaction possesses a strong driving force (Lexa *et al.*, 1990). Another is that a large acceleration of the reaction is observed upon passing from primary to secondary halides (Chock and Halpern, 1969). This is also predicted to happen, but presumably to a lesser extent, with the outer sphere, dissociative electron-transfer mechanism. Lastly, although this has not been specifically investigated, it seems that the application of the "kinetic advantage" method would point to the same conclusion. It is likely that the placing of rather positive standard potentials such as those of $Co(III)/Co(II)$ couples on the outer sphere line would give rise to rate constants definitely smaller than the experimental values. Other halogen-atom-transfer reactions, not leading to substitution, namely, halogen abstraction by $Re(CO_4)L$ radicals from several alkyl halides have been shown to have a smaller intrinsic barrier than outer sphere electron donors (Lee and Brown, 1987).

Similar investigations have been carried out and similar conclusions reached with the reaction of chromium(II) complexes with alkyl halides (Castro, 1963; Kochi and Davis, 1964; Kochi and Mocadlo, 1966; Kray and Castro, 1964). The main argument in favour of the halogen-atom-transfer mechanism in this case was the order of reactivity of the halides: tertiary > secondary > primary.

There is another substitution reaction, not involving transition-metal complexes, namely, reaction of trifluoromethyl bromide with sulphur dioxide anion radicals (165) (Andrieux *et al.*, 1990a) (this is an interesting route

$$CF_3Br + SO_2 + 2e^- \longrightarrow CF_3SO_2^- + Br^- \qquad (165)$$

to triflates through trifluoromethylsulphinates; Andrieux *et al.*, 1988d; Folest *et al.*, 1988; Tordeux *et al.*, 1988, 1989; Wakselman and Tordeux, 1986a,b) where halogen-atom transfer appears to be the rate-determining step. In aprotic solvents such as DMF, sulphur dioxide is reduced reversibly into its anion radical, similarly to aromatic hydrocarbons. One would thus expect them to react in a similar way (see Section 2) with CF_3Br, viz., along an outer sphere dissociative electron-transfer step followed by coupling of the ensuing $CF_3\cdot$ radical with $SO_2\cdot^-$ as shown in (166)–(168). If this were the

$$CF_3Br + SO_2\cdot^- \longrightarrow CF_3\cdot + SO_2 + Br^- \tag{166}$$

$$CF_3\cdot + SO_2\cdot^- \longrightarrow CF_3SO_2^- \tag{167}$$

$$SO_2 + Br^- \rightleftharpoons SO_2Br^- \tag{168}$$

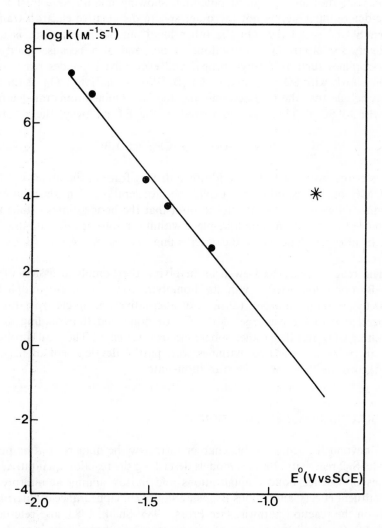

Fig. 20 Rate constants of the reaction of CF_3Br with electrochemically generated $SO_2\cdot^-$ (*) and aromatic anion radicals (●) as a function of their standard potentials. (Adapted from Andrieux *et al.*, 1990a.)

case, the rate constant of the reaction should be smaller than, or at most equal to, that of an aromatic anion radical of the same standard potential (the self-exchange intrinsic barrier is certainly larger for $SO_2 \cdot^-$ than for an aromatic anion radical, if only because the charge is more concentrated and solvent reorganization larger). It is in fact found (Fig. 20) that the reaction with $SO_2 \cdot^-$ is 4–5 orders of magnitude faster than with the aromatic anion radical of the same standard potential, showing unambiguously that the rate-determining step is not the outer sphere dissociative electron transfer from $SO_2 \cdot^-$ to CF_3Br. On the other hand, an S_N2 mechanism is quite unlikely for electronic distribution reasons and also because good S_N2 electrophiles such as methyl, propyl and even allyl bromides react much more slowly with $SO_2 \cdot^-$ than does CF_3Br (Wille et al., 1986). One is thus led to conclude that the rate determining step is bromine-atom transfer from CF_3Br to $SO_2 \cdot^-$ (169). As compared to the ET pathway, the Br-atom-

$$CF_3Br + SO_2 \cdot^- \longrightarrow CF_3 \cdot + SO_2Br^- \tag{169}$$

transfer reaction possesses an additional driving force of the order of 0.1 eV in DMF, but this is unlikely to explain the acceleration of the reaction fully. It must, however, be taken into account that the bonded interactions that stabilize the transition state take place within the solvent cage and are thus much stronger than when the intervening species are separated by the solvent.

The reaction may be viewed as involving the homolytic fission of the C—Br bond concertedly with the homolytic coupling of Br· and $SO_2 \cdot^-$ (strictly speaking an atom transfer), or alternatively as an electron transfer concerted with the cleavage of the C—Br bond and the coupling of the resulting SO_2 and Br^- (inner sphere electron transfer). The truth probably lies in between these two extremes with partial development of negative charge on the Br atom in the transition state.

MODELLING OF S_N2 SUBSTITUTION

It is beyond the scope of this chapter to review the numerous attempts to model S_N2 reactivity. Besides models describing the reaction qualitatively in terms of valence bond configurations and mostly aiming at interpreting substituent effects, anomalous Brønsted slopes and appearance of intermediates on the reaction pathway (see Pross, 1985; Shaik, 1985, and references therein), essentially two types of quantitative treatments have been developed. The more rigorous one, based on ab initio quantum chemical calculations (Berthier et al., 1969; Jorgensen, 1988) has reached quantitative success not only in modelling (Chandrasekhar and Jorgensen, 1985; Evanseck *et al.*, 1987) gas-phase nucleophilic displacement reactions (Riveros *et al.*, 1985) but has also provided quite interesting insights into the role of the solvent in liquid-phase reactions using Monte-Carlo techniques (Jorgensen, 1988, 1989) and more refined models of solvation (Aleman *et al.*, 1990; Bertran, 1989). This approach is necessarily restricted to a limited number of simple, model reactions justifying the continuous interest in the long standing quest for more general, albeit less rigorous, treatments. In this respect, the application of the Marcus cross-exchange relationship, originally developed for outer sphere electron-transfer reactions, to simple S_N2 reactions (methyl cation transfers) has met with remarkable success (Albery and Kreevoy, 1978; Lewis and Hu, 1984; Lewis *et al.*, 1985, 1987a,b; Lewis, 1986). The fact that the relationship applies, allowing the separation between extrinsic (driving force) and intrinsic factors and the expression of the intrinsic barrier as the average of the self-exchange intrinsic barriers, has been justified on the basis of a bond-energy–bond-order approximation (Marcus, 1968). How the intrinsic barriers are related to structural and environmental factors, a recurrent question (Bernasconi, 1987: Lewis, 1986), has not yet received a satisfactory, even if approximate, answer, unlike the case of purely outer sphere or outer sphere, dissociative electron-transfer reactions.

5 Concluding remarks

The weird "EC, CE, ECE ..." jargon of molecular electrochemists, a community well exposed to single electron transfer, draws a sharp distinction between electron-transfer ("E") and "chemical" reactions ("C"). Should they now totally abandon this dichotomy and see, together with all molecular chemists, single electron transfer in every chemical reaction, particularly in every nucleophilic substitution?

If, whatever the interest of conceiving electron-pair transfer reactions such as S_N2 substitution as an inner sphere electron-transfer process, single electron transfer is intended to qualify reactions in which the rate-determining step is an outer sphere, non-dissociative or dissociative electron-transfer step preceding the bond-formation step, then the answer is no. There are a number of cases where "true" S_N2 mechanisms (in which the bond-breaking and bond-formation steps are concerted) do occur, even with nucleophiles that are members of reversible one-electron reversible redox couples. In terms of activation energy, the S_N2 mechanism merges with the outer sphere, dissociative electron-transfer mechanism when the bonded interactions in the transition state vanish. Steric constraints at the electrophilic and nucleophilic reacting centres decrease the bonded interactions for obvious reasons and thus favour the outer sphere, dissociative electron-transfer mechanism. An increased thermodynamic affinity between R· and Nu· conversely favours the S_N2 mechanism in terms of driving force, but it remains to observe and rationalize in a wider range of experimental systems how intrinsic barriers are affected. This is related to the lack of a general model of S_N2 reactivity comparable to what already exists for outer sphere, non-dissociative and dissociative-electron transfer. Differences in stereo-chemistry and activation entropy are related to the fact that the S_N2 pathway requires a stricter alignment of the three reaction centres than the outer sphere, dissociative electron transfer pathway. This also allows the rationali-zation of the observation of partial inversion of configuration in borderline cases. It exemplifies the more general notion that selectivity, particularly stereoselectivity, is more likely to be reached in inner sphere than in outer sphere processes, an important point in the design of catalytic reactions. Gathering of more information about the possible geometrical preferences of outer sphere, dissociative electron-transfer pathways is desirable, how-ever.

The main mechanistic features of the $S_{RN}1$ reaction are now reasonably well understood. It is the realm of outer sphere, non-dissociative or dissocia-tive single electron transfers. They are responsible for the chain character of the reaction, and also appear in the dissociation of the substrate anion radical and in the formation of the anion radical of the substituted product, the key step of the reaction, that can be viewed as intramolecular single electron transfer concerted with bond breaking and bond formation, re-spectively. Quantitative modelling of the reactivity patterns for these two reactions is, however, still awaited. Single electron transfer is also respon-sible for major termination steps, at least in the aromatic case. Further investigation of the termination steps in other cases seems desirable and, more generally, there is still much to unravel at the quantitative level in the intricacies of the chain (possibly chains) propagation, initiation and termina-tion processes in all cases.

Besides its intrinsic interest, the investigation of outer sphere, non-dissociative or dissociative electron transfer is of key importance for a better understanding of inner sphere substitution processes and inner sphere processes in general. More is known in terms of quantitative modelling in the first case than in the second. The investigations have so far concerned mostly the carbon–halogen bond. Even if it is true that this is a particularly important example, other systems seem worth investigating. The transition between concerted and stepwise electron-transfer–bond breaking processes as a function of the driving force is an important fundamental problem that calls for a larger number of experimental exemplifications.

Acknowledgements

I would like to thank all my coworkers, whose names appear in the reference list, for their contribution to the resolution of the problems discussed in this chapter, particularly Christian Amatore, Claude P. Andrieux, Doris Lexa, Louis Nadjo, Jean Pinson and André Thiébault. A stay at the California Institute of Technology as a Scherman Fairchild Scholar in 1988 and 1989 gave me the opportunity of pleasant and useful conversations on related questions with Rudolph A. Marcus, Fred C. Anson and Harry B. Gray. Also related to the topics discussed in this chapter were frutiful scientific contacts between our group and the groups of Juan Bertran (Barcelona, Spain), Hans Schäfer (Münster, FRG) and Elio Vianello (Padova, Italy) suppported by the EEC "Science" program. I am indebted to Maurice Médebielle for his help in gathering the $S_{RN}1$ literature and to Jacques Moiroux and Annie Le Gorande for careful reading of the manuscript. Our own work in the field received the sustained support of the Centre National de la Recherche Scientifique.

References

Aalstad, B. and Parker, V. D. (1982). *Acta Chem. Scand., Ser. B* **36**, 47
Abe, T. (1973). *Chem. Lett.*, 1339
Abe, T. and Ikegamo, Y. (1976). *Bull. Chem. Soc. Jpn* **49**, 3227
Abe, T. and Ikegamo, Y. (1978). *Bull. Chem. Soc. Jpn* **51**, 196
Adam, W. and Arce, J. (1972). *J. Org. Chem.* **37**, 507
Alam, N., Amatore, C., Combellas, C., Thiébault, A. and Verpeaux, J. N. (1987). *Tetrahedron Lett.* **49**, 6171
Alam, N., Amatore, C., Combellas, C., Pinson, J., Savéant, J.-M., Thiébault, A. and Verpeaux, J. N. (1988). *J. Org. Chem.* **53**, 1493
Albery, W. J. and Kreevoy, M. M. (1978). *Adv. Phys. Org. Chem.* **16**, 87
Albright, T. A., Burdett, J. K. and Whangbo, M. H. (1985). "Orbital Interactions in Chemistry". Wiley, New York
Aleman, C., Maseras, F., Lledos, A., Duran, M. and Bertran, J. (1990). *J. Phys. Org. Chem.*, in press
Alwair, K. and Grimshaw, J. (1973a). *J. Chem. Soc., Perkin Trans. 2*, 1150
Alwair, K. and Grimshaw, J. (1973b). *J. Chem. Soc., Perkin Trans. 2*, 1811
Amatore, C., Chaussard, J., Pinson, J., Savéant, J.-M. and Thiébault, A. (1979a). *J. Am. Chem. Soc.* **101**, 6012
Amatore, C., Savéant, J.-M. and Thiébault, A. (1979b). *J. Electroanal. Chem.* **103**, 303
Amatore, C., Pinson, J., Savéant, J.-M. and Thiébault, A. (1980a). *J. Electroanal. Chem.* **107**, 59
Amatore, C., Pinson, J., Savéant, J.-M. and Thiébault, A. (1980b). *J. Electroanal. Chem.* **107**, 75
Amatore, C., Pinson, J., Savéant, J.-M. and Thiébault, A. (1981a). *J. Electroanal. Chem.* **123**, 231

Amatore, C., Pinson, J., Savéant, J.-M. and Thiébault, A. (1981b). *J. Am. Chem. Soc.* **103**, 6930

Amatore, C., Badoz-Lambling, J., Bonnel-Huyghes, C., Pinson, J., Savéant, J.-M. and Thiébault, A. (1982a). *J. Am. Chem. Soc.* **104**, 1979

Amatore, C., Pinson, J., Savéant, J.-M. and Thiébault, A. (1982b). *J. Am. Chem. Soc.* **104**, 817

Amatore, C., Gareil, M. and Savéant, J.-M. (1984a). *J. Electroanal. Chem.* **176**, 377

Amatore, C., Oturan, M. A., Pinson, J., Savéant, J.-M. and Thiébault, A. (1984b). *J. Am. Chem. Soc.* **106**, 6318

Amatore, C., Savéant, J.-M., Combellas, C., Robvielle, S. and Thiébault, A. (1985a). *J. Electroanal. Chem.* **184**, 25

Amatore, C., Oturan, M. A., Pinson, J., Savéant, J.-M. and Thiébault, A. (1985b). *J. Am. Chem. Soc.* **107**, 3451

Amatore, C., Combellas, C., Robvielle, S., Savéant, J.-M. and Thiébault, A. (1986a). *J. Am. Chem. Soc.* **108**, 4754

Amatore, C., Gareil, M., Oturan, M. A., Pinson, J., Savéant, J.-M. and Thiébault, A. (1986b). *J. Org. Chem.* **51**, 3757

Amatore, C., Jutand, A. and Pflüger, F. (1987). *J. Electroanal. Chem.* **218**, 361

Amatore, C., Combellas, C., Pinson, J., Savéant, J.-M. and Thiébault, A. (1988). *J. Chem. Soc., Chem. Commun.*, 7

Amatore, C., Beugelmans, R., Bois-Choussy, M., Combellas, C. and Thiébault, A. (1989). *J. Org. Chem.* **54**, 5688

Andrieux, C. P. and Savéant, J.-M. (1986a). *In* "Investigation of Rates and Mechanisms of Reactions", *Techniques of Chemistry* (ed. C. F. Bernasconi), Vol. VI/4E, Part 2, pp. 305–390. Wiley, New York

Andrieux, C. P. and Savéant, J.-M. (1986b). *J. Electroanal. Chem.* **205**, 43

Andrieux, C. P. and Savéant, J.-M. (1988). Unpublished results

Andrieux, C. P. and Savéant, J.-M. (1989). *J. Electroanal. Chem.* **267**, 15

Andrieux, C. P., Dumas-Bouchiat, J.-M. and Savéant, J.-M. (1978). *J. Electroanal. Chem.* **87**, 39

Andrieux, C. P., Blocman, C., Dumas-Bouchiat, J.-M. and Savéant, J.-M. (1979). *J. Am. Chem. Soc.* **101**, 3431

Andrieux, C. P., Blocman, C., M'Halla, F., Dumas-Bouchiat, J.-M. and Savéant, J.-M. (1980). *J. Am. Chem. Soc.* **102**, 3806

Andrieux, C. P., Savéant, J.-M. and Zann, D. (1984a). *Nouv. J. Chim.* **8**, 107

Andrieux, C. P., Merz, A., Savéant, J.-M. and Tomahogh, R. (1984b). *J. Am. Chem. Soc.* **106**, 1957

Andrieux, C. P., Merz, A. and Savéant, J.-M. (1985). *J. Am. Chem. Soc.* **107**, 6097

Andrieux, C. P., Gallardo, I., Savéant, J.-M. and Su, K. B. (1986a). *J. Am. Chem. Soc.* **108**, 638

Andrieux, C. P., Savéant, J.-M. and Su, K. B. (1986b). *J. Phys. Chem.* **90**, 3815

Andrieux, C. P., Badoz-Lambling, J., Combellas, C., Lacombe, D., Savéant, J.-M., Thiébault, A. and Zann, D. (1987). *J. Am. Chem. Soc.* **109**, 1518

Andrieux, C. P., Hapiot, P. and Savéant, J.-M. (1988a). *J. Phys. Chem.* **92**, 5987

Andrieux, C. P., Garreau, D., Hapiot, P., Pinson, J. and Savéant, J.-M. (1988b). *J. Electroanal. Chem.* **243**, 321

Andrieux, C. P., Garreau, D., Hapiot, P. and Savéant, J.-M. (1988c). *J. Electroanal. Chem.* **248**, 447

Andrieux, C. P., Gelis, L., Jaccaud, M., Leroux, F. and Savéant, J.-M. (1988d). French Patent 88.09336

Andrieux, C. P., Gallardo, I. and Savéant, J.-M. (1989). *J. Am. Chem. Soc.* **111**, 1620
Andrieux, C. P., Gelis, L. and Savéant, J.-M. (1990a). *J. Am. Chem. Soc.* **112**, 786
Andrieux, C. P., Gelis, L., Médebielle, M., Pinson, J. and Savéant, J.-M. (1990b). *J. Am. Chem. Soc.*, **112**, 3509
Andrieux, C. P., Hapiot, P. and Savéant, J.-M. (1990c). *Electroanalysis*, **2**, 183
Andrieux, C. P., Hapiot, P. and Savéant, J.-M. (1990d). *Chem. Rev.*, **90**, 723
Andrieux, C. P., Le Gorande, A. and Savéant, J.-M. (1990e). Submitted
Angel, D. H. and Dickinson, T. (1972). *J. Electroanal. Chem.* **35**, 55
Anson, F. C., Rathjen, N. and Frisbee, R. D. (1970). *J. Electrochem. Soc.* **117**, 477
Archibald, T. G., Taran, C. and Baum, K. (1989). *J. Fluorine Chem.* **43**, 243
Arevalo, M. C., Farnia, G., Severin, M. G. and Vianello, E. (1987). *J. Electroanal. Chem.* **220**, 201
Ashby, E. C. (1988). *Acc. Chem. Res.* **21**, 414
Ashby, E. C. and DePriest, R. (1982). *J. Am. Chem. Soc.* **104**, 6146
Ashby, E. C., DePriest, R. and Su, W. Y. (1984). *Organometallics* **3**, 1718
Ashby, E. C., Su, W. Y. and Pham, T. N. (1985). *Organometallics* **4**, 1493
Baizer, M. M. and Lund, H. (1983). "Organic Electrochemistry". Dekker, New York
Bank, S. and Juckett, D. A. (1975). *J. Am. Chem. Soc.* **97**, 7742
Bard, A. J. and Faulkner, L. R. (1980). "Electrochemical Methods, Fundamentals and Applications". Wiley, New York
Bard, R. R., Bunnett, J. F., Creary, X. and Tremelling, M. J. (1980). *J. Am. Chem. Soc.* **102**, 2852
Becker, J. Y. (1983). *In* "The Chemistry of Functional Groups" (ed. S. Patai and Z. Rappoport), Supplement D, Ch. 6, pp. 203–285. Wiley, New York
Beland, F. A., Farwell, S. O., Callis, P. R. and Geer, R. D. (1977). *J. Electroanal. Chem.* **78**, 145
Benson, S. W. (1976). "Thermodynamical Kinetics", 2nd edn. Wiley, New York
Bernasconi, C. F. (1987). *Acc. Chem. Res.* **20**, 301
Berthier, G., David, D. J. and Veillard, A. (1969). *Theor. Chim. Acta* **14**, 329
Bertran, J. (1989). "New Theoretical Concepts for Understanding Chemical Reactions", p. 231. Kluwer Academic
Beugelmans, R. (1984). *Bull. Soc. Chim. Belg.* **93**, 547
Beugelmans, R., Bois-Choussy, M. and Boudet, B. (1982). *Tetrahedron* **38**, 3479
Bindra, P., Brown, A. P., Fleischmann, M. and Pletcher, D. (1975). *J. Electroanal. Chem.* **158**, 39
Boiko, V. N., Schupak, G. M. and Yagupol'skii, L. M. (1977). *J. Org. Chem. USSR (Engl. Transl.)* **13**, 1866
Bordwell, F. G. and Hughes, D. L. (1983). *J. Org. Chem.* **48**, 2206
Bordwell, F. G. and Harrelson, J. A., Jr (1987). *J. Am. Chem. Soc.* **109**, 8112
Bordwell, F. G. and Harrelson, J. A., Jr (1989). *J. Am. Chem. Soc.* **111**, 1052
Bordwell, F. G. and Wilson, C. A. (1987). *J. Am. Chem. Soc.* **109**, 5470
Bordwell, F. G., Bausch, M. J. and Wilson, C. A. (1987). *J. Am. Chem. Soc.* **109**, 5465
Bowman, W. R. (1988a). *In* "Photoinduced Electron Transfer" (ed. M. A. Fox and M. Channon), Part C, pp. 487–552. Elsevier, Amsterdam
Bowman, W. R. (1988b). *Chem. Soc. Rev.* **17**, 283
Bowyer, W. J. and Evans, D. H. (1988). *J. Electroanal. Chem.* **224**, 227
Brickenstein, E. K. and Khairutdinov, R. F. (1985). *Chem. Phys. Lett.* **115**, 176
Bridger, R. F. A. and Russell, G. A. (1963). *J. Am. Chem. Soc.* **85**, 375

Britton, W. E. and Fry, A. J. (1975). *Anal. Chem.* **47**, 9
Brunet, J. J., Sidot, C. and Caubère, P. (1983). *J. Org. Chem.* **48**, 1166
Bunnett, J. F. (1978). *Acc. Chem Res.* **11**, 413
Butler, J. A. V. (1924a). *Trans. Faraday Soc.* **19**, 724
Butler, J. A. V. (1924b). *Trans. Faraday Soc.* **19**, 73
Cabaret, D., Maigrot, N. and Welvart, Z. (1985). *Tetrahedron* **41**, 5357
Calef, D. F. and Wolynes, P. G. (1983a). *J. Phys. Chem.* **87**, 3387
Calef, D. F. and Wolynes, P. G. (1983b). *J. Chem. Phys.* **78**, 470
Casado, J., Gallardo, I. and Moreno, M. (1987). *J. Electroanal. Chem.* **219**, 197
Casanova, J. J. and Rogers, H. R. (1974). *J. Org. Chem.* **39**, 2408
Castro, C. E. and Kray, W. C. (1963). *J. Am. Chem. Soc.* **85**, 2768
Chami, Z., Gareil, M., Pinson, J., Savéant, J.-M. and Thiébault, A. (1990). *J. Org. Chem.*, in press
Chandrasekhar, J. and Jorgensen, W. L. (1985). *J. Am. Chem. Soc.* **107**, 154
Chase, M. W., Davies, C. A., Downey, J. R., Frurip, D. J., McDonald, R. A. and Syverud, A. N. (1985). *J. Phys. Chem. Ref. Data* **14**, Suppl. 1
Chen, Q. Y. and Qiu, Z. M. (1986). *J. Fluorine Chem.* **31**, 301
Chen, Q. Y. and Qiu, Z. M. (1987a). *J. Chem. Soc., Chem. Commun.*, 1240
Chen, Q. Y. and Qiu, Z. M. (1987b). *J. Fluorine Chem.* **35**, 343
Chock, P. B. and Halpern, J. (1969). *J. Am. Chem. Soc.* **91**, 582
Ciminale, F., Bruno, G., Testafari, L. and Tiecco, M. (1978). *J. Org. Chem.* **43**, 4509
Cleary, J. A., Mubarak, M. S., Viera, K. L., Anderson, K. L. and Peters, D. G. (1986). *J. Electroanal. Chem.* **198**, 107
Closs, G. L. and Miller, J. R. (1988). *Science* **240**, 440
Compton, R. N., Reinhart, P. W. and Cooper, C. C. (1978). *J. Chem. Phys.* **68**, 4360
Conway, B. E. (1985). *In* "Modern Aspects of Electrochemistry" (ed. B. E. Conway, R. E. White and J. O'M. Bockris), Vol. 16, p. 103. Plenum, New York
Corrigan, D. A. and Evans, D. H. (1980). *J. Electroanal. Chem.* **106**, 287
Daasbjerg, K., Lund, T. and Lund, H. (1989). *Tetrahedron Lett.* **30**, 493
Danen, W. C., Kensler, T. T., Lawless, J. G., Marcus, M. F. and Hawley, M. D. (1969). *J. Phys. Chem.* **89**, 2787
Debye, P. (1942). *Trans. Electrochem. Soc.* **82**, 265
Degrand, C. (1986). *J. Chem. Soc., Chem. Commun.*, 1113
Degrand, C. (1987a). *J. Org. Chem.* **52**, 1421
Degrand, C. (1987b). *J. Electroanal. Chem.* **238**, 239
Degrand, C. and Prest, R. (1990a). *J. Electroanal. Chem.*, **282**, 281
Degrand, C. and Prest, R. (1990b). *J. Org. Chem.*, **55**, 5242
Degrand, C., Prest, R. and Compagnon, J. P. (1987). *J. Org. Chem.* **52**, 5229
Delahay, P. (1965). "Double Layer and Electrode Kinetics". Wiley, New York
Dietz, R. and Peover, M. E. (1968). *Discuss. Faraday Soc.* **45**, 155
Eberson, L. (1982). *Acta Chem. Scand., Ser. B* **36**, 533
Eberson, L. (1987). "Electron Transfer Reactions in Organic Chemistry". Springer-Verlag, Berlin
Eberson, L. and Ekström, M. (1988a). *Acta Chem. Scand. Ser. B* **42**, 101
Eberson, L. and Ekström, M. (1988b). *Acta Chem. Scand. Ser. B* **42**, 113
Eberson, L., Ekström, M., Lund, T. and Lund, H. (1989). *Acta Chem. Scand. Ser. B* **43**, 101
Erdey-Gruz, T. and Volmer, M. (1930). *Z. Physik. Chem.* **150A**, 203
Espenson, J. H. (1986). *In* "Techniques of Chemistry" (ed. C. F. Bernasconi), Vol. VI/4E, Part 2, pp. 487–563. Wiley, New York

Evans, D. H. and O'Connell, K. M. (1986). *Electroanal. Chem.* **14**, 113
Evanseck, J. D., Blake, J. F. and Jorgensen, W. L. (1987). *J. Am. Chem. Soc.* **109**, 2349
Falsig, M., Lund, H., Nadjo, L. and Savéant, J.-M. (1980). *Nouv. J. Chim.* **4**, 445
Feiring, A. E. (1983). *J. Org. Chem.* **48**, 347
Feiring, A. E. (1984). *J. Fluorine Chem.* **25**, 151
Fitch, A. and Evans, D. H. (1986). *J. Electroanal. Chem.* **202**, 83
Folest, J.-C., Nedelec, J.-Y. and Perichon, J. (1988). *J. Synth. Comm.* **218**, 1491
Fox, L. S. and Gray, H. B. (1990). *Science* **247**, 1069
Fox, M. A., Younathan, J. and Fryxell, G. E. (1983). *J. Org. Chem.* **26**, 3109
Fraunfelder, H. and Wolynes, P. G. (1985). *Science* **229**, 337
Fukusumi, S., Koumitsu, S., Hironaka, S. and Tanaka, K. (1987). *J. Am. Chem. Soc.* **109**, 305
Galli, C. and Bunnett, J. F. (1981). *J. Am. Chem. Soc.* **103**, 7140
Garreau, D., Hapiot, P. and Savéant, J.-M. (1989). *J. Electroanal. Chem.* **272**, 1
Garreau, D., Hapiot, P. and Savéant, J.-M. (1990). *J. Electroanal. Chem.*, **281**, 73
Garst, J. F. (1971). *Acc. Chem. Res.* **4**, 400
Garst, J. F. and Abel, B. N. (1975). *J. Am. Chem. Soc.* **97**, 4926
Garst, J. F. and Barton, F. E. (1974). *J. Am. Chem. Soc.* **96**, 523
Garst, J. F., Barbas, J. T. and Barton, F. E. (1968). *J. Am. Chem. Soc.* **90**, 7159
Garst, J. F., Pacifici, J. A., Singleton, V. D., Ezzel, M. F. and Morris, J. I. (1975). *J. Am. Chem. Soc.* **97**, 5242
Gatti, N., Jugelt, W. and Lund, H. (1987). *Acta Chem. Scand. Ser. B* **41**, 646
Gennet, T., Milner, D. F. and Weaver, M. J. (1985). *J. Phys. Chem.* **89**, 2787
Glasstone, S., Laidler, K. J. and Eyring, H. (1941). "The Theory of Rate Processes". McGraw-Hill, New York
Griggio, L. (1982). *J. Electroanal. Chem.* **140**, 155
Griggio, L. and Severin, M. G. (1987). *J. Electroanal. Chem.* **223**, 185
Grimshaw, J. and Thompson, N. (1986). *J. Electroanal. Chem.* **205**, 35
Grimshaw, J. and Trocha-Grimshaw, J. (1974). *J. Electroanal. Chem.* **56**, 443
Grimshaw, J., Langan, J. R. and Salmon, G. A. (1988). *J. Chem. Soc., Chem. Commun.*, 1115
Gould, I. R., Mattes, S. L. and Farid, S. (1987). *J. Am. Chem. Soc.* **109**, 3675
Gueutin, C., Lexa, D., Savéant, J.-M. and Wang, D. L. (1989). *Organometallics* **8**, 1607
Gunner, M. R., Robertsson, D. E. and Dutton, P. L. (1986). *J. Phys. Chem.* **90**, 3783
Halpern, J. (1974). *Ann. N.Y. Acad. Sci.* **239**, 2
Halpern, J. and Maher, J. P. (1964). *J. Am. Chem. Soc.* **86**, 5361
Halpern, J. and Maher, J. P. (1965). *J. Am. Chem. Soc.* **87**, 5361
Halpern, J. and Phelan, P. F. (1972). *J. Am. Chem. Soc.* **94**, 1881
Hapiot, P., Moiroux, J. and Savéant, J.-M. (1990). *J. Am. Chem. Soc.* **112**, 1337
Hasegawa, A. and Williams, S. (1977). *Chem. Phys. Lett.* **46**, 66
Hasegawa, A., Shiotani, M. and Williams, S. (1977). *Faraday Discuss. Chem. Soc.* 157
Hawley, M. D. (1980). *In* "Encyclopedia of the Electrochemistry of the Elements" (ed. A. J. Bard and H. Lund), Vol. XIV, Organic Section. Wiley, New York
Hebert, E., Mazaleyrat, J. P., Nadjo, L., Savéant, J.-M. and Welvart, Z. (1985). *Nouv. J. Chem.* **9**, 75
Hegedus, L. S. and Miller, L. L. (1975). *J. Am. Chem. Soc.* **97**, 459
Heinze, J. and Schwart, J. (1981). *J. Electroanal. Chem.* **126**, 283

Houser, K. J., Bartak, D. E. and Hawley, M. D. (1973). *J. Am. Chem. Soc.* **95**, 6033
Howell, J. O. and Wightman, R. M. (1984). *Anal. Chem.* **56**, 524
Hoz, S. and Bunnett, J. F. (1977). *J. Am. Chem. Soc.* **99**, 4690
Hupp, J. T. and Weaver. M. J. (1984). *J. Phys. Chem.* **88**, 1463
Hupp, J. T. and Weaver, M. J. (1985). *J. Phys. Chem.* **89**, 2795
Hupp, J. T., Weaver, T. and Weaver, M. J. (1983). *J. Electroanal. Chem.* **153**, 1
Hush, N. S. (1957). *Z. Elektrochem.* **61**, 734
Hush, N. S. (1958). *J. Chem. Phys.* **28**, 962
Hush, N. S. (1961). *Trans. Faraday Soc.* **57**, 557
Inesi, A. and Rampazzo, L. (1974). *J. Electroanal. Chem.* **54**, 289
Jaun, B., Schwartz, J. and Breslow, R. (1980). *J. Am. Chem. Soc.* **102**, 5741
Jorgensen, W. L. (1988). *Adv. Chem. Phys.* **70**, Part II, 469
Jorgensen, W. L. (1989). *Acc. Chem. Res.* **22**, 184
Keefe, J. R. and Kresge, A. J. (1986). *In* "Investigation of Rates and Mechanisms of
    Reactions", Techniques of Chemistry, (ed. C. F. Bernasconi), Vol. VI/4E, Part, 1,
    pp. 747–790. Wiley, New York
Kim, J. K. and Bunnett, J. F. (1970). *J. Am. Chem. Soc.* **92**, 7463
King, K. D., Golden, D. M. and Benson, S. W. (1971). *J. Phys. Chem.* **75**, 987
Kitching, W., Olszowy, H. A. and Harvey, K. (1981). *J. Org. Chem.* **46**, 2423
Kitching, W., Olszowy, H. A. and Harvey, K. (1982). *J. Org. Chem.* **47**, 1893
Klein, A. J. and Evans, D. H. (1979). *J. Am. Chem. Soc.* **101**, 757
Klinger, R. J. and Kochi, J. K. (1981). *J. Am. Chem. Soc.* **103**, 5839
Klinger, R. J. and Kochi, J. K. (1982). *J. Am. Chem. Soc.* **104**, 4186
Kochi, J. K. and Davis, D. D. (1964). *J. Am. Chem. Soc.* **86**, 5264
Kochi, J. K. and Mocadlo, P. M. (1966). *J. Am. Chem. Soc.* **88**, 4094
Kojima, H. and Bard, A. J. (1975). *J. Am. Chem. Soc.* **97**, 6317
Kondratenko, N. V., Popov, V. I., Boiko, V. N. and Yagupol'skii, L. M. (1977).
    J. Org. Chem. USSR (Engl. Transl.) **13**, 2086
Kornblum, N. (1971). *Pure Appl. Chem.* **4**, 67
Kornblum, N. (1975). *Angew. Chem. Internat. Ed. Engl.* **14**, 734
Kornblum, N. (1982). *In* "The Chemistry of Functional Groups" (ed. S. Patai),
    Suppl. F., p. 361. Wiley, New York
Kornblum, N. and Fifold, M. J. (1989). *Tetrahedron* **45**, 1311
Kornblum, N. and Wade, P. A. (1987). *J. Org. Chem.* **52**, 5301
Kornblum, N., Michel, R. E. and Kerber, R. C. (1966). *J. Am. Chem. Soc.* **88**, 5662
Kornblum, N., Earl, G. W., Greene, G. S., Holy, N. L., Manthey, J. W., Musser,
    M. T. and Snow, D. H. (1967). *J. Am. Chem. Soc.* **89**, 5714
Kornblum, N., Cheng, L., Davies, T. M., Earl, G. W., Holy, N. L., Kerber, R. C.,
    Kestern, M. M., Manthey, J. W., Musser, M. T., Pinnick, H. W., Snow, D. H.,
    Stuchal, F. W. and Swiger, R. T. (1987). *J. Org. Chem.* **52**, 196
Kornblum, N., Ackerman, P., Manthey, J. W., Musser, M. T., Pinnick, H. W.,
    Singaram, S. and Wade, P. A. (1988). *J. Org. Chem.* **53**, 1475
Kray, W. C. and Castro, C. E. (1964). *J. Am. Chem. Soc.* **86**, 4603
Kreevoy, M. M. and Truhlar, D. G. (1986). *In* "Investigation of Rates and
    Mechanisms of Reactions", Techniques of Chemistry (ed. C. F. Bernasconi), Vol.
    VI/4E, Part 1, pp. 13–96. Wiley, New York
Kresge, A. J. (1975). *Acc. Chem. Res.* **8**, 354
Kuivila, H. G. and Alnajjar, M. S. (1982). *J. Am. Chem. Soc.* **104**, 6146
Kwiatek, J. (1967). *Catalysis Rev.* **1**, 437
Kwiatek, J. and Seyler, J. K. (1965). *J. Organomet. Chem.* **3**, 421

Kwiatek, J. and Seyler, J. K. (1968). *Adv. Chem. Ser.* **70**, 207
Lablache-Combier, A. (1988a). *In* "Photoinduced Electron Transfer" (ed. M. A. Fox and M. Channon), Part C, pp. 134–312. Elsevier, Amsterdam
Lawless, J. G. and Hawley, M. D. (1969). *J. Electroanal. Chem.* **21**, 365
Lee, K. W. and Brown, T. L. (1987). *J. Am. Chem. Soc.* **109**, 3269
Lewis, E. S. (1986). *J. Phys. Chem.* **90**, 3756
Lewis, E. S. (1989). *J. Am. Chem. Soc.* **111**, 7576
Lewis, E. S. and Hu, D. D. (1984). *J. Am. Chem. Soc.* **106**, 3292
Lewis, E. S, MacLaughlin, M. L. and Douglas, T. A. (1985). *J. Am. Chem. Soc.* **107**, 6668
Lewis, E. S., Douglas, T. A. and MacLaughlin, M. L. (1987a). *Adv. Chem. Ser.* **215**, 35
Lewis, E. S., Yousaf, T. I. and Douglas, T. A. (1987b). *J. Am. Chem. Soc.* **109**, 6137
Lexa, D. and Savéant, J.-M. (1982). *J. Am. Chem. Soc.* **104**, 3503
Lexa, D., Mispelter, J. and Savéant, J.-M. (1981). *J. Am. Chem. Soc.* **103**, 6806
Lexa, D., Savéant, J.-M., Su, K. B. and Wang, D. L. (1987). *J. Am. Chem. Soc.* **109**, 6464
Lexa, D., Savéant, J.-M., Su, K. B. and Wang, D. L. (1988). *J. Am. Chem. Soc.* **110**, 7617
Lexa, D., Savéant, J.-M., Schäfer, H., Su, K. B., Vering, B. and Wang, D. L. (1990). *J. Am. Chem. Soc.*, **112**, 6162
Lund, T. and Lund, H. (1986). *Acta Chem. Scand., Ser. B* **40**, 470
Lund, T. and Lund, H. (1987). *Acta Chem. Scand., Ser. B* **41**, 93
Lund, T. and Lund, H. (1988). *Acta Chem. Scand., Ser. B* **42**, 269
Lund, H. and Pedersen, S. U. (1989). Personal communication
Lund, T., Pedersen, S. U., Lund, H., Cheung, K. W. and Utley, J. H. P. (1987). *Acta Chem. Scand. B* **41**, 285
McLendon, G. (1988). *Acc. Chem. Res.* **21**, 16
McManis, G. E., Golvin, M. N. and Weaver, M. J. (1986). *J. Phys. Chem.* **90**, 6563
Malissard, M., Mazaleyrat, J. P. and Welvart, Z. (1977). *J. Am. Chem. Soc.* **99**, 6933
Marcus, R. A. (1956). *J. Chem. Phys.* **24**, 4966
Marcus, R. A. (1963). *J. Chem. Phys.* **39**, 1734
Marcus, R. A. (1964). *Ann. Rev. Phys. Chem.* **15**, 155
Marcus, R. A. (1965). *J. Chem. Phys.* **43**, 669
Marcus, R. A. (1968). *J. Phys. Chem.* **72**, 891
Marcus, R. A. (1969). *J. Am. Chem. Soc.* **91**, 7224
Marcus, R. A. (1977). *In* "Special Topics in Electrochemistry" (ed. P. A. Rock), pp. 161–179. Elsevier, New York
Marcus, R. A. (1982). *Faraday Discuss. Chem. Soc.* **74**, 7
Marcus, R. A. and Sumi, H. (1986a). *J. Chem. Phys.* **84**, 4272
Marcus, R. A. and Sumi, H. (1986b). *J. Chem. Phys.* **84**, 4894
Marcus, R. A. and Sumi, H. (1986c). *J. Electroanal. Chem.* **204**, 59
Marcus, R. A. and Sutin, N. (1985). *Biophys. Biochim. Acta* **811**, 265
Margel, S. and Levy, M. (1974). *J. Electroanal. Chem.* **56**, 259
Marzilli, L. G., Marzilli, P. A. and Halpern, J. (1970). *J. Am. Chem. Soc.* **92**, 5752
Marzilli, L. G., Marzilli, P. A. and Halpern, J. (1971). *J. Am. Chem. Soc.* **93**, 1374
Médebielle, M., Pinson, J. and Savéant, J.-M. (1990). *Tetrahedron Lett.*, **31**, 1275
Mendenhall, G. D., Golden, D. M. and Benson, S. W. (1973). *J. Phys. Chem.* **77**, 2707
Merz, A. and Tomahogh, R. (1979). *Angew. Chem., Internat. Ed. Engl.* **18**, 938

M'Halla, F., Pinson, J. and Savéant, J.-M. (1978). *J. Electroanal. Chem.* **89**, 347
M'Halla, F., Pinson, J. and Savéant, J.-M. (1980). *J. Am. Chem. Soc.* **102**, 4120
Miller, J. R., Beitz, J. V. and Huddleston, R. K. (1986). *J. Am. Chem. Soc.* **106**, 5057
Mishra, S. P. and Symons, M. C. R. (1973). *J. Chem. Soc., Chem. Commun.* 577
Mohilner, D. M. (1969). *J. Phys. Chem.* **73**, 2652
Momot, E. and Bronoel, G. (1974). *C.R. Acad. Sci.* **278**, 319
Montenegro, M. I. and Pletcher, D. (1986). *J. Electroanal. Chem.* **200**, 371
Nadjo, L. and Savéant, J.-M. (1971). *J. Electroanal. Chem.* **30**, 41
Nadjo, L. and Savéant, J.-M. (1973). *J. Electroanal. Chem.* **48**, 113
Nadjo, L., Savéant, J.-M. and Su, K. B. (1985). *J. Electroanal. Chem.* **196**, 23
Nadler, W. and Marcus, R. A. (1987). *J. Chem. Phys.* **86**, 3906
Nelson, R. F., Carpenter, A. K. and Seo, E. T. (1973). *J. Electrochem. Soc.* **120**, 206
Neta, P. and Behar, D. (1981). *J. Am. Chem. Soc.* **103**, 103
Newcomb, M. and Curran, D. P. (1988). *Acc. Chem. Res.* **21**, 206
Newcomb, M., Thomas, R. V. and Goh, S. H. (1990). *J. Am. Chem. Soc.*, in press.
Newton, M. D. and Sutin, N. (1984). *Ann. Rev. Phys. Chem.* **35**, 437
Nielson, R. M., McManis, G. E., Golvin, M. N. and Weaver, M. J. (1988). *J. Phys. Chem.* **92**, 3441
Norris, R. K. (1983). "The Chemistry of Halides, Pseudohalides and Azides" *in* "The Chemistry of Functional Groups" (ed. S. Patai and Z. Rappoport), Suppl. D, Ch. 16, p. 681. Wiley, New York.
Norris, R. K. (1990). *Compr. Org. Synth.*, in press
O'Connell, K. M. and Evans, D. H. (1983). *J. Am. Chem. Soc.*, **105**, 1473
Ohno, T., Yoshimura, A. and Mataga, N. (1986). *J. Phys. Chem.* **90**, 3295
Ohno, T., Yoshimura, A., Shioyama, H. and Mataga, N. (1987). *J. Phys. Chem.* **91**, 4365
Oturan, M. A., Pinson, J., Savéant, J.-M. and Thiébault, A. (1989). *Tetrahedron Lett.* **30**, 1373
Palacios, S. M., Alonso, R. A. and Rossi, R. A. (1985). *Tetrahedron.* **41**, 4147
Parker, V. D. (1981a). *Acta Chem. Scand., Ser. B* **35**, 595
Parker, V. D. (1981b). *Acta Chem. Scand., Ser. B.* **35**, 655
Parsons, R. and Passeron, E. (1966). *J. Electroanal. Chem.* **12**, 525
Peover, M. E. and Powell, J. S. (1969). *J. Electroanal. Chem.* **20**, 427
Perrin, C. L. (1984). *J. Phys. Chem.* **88**, 3611
Pierini, A. B., Penenory, A. B. and Rossi, R. A. (1985). *J. Org. Chem.* **50**, 2739
Pierini, A. B., Baumgartner, M. T. and Rossi, R. A. (1988). *Tetrahedron Lett.* **29**, 342
Pinson, J. and Savéant, J.-M. (1974). *J. Chem. Soc., Chem. Commun.*, 933
Pinson, J. and Savéant, J.-M. (1978). *J. Am. Chem. Soc.* **100**, 1506
Popov, V. I., Boiko, V. N., Kondratenko, N. V., Sembur, V. P. and Yagupol'skii, L. M. (1977). *J. Org. Chem. USSR (Engl. Transl.)* **13**, 1985
Popov, V. I., Boiko, V. N. and Yagupol'skii, L. M. (1982). *J. Fluorine Chem.* **21**, 363
Pross, A. (1977). *Adv. Phys. Org. Chem.* **14**, 69
Pross, A. (1985). *Adv. Phys. Org. Chem.* **21**, 99
Pross, A. and McLennan, D. J. (1984). *J. Chem. Soc., Perkin Trans, 2*, 981
Pross, A. and Shaik, S. S. (1982a). *J. Am. Chem. Soc.* **104**, 187
Pross, A. and Shaik, S. S. (1982b). *J. Am. Chem. Soc.* **104**, 2708
Pross, A. and Shaik, S. S. (1983). *Acc. Chem. Res.* **16**, 363
Rajan, S. and Sridaran, P. (1977). *Tetrahedron Lett.*, 2177
Riveros, J. M., José, S. M. and Takashima, K. (1985). *Adv. Phys. Org. Chem.* **21**, 197
Rossi, R. A. (1982). *Acc. Chem. Res.* **15**, 164

Rossi, R. A. and Bunnett, J. F. (1973). *J. Org. Chem.* **38**, 3020
Rossi, R. A. and Pierini, A. B. (1980). *J. Org. Chem.* **45**, 2914
Rossi, R. A. and Rossi, R. H. (1983). "ACS Monograph 178", The American Chemical Society, Washington DC
Rossi, R. A., Palacios, S. M. and Santiago, A. N. (1982). *J. Org. Chem.* **47**, 4655
Rossi, R. A., Palacios, S. M. and Santiago, A. N. (1984a). *J. Org. Chem.* **49**, 4609
Rossi, R. A., Santiago, A. N. and Palacios, S. M. (1984b). *J. Org. Chem.* **49**, 3387
Russell, G. A. (1970). *Spec. Publ. Chem. Soc.* **24**, 271
Russell, G. A. (1987). *Adv. Phys. Org. Chem.* **24**, 271
Russell, G. A. (1989). *Acc. Chem. Res.* **22**, 1
Russell, G. A. and Danen, W. C. (1966). *J. Am. Chem. Soc.* **88**, 5663
Russell, G. A., Norris, R. K. and Panek, E. J. (1971). *J. Am. Chem. Soc.* **93**, 5839
Russell, G. A., Mudryk, B. and Jawdosiuk, M. (1981). *J. Am. Chem. Soc.* **103**, 4610
Russell, G. A., Mudryk, B., Jawdosiuk, M. and Wrobel, Z. (1982a). *J. Org. Chem.* **38**, 1059
Russell, G. A., Mudryk, B., Ros, F. and Jawdosiuk, M. (1982b). *Tetrahedron* **103**, 4610
Savéant, J.-M. (1980a). *Acc. Chem. Res.* **13**, 323
Savéant, J.-M. (1980b). *J. Electroanal. Chem.* **112**, 1975
Savéant, J.-M. (1983). *J. Electroanal. Chem.* **143**, 447
Savéant, J.-M. (1986). "Advances in Electrochemistry", Ch. IV, pp. 289–336. The Robert A. Welsh Foundation, Houston
Savéant, J.-M. (1987). *J. Am. Chem. Soc.* **109**, 6788
Savéant, J.-M. (1988). *Bull. Soc. Chim. Fr.*, 225
Savéant, J.-M. and Tessier, D. (1975). *J. Electroanal Chem.* **65**, 57
Savéant, J.-M. and Tessier, D. (1977). *J. Phys. Chem.* **81**, 2192
Savéant, J.-M. and Tessier, D. (1978a). *J. Phys. Chem.* **82**, 1723
Savéant, J.-M. and Tessier, D. (1982). *Faraday Discuss. Chem. Soc.* **74**, 57
Savéant, J.-M. and Thiébault, A. (1978b). *J. Electroanal. Chem.* **89**, 335
Scamehorn, R. G. and Bunnett, J. F. (1977). *J. Org. Chem.* **42**, 1449
Schlesener, G. J., Amatore, C. and Kochi, J. K. (1986). *J. Phys. Chem.* **90**, 3747
Schneider, P. W., Phelan, P. F. and Halpern, J. (1969). *J. Am. Chem. Soc.* **91**, 77
Scouten, C. G., Barton, F. E., Burgess, J. R., Story, P. R. and Garst, J. F. (1969). *J. Chem. Soc., Chem. Commun.* 78
Sease, W. J. and Reed, C. R. (1975). *Tetrahedron Lett.*, 393
Semmelhack, M. F. and Bargar, T. (1980). *J. Am. Chem. Soc.* **102**, 7765
Severin, M. G., Arevalo, M. C., Farnia, G. and Vianello, E. (1987). *J. Phys. Chem.* **91**, 466
Severin, M. G., Farnia, G., Vianello, E. and Arevalo, M. C. (1988). *J. Electroanal. Chem.* **251**, 369
Shaik, S. S. (1985). *Prog. Phys. Org. Chem.* **15**, 197
Simonet, J., Michel, M. A. and Lund, H. (1975). *Acta Chem. Scand., Ser. B.* **29**, 489
Smoluchovski, M. (1916a). *Phys. Z.* **17**, 557
Smoluchovski, M. (1916b). *Phys. Z.* **17**, 585
Smoluchovski, M. (1917). *Z. Phys. Chem. Stoechiom. Verwandtschaftsl.* **92**, 129
Snadrini, D., Maestri, M., Belser, P., von Zelewsky, A. and Balzani, V. (1985). *J. Phys. Chem.* **89**, 3075
Steelhammer, J. C. and Wentworth, W. E. (1969). *J. Chem. Phys.* **51**, 1802
Suga, K., Mizota, H., Kanzaki, Y. and Aoyagui, S. (1973). *J. Electroanal. Chem.* **41**, 313

Swartz, J. E. and Tenzel, T. T. (1984). *J. Am. Chem. Soc.* **106**, 2520

Symons, M. C. R. (1981). *Pure Appl. Chem.* **53**, 223

Taube, H. (1970). "Electron Transfer Reactions of Complex Ions in Solution". Academic Press, New York

Tordeux, M., Langlois, B. and Wakselman, C. (1988). French Patent 2593808; *Chem. Abstr.* **108**, 166975v

Tordeux, M., Langlois, B. and Wakselman, C. (1989). *J. Org. Chem.* **54**, 2452

Tremelling, M. J. and Bunnett, J. F. (1980). *J. Am. Chem. Soc.* **102**, 7375

Van der Zawn, G. and Hynes, J. T. (1982). *J. Chem. Phys.* **76**, 2993

Van der Zawn, G. and Hynes, J. T. (1985). *J. Phys. Chem.* **89**, 4181

Voloschhuk, V. G., Boiko, V. N. and Yagupol'skii, L. M. (1977). *J. Org. Chem. USSR (Engl. Transl.)* **13**, 1866

Von Stackelberg, M. and Stracke, W. (1949). *Z. Elektrochem.* **53**, 118

Wade, P. A., Morrison, H. A. and Kornblum, N. (1987). *J. Org. Chem.* **52**, 3102

Waisman, E., Worry, G. and Marcus, R. A. (1977). *J. Electroanal. Chem.* **82**, 9

Wakselman, C. and Tordeux, M. (1984). *J. Chem. Soc., Chem. Commun.*, 793

Wakselman, C. and Tordeux, M. (1985). *J. Org. Chem.* **50**, 4047

Wakselman, C. and Tordeux, M. (1986a). *Bull. Soc. Chim. Fr.*, 1868

Wakselman, C. and Tordeux, M. (1986b). French Patent 2564829; *Chem. Abstr.* **105**, 171845v

Walder, L. (1989). Personal communication

Wasielevski, M. R., Niemczyk, M. R., Svec, W. A. and Pewitt, E. B. (1985). *J. Am. Chem. Soc.* **107**, 1080

Weaver, M. J. and Anson, F. C. (1976). *J. Phys. Chem.* **80**, 1861

Weaver, M. J. and Gennet, T. (1985). *Chem. Phys. Lett.* **113**, 213

Wentworth, W. E., Becker, R. S. and Tung, R. (1967). *J. Phys. Chem.* **71**, 1652

Wentworth, W. E., George, R. and Keith, H. (1969). *J. Chem. Phys.* **51**, 1791

Wightman, R. M. and Wipf, D. O. (1988). *Electroanal. Chem.* **15**, 267

Wille, H. J., Kastening, B. and Knittel, D. (1986). *J. Electroanal. Chem.* **214**, 221

Wolfe, J. F. and Carver, D. R. (1978). *Org. Prep. Proc. Int.* **10**, 225

Wong, C. L. and Kochi, J. K. (1979). *J. Am. Chem. Soc.* **101**, 5593

Zavada, J., Krupicka, J. and Sicher, J. (1963). *Coll. Czech. Chem. Comm.* **28**, 1664

Zhang, X., Leddy, J. and Bard, A. J. (1985). *J. Am. Chem. Soc.* **107**, 3719

Zhang, X., Yang, H. and Bard, A. J. (1987). *J. Am. Chem. Soc.* **109**, 1916

Zusman, L. D. (1980). *J. Chem. Phys.* **49**, 295

The Captodative Effect

REINER SUSTMANN AND HANS-GERT KORTH

Institut für Organische Chemie der Universität Essen, Postfach 103 764, D-4300 Essen 1, FRG

1 Introduction

Substituent effects have fascinated organic chemists for generations and their study is still an active area of research. The generalization of the influence of substituents is expected to lead to an understanding of physical properties, structures, equilibria and reactions in organic chemistry (Schleyer, 1987). Substituents can be considered as perturbations of a given standard system and it is often believed that their character remains basically unaltered from one molecular situation to another, i.e. an invariable universal nature of a substituent is assumed.

It is general practice to divide the influence of substituents into steric effects (for a review, see Gallo, 1983) and polar effects (Reynolds, 1983;

ADVANCES IN PHYSICAL ORGANIC CHEMISTRY
VOLUME 26   ISBN 0-12-033526-3

Topsom, 1976) and to treat them separately. In this review we will be concerned solely with polar or electronic substituent effects. Although it is possible to define a number of different electronic effects (field effects, σ-inductive effects, π-inductive effects, π-field effects, resonance effects), it is customary to use a dual substituent parameter scale, in which one parameter describes the polarity of a substituent and the other the charge transfer (resonance) (Topsom, 1976). In terms of molecular orbital theory, particularly in the form of perturbation theory, this corresponds to a separate evaluation of charge (inductive) and overlap (resonance) effects. This is reflected in the Klopman–Salem theory (Devaquet and Salem, 1969; Klopman, 1968; Salem, 1968) and in our theory (Sustmann and Binsch, 1971, 1972; Sustmann and Vahrenholt, 1973). A related treatment of substituent effects has been proposed by Godfrey (Duerden and Godfrey, 1980).

Effects of single, isolated substituents on physical properties, reactions, or equilibria can be treated successfully on the basis of dual substituent parameter (DSP) approaches. To deal with di- or poly-substituted systems is less straightforward. Here the question arises of how individual substituents combine their respective actions. It is in this context that the captodative substituent effect has to be seen.

2 Capto and dative substituents

The concept of captodative substitution implies the simultaneous action of a *captor* (acceptor) and a *donor* substituent on a molecule. Furthermore, in the definition of Viehe *et al.* (1979), which was given for free radicals, both substituents are bonded to the same or to two vinylogous carbon atoms, i.e. 1,1- and 1,3-substitution, and so forth is considered. One might, however, also include 1,2-, 1,4-, ... disubstitution, a situation which is more often referred to as "push–pull" substitution. Before discussing captodative substituent effects it might be helpful to analyse the terms *capto* and *dative* in more detail.

At first glance the significance of these two terms seems to be obvious. A substituent which withdraws electron density is a *captor* and a substituent which donates electron density is a *donor*. However, both properties cannot be discussed independently from a partner from which they accept, or to which they donate, electrons. This raises the question of whether it is in principle possible to define a universal donor or acceptor character for a substituent.

The nature of the atom to which the substituent is attached plays a decisive role in the behaviour of the substituent. The relative electronegativity of the atoms concerned determines whether electron density is withdrawn or donated by the polar effect. In organic chemistry, substituents are preferentially bonded to carbon atoms. Even though the central carbon atoms may be of different hybridization and charge character, being therefore of different effective electronegativity, more or less general substituent scales can nevertheless be defined. The different nature of a carbon atom and its immediate neighbourhood causes a fine tuning, but does not alter the character of a substituent markedly. However, if a substituent is attached to an atom of an electronegativity which differs greatly from carbon then it cannot be expected that identical substituent scales will be valid.

Even though the classification of substituent effects into polar and resonance contributions seems reasonable, it will be difficult to treat them completely independently, particularly in those cases where both effects can operate simultaneously. In molecular orbital terms, resonance effects derive from the interaction of occupied with unoccupied molecular orbitals. The extent of stabilization, however, is a function of the energetic separation of the interacting orbitals. The latter is strongly influenced by the nature of the atoms involved in the interacting molecular orbitals, i.e. by polar effects. The amount of overlap between the orbitals is also important. The polar effect polarizes the electron distribution in the occupied molecular orbitals due to electronegativity differences. Thus, the resulting action of a substituent, which is electron withdrawing through a polar effect and which is also capable of resonance, represents the sum of these two contributions.

Despite these difficulties, it appears that the most potent substituent scales are those where a DSP set is used (Topsom, 1976) instead of a single σ-value for the electronic effect. While one σ-inductive (σ_I) parameter is used in all molecular situations, it seems preferable to apply several σ-resonance (σ_R) parameter sets. Here, the system to which the substituent is attached is taken into account. This corresponds to the fine tuning above mentioned. Some values for common substituents are given in Table 1 (Topsom 1976).

Table 1 σ_I- and σ_R-Values for some common substituents.[a]

	σ_I	σ_R^0	σ_R^-	σ_R^+
$N(CH_3)_2$	0.06	−0.52	−0.34	−1.75
NH_2	0.12	−0.48	−0.48	−1.61
OCH_3	0.27	−0.45	−0.45	−1.02
F	0.50	−0.34	−0.45	−0.57
Cl	0.46	−0.23	−0.23	−0.36
C_6H_5	0.10	−0.11	0.04	−0.30
CH_3	−0.04	−0.11	−0.11	−0.25
CF_3	0.45	0.08	0.12	0.08
CN	0.56	0.13	0.33	0.13
$COCH_3$	0.28	0.16	0.47	0.16
NO_2	0.65	0.15	0.46	0.15

[a] Topsom (1976).

According to the σ_I-values all substituents, except methyl, appear to be electron captors even though some values are very small, for instance the dimethylamino- and phenyl groups. The σ_I-scale reflects mainly the difference in electronegativity of the atoms concerned and, as most substituents in organic chemistry incorporate heteroatoms of higher electronegativity than carbon, the σ_I-constants are practically all of identical sign. Thus, in the sense of the polar effect, most common substituents have to be classified as captors. The situation could be different, for example, if heteroatoms like boron or silicon are the central atoms of a substituent. The σ_R-values disclose two classes of substituents, those with electron-donating (negative σ_R) and those with electron-accepting (positive σ_R) properties. This is the basis for the definition of the term "captodative". It has been given intuitively and concentrates on resonance (charge-transfer) effects. The deciding classification property thus derives from resonance, but the direction of charge transfer is connected closely to the polar properties of the atoms of a substituent. Where both σ-values carry the same sign no ambiguity will arise as to the total substituent character; conflicts may arise in those examples where σ_I and σ_R have opposite sign.

Table 2 $\sigma_\alpha^{\bullet}$-Values.[a]

Substituent	$\sigma_\alpha^{\bullet}$	Substituent	$\sigma_\alpha^{\bullet}$
4-SCH$_3$	0.063	4-Cl	0.011
4-COCH$_3$	0.060	4-i-C$_3$H$_7$	0.009
4-SPh	0.058	4-t-C$_4$H$_9$	0.008
4-COPh	0.055	4-S(O)$_2$CH$_3$	0.005
4-COOCH$_3$	0.043	3-CH$_3$	0.002
4-CN	0.040	4-OCOPh	0.000
4-SCOCH$_3$	0.029	H	0.000
4-S(O)Ph	0.026	3-OCH$_3$	−0.001
4-OCH$_3$	0.018	3-OPh	−0.002
4-OPh	0.018	4-OCOCH$_3$	−0.005
4-S(O)CH$_3$	0.018	2-F	−0.007
4-S(O$_2$)Ph	0.018	3-Cl	−0.007
4-Si(CH$_3$)$_3$	0.017	3-F	−0.009
4-S(O)OMe	0.016	4-CF$_3$	−0.009
4-CH$_3$	0.015	4-F	−0.011
4-SO$_3$CH$_3$	0.013	3-COOCH$_3$	−0.014
4-C$_2$H$_5$	0.012	3-CF$_3$	−0.017
		3-CN	−0.026

[a] Arnold (1986).

The σ-scales in Table 1 were derived for molecules with a closed shell of electrons and not for free radicals. However, the concept of captodative substituent effects was proposed for free radicals. Several attempts have been made to define σ-values for free radicals; for a critical discussion see Jackson (1986). The largest parameter set was delineated from esr hyperfine coupling constants for benzylic radicals by Arnold (Dust and Arnold, 1983; Wayner and Arnold, 1984). Values are listed in Table 2. This scale makes no distinction between polar and resonance effects, i.e. it combines both contributions in a single number, as does the original Hammett σ-scale. It is not obvious how the separation into a set of donor or acceptor substituents could be done on the basis of this scale. The $\sigma_\alpha^{\cdot}$-values were proposed as a measure of the radical stabilizing power of the substituents even though the observed changes in the hyperfine coupling constants of the α-hydrogens might not only reflect the changes in spin delocalization on introduction of the substituent (see below).

As a summary of these considerations we must conclude that on the basis of polar effects most substituents are *captors* and that it is the resonance effect which leads to the discrimination of two classes of substituents.

3   Historical aspects

In 1952, Dewar devoted one of the papers in his famous series on perturbational molecular orbital theory of organic chemistry to free radical chemistry (Dewar, 1952). In Theorem 65 he states:

A +E substituent R and a −E substituent T can conjugate mutually through an odd alternant hydrocarbon radical S if R, T are both attached to active atoms in S.

Mutual conjugation of, as we describe it nowadays, +M and −M substituents is equivalent to an *extra* stabilization of the system. Thus, we can interpret this statement as the first formulation of the captodative effect even though the term was coined much later. The difficulty of organic chemists to comprehend the rather mathematically formulated theorem must have hindered its wider recognition and seems to be the reason that the phenomenon of interaction of +M and −M substituents has been reinvented several times since then. It is remarkable that these rediscoveries were always initiated by experimental studies in free radical chemistry.

Balaban (1971; Balaban et al., 1977) investigated radicals of type [1] by esr spectroscopy and noted their long lifetime, which he attributed to the "push–pull" character of the substituents involved and their mutual conjugation. Katritzky (Baldock et al., 1973, 1974; Katritzky and Soti, 1974) recognized in an analysis of merocyanines that there should be a related class of free radicals [2] which, in accordance with the stability of merocyanines, should exhibit high stability as a consequence of the interaction of $+M$ and $-M$ substituents. He called this "merostabilization" and provided a number of examples. He also pointed out the merostabilized character of radicals [3]–[7], examples which had been known for some time. Balaban's and Katritzky's ideas seem to have evolved in parallel and independently of each other.

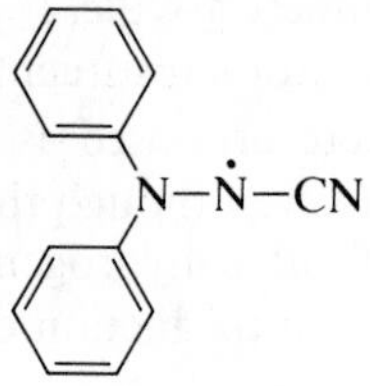

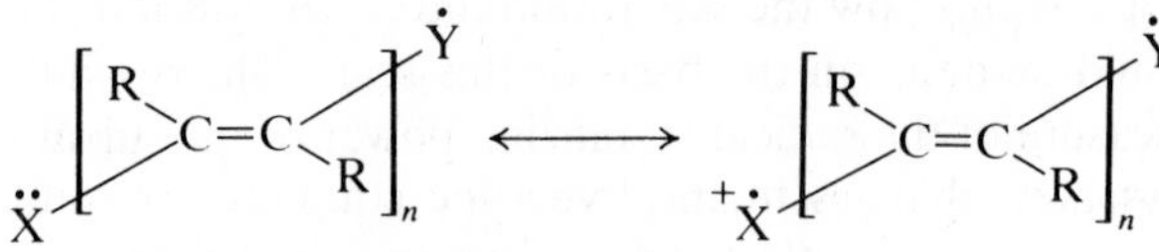

$$X = R_2N—, RS—, RO—$$
$$Y = N—R, —O$$

[1]   [2]

$$R_2\overset{\cdot+}{N}—\overset{-}{O} \longleftrightarrow R_2N—\overset{\cdot}{O}$$

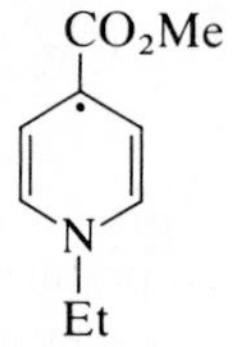

[3] 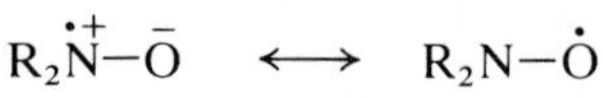   [4] 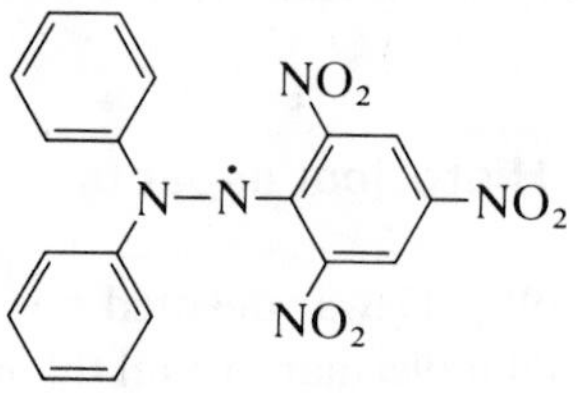   [5]

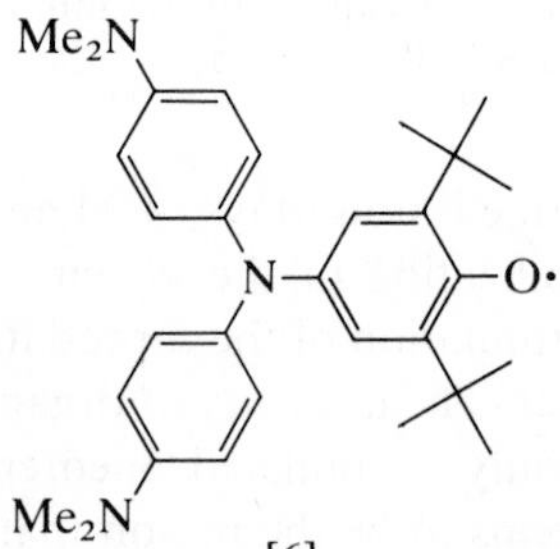

[6] 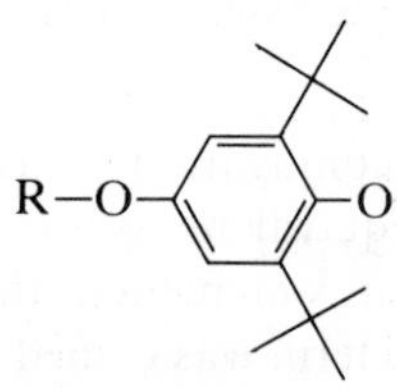   [7]

The familiarity with qualitative valence bond descriptions of substituent effects in combination with the known substituent effects in carbocations and carbanions led Viehe and his group to the postulate of a captodative effect for free radicals (Stella et al., 1978; Viehe et al., 1979). They did not seem to be aware of the earlier work which was of a more physical organic character. The fact that carbocations [8] are stabilized by $+M$ substituents, and carbanions [9] by $-M$ substituents, raised the idea that free radicals, as the intermediate class of compounds, should be particularly stabilized if both types of substituents were attached to the radical centre or to vinylogous positions [10]. Formulae [11a, b, c] for the alkoxy alkoxycarbonyl-substituted methyl radical make clear that polar resonance structures can be drawn which qualitatively reflect an enhancement of the stability of these radicals. As a consequence of the contribution of these polar structures to the ground state of the radicals, it is believed that polar solvents will exert a favourable influence on their stability (see below). This conceptually simple approach was immediately attractive to organic chemists and initiated numerous studies in the field of free radical chemistry. The implication of Viehe's work is that captodative substitution causes a stabilization which surpasses an additive effect of the two substituents or is at least greater than the stabilization by two substituents of the same kind and that this should have chemical consequences (Viehe *et al.*, 1979, 1985, 1988, 1989). Captodative, as originally defined by Viehe (Viehe *et al.*, 1979) meant a synergetic (sometimes also called synergistic) action of the substituents, i.e. more than an additive substituent effect. A continuing controversy has evolved as to the existence of such a synergetic substituent effect.

Before discussing experimental studies to determine quantitatively the influence of substituents on radical centres some theoretical approaches will be discussed.

$$
\begin{array}{ccc}
d{-}\overset{\displaystyle|}{\underset{\displaystyle d}{C}}{}^{+} & c{-}\overset{\displaystyle|}{\underset{\displaystyle c}{C}}{}^{-} & c{-}\overset{\displaystyle|}{\underset{\displaystyle d}{C}}{}^{\displaystyle\cdot}
\end{array}
$$

d = donor   c = captor

[8]   [9]   [10]

$$
\begin{array}{ccc}
O{=}\overset{\displaystyle\cdot}{C}{-}\overset{}{C}H{-}OR & O{=}\overset{\displaystyle-}{C}{-}\overset{\displaystyle\cdot+}{C}H{-}\overset{}{O}R & \overset{\displaystyle-}{O}{-}C{=}CH{-}\overset{\displaystyle\cdot+}{O}R \\
\ \ \ |\ \ \ & \ \ \ |\ \ \ & \ \ \ |\ \ \ \\
\ \ OR & \ \ OR & \ \ OR
\end{array}
$$

[11a]   [11b]   [11c]

4 Theoretical studies

The first indication of the existence of a captodative substituent effect by Dewar (1952) was based on π-molecular orbital theory. The combined action of the π-electrons of a donor and a captor substituent on the total π-electron energy of a free radical was derived by perturbation theory. Besides the formulation of this special stabilizing situation and the quotation of a literature example [5] (Goldschmidt, 1920, 1929) as experimental evidence, the elaboration of the phenomenon was not pursued further, neither theoretically nor experimentally.

When Viehe described the captodative effect, frontier molecular orbital (FMO) theory had become fashionable and an interpretation of the effect was attempted on this basis (Viehe *et al.*, 1979). The qualitative discussion, however, does not provide a satisfying picture. A more thorough derivation of the influence of a captor and a donor substituent on the π-electron energy of a radical in terms of one-electron theory was given by Klessinger (1980). He showed that the effect of two substituents with opposing electronic influences on the singly occupied molecular orbital (SOMO) is not the reason for a special substituent effect. As is evident from Fig. 1, the energy of the SOMO closely resembles that of an unsubstituted carbon radical. Rather, the stabilization of the non-bonding electrons of the donor substituent causes the captodative effect. However, this derivation is also of a qualitative nature and does not allow a quantitative assessment of the amount of stabilization. The next step in the development of these ideas was the attempt to obtain a quantitative insight into the amount of captodative stabilization.

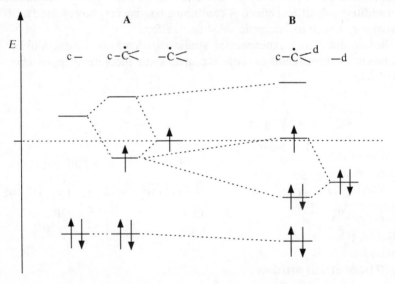

Fig. 1 FMO diagram for the formation of a captodative-substituted radical c—Ċ—d by successive interaction of (A) a carbon radical with a captor (c) and of (B) a captor-substituted carbon radical with a donor (d) substituent.

A first quantitative assessment of the effect by *ab initio* calculations was presented by Schleyer's group (Crans *et al.* 1980). They used unrestricted Hartree–Fock (UHF) calculations at the fully geometry optimized, split valence, 4-31G basis-set level to investigate the interaction of a π-donor and a π-acceptor both bonded to a CH radical centre. The existence of a synergetic captodative effect was indicated, in particular for the case of the amino-group as donor and boryl (BH_2) as acceptor (12.3 kcal mol^{-1} extra stabilization). Leroy and his group were the first to undertake a systematic *ab initio* quantum mechanical study of this problem. The topic has been treated in a number of publications (Leroy *et al.*, 1982, 1985, 1986, 1987, 1989; Leroy, 1983, 1985a,b, 1986, 1988). In order to evaluate critically the theoretical results, we briefly outline their definition of stabilization. Following earlier work (see Cox and Pilcher, 1970, for a discussion), stabilization/destabilization is defined as in (1) as the difference of the heat of atomization

$$SE^0 = \Delta H_a^0 - \sum N_{ab} E_{ab}^0 \tag{1}$$

(ΔH_a^0) and the sum of bond-energy terms (E_{ab}^0) of the species under consideration. Here, N_{ab} stands for the number of equivalent bonds having E_{ab}^0. The bond-energy terms are obtained from a set of reference compounds. The heat of atomization is evaluated from the difference of the heat of formation of the constituent atoms and the species under consideration (2). The procedure depends on the reliability of the heat of formation of the individual molecules. In cases where experimental values exist it is not necessary to have recourse to computational methods. However, the heats of formation of many molecules which are of interest in the context of captodative substitution are not known experimentally.

$$\Delta H_a^0 = \Delta H_f^0 \text{ (atoms)} - \Delta H_f^0 \text{ (molecule)} \tag{2}$$

A semiempirical procedure based on isodesmic reactions is used to determine unknown heats of formation. The choice of the isodesmic reaction is critical for the validity of the approach. By definition, all the bonds are conserved in number and nature in this type of process. Also, the bond lengths of equivalent bonds in reactants and products should be as similar as possible. The stabilization energies are based on the type of isodesmic reaction which is given for the influence of amino- and cyano-groups in a captodative radical in equation (3). Only radicals appear as reactants and

$$H_2N-\overset{\bullet}{C}H-CN + \overset{\bullet}{C}H_3 \longrightarrow H_2N-\overset{\bullet}{C}H_2 + \overset{\bullet}{C}H_2-CN \tag{3}$$

products, i.e. the requirements for a minimal deviation from the prerequisites for isodesmic reactions are fulfilled in this case. Using heats of formation of reference compounds, it is possible to determine the heat of formation of a given compound if the reaction enthalpy of the isodesmic reaction is known. Thus, from a knowledge of the heat of formation of three radicals in (3) the value of ΔH_f^0 for the fourth radical can be determined.

The combination of *ab initio* quantum mechanical calculations and experimental ΔH_f^0 values of reference compounds leads to a well-defined procedure for obtaining heats of formation. Possible errors might derive from inaccurate experimental values or from deficiencies in the *ab initio* computational methods. The calculations were carried out with the Gaussian series of programs using gradient techniques for full geometry optimizations and a 4-31G or 6-31G basis set. It should be recalled at this stage that the stabilization as defined in (1) differs from the original one of Dewar because it corresponds to a global stabilization of the species and is not restricted to the π-electron energy alone.

Some stabilization energies are collected in Table 3. The presentation provides an easy overview of the stabilization in singly and doubly substituted methyl radicals. Values in parentheses for doubly substituted radicals represent the sum of the stabilization energies derived from monosubstituted radicals. By comparing the values calculated directly for the doubly substituted radical, information is obtained on antagonistic, additive or synergetic substituent effects. Apart from methyl and ethyl radicals, all other radicals are stabilized. Some points merit comment.

Table 3 Influence of substituents on the stabilization energies (kcal mol^{-1}) of ĊHXY radicals.[a]

X Y	H	CH₃	CN	C≡CH	COOH	OH	NH₂	CH=CH₂
H	−2.65	−0.50	6.60	10.56	3.43	2.53	7.53	13.21
CH₃		1.82	8.35	13.18	—	—	7.89	15.43
		(−1.00)	(6.10)	(10.06)	(2.93)	(2.03)	(7.03)	(12.71)
CN			3.34	—	—	5.83	14.96	16.94[b]
			(13.20)	(17.16)	(10.03)	(9.13)	(14.13)	(19.81)
C≡CH				—	—	—	17.81	—
				(21.12)	(13.99)	(13.09)	(18.09)	(23.77)
COOH					—	10.82	19.95	—
					(6.86)	(5.96)	(10.96)	(16.64)
OH						8.73	—	16.64[b]
						(5.06)	(10.06)	(15.74)
NH₂							10.21	—
							(15.06)	(20.74)
CH=CH₂								20.31
								(26.42)

[a] Leroy *et al.* (1986).
[b] Mean value calculated from the *trans*- and *cis*-configurations.

Alkyl-substituted radicals.   In the series of alkylated methyl radicals, stabilization increases with increasing degree of methylation, going parallel to expectation. The first methyl group (ethyl radical) reduces the destabilization of the methyl radical. The ethyl radical, however, still remains slightly destabilized. On adding the second and third methyl groups (i-propyl radical and t-butyl radical) a net stabilization is obtained.

Monosubstituted radicals.   A cyano-group stabilizes better than a carboxyl group, and the stabilization by a hydroxyl group is less than that by an amino-group, the latter being indeed more stabilizing than a cyano group. The strongest effects are observed for vinyl and alkynyl substituents. Introduction of a vinyl group produces the allyl radical and the calculated stabilization should be related to the resonance energy of the allyl group. Indeed, the stabilization of 13.21 kcal mol^{-1} comes close to experimentally determined values for this quantity (for a discussion see Korth *et al.*, 1981).

Disubstituted radicals.   No general statement is possible on how a second substituent will influence the global stabilization. A second captor group $(\overset{\cdot}{C}H_2CN \rightarrow \overset{\cdot}{C}H(CN)_2)$ decreases the stabilization; not only is the total stabilization not additive but, even worse, it is only about 50% of the stabilization in the cyanomethyl radical. On the other hand, the stabilization by a second hydroxyl group $(\overset{\cdot}{C}H_2OH \rightarrow \overset{\cdot}{C}H(OH)_2)$ is higher than that of the first, in total being higher than twice the value of the monohydroxyl radical. Here, a peculiarity of the approach becomes apparent. The radical chemist would like to know the influence of the substituent on the radical centre, isolated from other contributions of the substituent to the stability of the radical species as a whole. The present procedure provides total substituent effects on the energy of the molecule. The high stabilization by the second hydroxyl group originates from the strong closed-shell anomeric stabilization which exists in the *molecule* and is not the consequence of a substituent effect on the *radical centre*; thus, it cannot be taken as evidence for a synergetic stabilization of the radical centre. Captodative-substituted radicals do not behave uniformly. The cyanohydroxymethyl radical shows less than additive substituent stabilization, and carboxyhydroxy-, aminocarboxy- and aminocyano-methyl radicals display stabilization energies which are higher than expected from the sum of the individual substituent effects. The computed high stabilization in the case of the carboxyl-substituted captodative radicals is surprising. Qualitatively, the organic chemist would expect that the combination of a cyano-group as captor and a hydroxyl or amino-group as donor should be better. Here, again, it is likely that the total stabilization contains other than π-electron contributions. The hydrogen atom of the carboxyl group or those of the hydroxyl or amino-groups could participate in intramolecular hydrogen bonding, which would give additional stability to the system.

Polysubstituted radicals. In these cases no additivity of substituent effects is found.

The foregoing discussion shows that the approach taken does not necessarily provide the organic chemist with an answer to the question of special effects on the radical centre in captodative-substituted radicals. Stabilization of the radical centre and stabilization of the complete radical structure must be considered separately. It is only the latter situation which can be dealt with by the approach of Leroy and coworkers.

As mentioned above, different approaches to the evaluation of the stabilization in free radicals do not necessarily lead to identical results. Pasto (Pasto *et al.*, 1987; Pasto, 1988), also performing *ab initio* calculations, used the isodesmic reaction (4) for the evaluation of substituent effects. The heat

$$X_n\dot{C}H_{3-n} + CH_4 \longrightarrow X_nCH_{4-n} + \dot{C}H_3 \tag{4}$$

of this reaction is called the resonance-stabilization energy (*RSE*). In our view, the isodesmic reaction selected by Leroy fulfills the requirements of cancellation of correlation effects better than (4). Also, the *ab initio* procedures used were not identical. Calculations on the various species were carried out at the ROHF 4-31G level with full geometry optimization. It is therefore of interest to see whether the *RSE* values obtained by Pasto are close to the *SE* values of Leroy (see Table 4).

Some of the results will be discussed in detail.

Monosubstituted radicals. All substituents are stabilizing. The amount of stabilization for cyano-, ethynyl and, in particular, vinyl groups is smaller than in the Leroy scheme, but higher for carboxyl, hydroxyl and amino-groups. The total range of stabilization energies is compressed compared with Leroy's data.

Disubstituted radicals. Whereas dihydroxyl substitution at the radical centre gave more than additive stabilization in Leroy's results, just the opposite is true in this case. This could result from a different geometry which does not account for the anomeric effect. Contrasting results are also obtained for the dicyanomethyl radical, which demonstrates a stabilization slightly more than additivity. Cyanohydroxyl captodative substitution also shows slightly more than additive stabilization. This statement differs from that in the original literature (Pasto, 1988). In the evaluation of Δ*RSE* (the sum of the effects in the monosubstituted radicals minus *RSE* for the disubstituted radical) the value for the formyl group has probably been taken erroneously for that of the cyano-group. Aminocarbonyl substitution is an example where the postulate of a synergetic substituent effect is also fulfilled. An alkynyl or vinyl group in combination with a cyano-group also leads to a stabilization which is more than additive. Hydroxyl and alkynyl groups show stabilization which is less than additive, and hydroxyl and vinyl groups show a stabilization that is more than additive. As can be deduced from the ΔRSE values there are cases of near additivity, and also of less or more than additivity. However, the results show that no general statement can be made.

Table 4 Resonance-stabilization energies (RSE) (kcal mol^{-1}) for substituted methyl radicals ĊHXY.[a]

Substituents			
X	Y	RSE	ΔRSE[b]
CH_3	H	3.27	
CN	H	5.34	
CHO	H	7.66	
C≡CH	H	8.00	
COOH	H	5.72	
OH	H	5.73	
NH_2	H	10.26	
CH=CH$_2$	H	7.80	
CH_3	CH_3	5.81	−0.73
CN	CN	11.17	+0.49
C≡CH	C≡CH	16.70	+0.70
CHO	CHO	14.73	−0.59
OH	OH	4.70	−6.76
CN	C≡CH	14.30	+0.96
CN	CHO	12.35	−0.65
CN	OH	11.35	+0.26
CN	NH_2	16.73	+1.13
CN	CH=CH$_2$	15.81	+2.67
OH	C≡CH	13.26	−0.47
OH	CH=CH$_2$	14.37	+0.84

[a] Pasto et al. (1987); Pasto (1988).
[b] ΔRSE are the differences between the sum of stabilization of single substituents and the value for the disubstituted radical. A positive sign indicates more than additivity.

A special property of captodative radicals can neither be recognized in Leroy's nor in Pasto's work. The substituents may show additivity, less than additivity or, even, more than additivity in the calculated stabilization. If we combine the results in both schemes it has to be concluded that captodative radicals are not a distinguishable class of radicals. They should not obey different rules from radicals stabilized by non-captodative substituents. Radicals can be stabilized by all kinds of substituents, although to a different extent. The $\sigma^{\cdot}$-values of Arnold (Dust and Arnold, 1983; Wayner and Arnold, 1984) provide good support. According to the σ_R-parameters (Table 1) one might anticipate the highest captodative substituent effect if substituents with high positive σ_R-values are combined with those having large negative values. Thus, the best combination should be dimethylamino- and nitro-groups as substituents. This example, however, has not been treated theoretically. The observation that the aminocyanomethyl radical exhibits a slightly more than additive substituent effect in both schemes might be a demonstration of the presumption that the combination of substituents with widely differing σ_R-values of opposite sign are the best choice for the realization of a captodative substituent effect.

As no clear-cut conclusion can be drawn from the analysis of these *ab initio* calculations it is not surprising that further attempts have been made to answer the question of the existence of a special captodative effect by improving the calculational methods. Clark (Clark, 1988) has carried out extensive calculations for the cyanomethyl-, aminomethyl- and aminocyanomethyl-radical system by perturbational methods including different amounts of correlation . The results using the isodesmic reactions (5)–(8) are shown in Table 5.

Table 5 Heats of reaction (kcal mol^{-1}) for the isodesmic reactions (5)–(8) for different calculational levels.[a]

ΔH_R	UHF	AUHF	UMP2	AUMP2	UMP3	AUMP3	UMP4	AUMP4
(5)	−8.8	−8.9	−11.4	−11.7	−10.9	−10.9	−11.4	−11.6
(6)	−9.8	−4.8	−1.0	−6.3	−4.6	−6.2	−5.1	−6.9
(7)	−16.8	−12.9	−14.0	−19.2	−16.1	−18.0	−17.1	−19.2
(5)+(6)−								
(7)	1.8	0.8	−1.6	−1.2	−0.6	−0.9	−0.6	−0.7
(8)	−0.1	−1.0	−4.0	−3.7	−2.7	−2.9	−3.3	−3.3

[a] Clark (1988).

$$\dot{C}H_3 + CH_3NH_2 \longrightarrow CH_4 + \dot{C}H_2NH_2 \tag{5}$$

$$\dot{C}H_3 + CH_3CN \longrightarrow CH_4 + \dot{C}H_2CN \tag{6}$$

$$\dot{C}H_3 + NH_2CH_2CN \longrightarrow CH_4 + H_2N\dot{C}HCN \tag{7}$$

$$\dot{C}H_2NH_2 + \dot{C}H_2CN \longrightarrow \dot{C}H_3 + H_2N\dot{C}HCN \tag{8}$$

The isodesmic reactions (5)–(7) follow Pasto's approach, the last one corresponding to Leroy's scheme. On the basis of the first three equations, a non-additive substituent effect is found in the UHF and the AUHF calculations. Inclusion of correlation attributes a small extra stabilization to the captodative system, between -0.6 and -1.6 kcal mol^{-1}, depending on the amount of correlation included. Considering the isodesmic reaction (8), a net extra stabilization is observed in all cases, being smallest for the calculation without correlation corrections. Inclusion of correlation of a different level by perturbational methods yields an extra stabilization between -3 and -4 kcal mol^{-1}. If one adopts Leroy's proposal to choose an isodesmic reaction where the number, the nature and the lengths of the bonds in reactants and products are as similar as possible, then the last reaction seems to be best suited to account for correlation effects. It is interesting to note that Clark's calculations demonstrate that even the best choice of isodesmic reaction does not provide the certainty that enough care has been taken of correlation effects. Assuming that these results constitute the best theoretical answer to the problem of captodative stabilization presently available, it has to be concluded that there is a small amount of extra stabilization in the aminocyanomethyl radical. However, the extra stabilization is not so large that unique properties are to be expected for these radicals.

The results also cast a shadow on the results of Leroy and Pasto. It seems that correlation corrections are indispensable. It is, however, difficult to carry out such elaborate calculations for all the systems considered by Leroy and Pasto. The point which has to be kept in mind is that each captodative or otherwise substituted radical is unique. Therefore, one cannot generalize to a whole class of similarly substituted radicals.

5 Experimental studies

ESR-SPECTROSCOPIC MEASUREMENTS

Esr spectroscopy has been used extensively in connection with the problem of radical stabilization. Two properties of the radicals are analysed to obtain information on their stability: spin-density distribution and lifetime. The former has a solid physical basis in hyperfine splitting and the universally accepted hypothesis that spin delocalization accords stability to a free radical (for a discussion, see Walton, 1984). The more the unpaired spin density is delocalized, the higher is, supposedly, the stability of the radical. This argument, as we shall see, is mainly used on a qualitative basis (see, however, Walton, 1984). The lifetime of a radical has to be defined more precisely in order to give a meaningful measure of stability. Griller and Ingold (1976) proposed the term "persistency" to describe the kinetic stability of a free radical in relation to the lifetime of the methyl radical under identical conditions. This, however, is less important in our discussion than the thermodynamic stability of free radicals. A non-persistent (transient) radical can be stabilized, e.g. the allyl radical, and a persistent radical, e.g. the 2,4,6-tri-t-butylphenyl radical, may not be stabilized. In general, it is the magnitude of the activation energy of possible termination reactions which determines the lifetime of a radical. Bimolecular reactions having low activation barriers reduce the lifetime of a given radical. Even if the radical can be observed by esr spectroscopy over a longer period of time, one has to be careful in interpreting this in terms of radical stability. It does not necessarily mean that an *individual* radical has a long lifetime. The free radical may be in equilibrium with its diamagnetic dimer and esr spectroscopy will measure the equilibrium concentration. Under these circumstances only the rate constant for combination would give a true measure of the kinetic stability. There is, however, no doubt that in many cases the intrinsic thermodynamic stability of a free radical contributes to the radical/dimer equilibrium. The bond-dissociation energy (BDE, see below) is a more appropriate measure of stability here.

In order to find out whether captodative substitution of a methyl radical can lead to persistency, the rate of disappearance by bimolecular self-reaction was measured for typical sterically unhindered captodative radicals (Korth *et al.*, 1983). The t-butoxy(cyano)methyl radical, t-butylthio(cyano)-methyl radical and methoxy(methoxycarbonyl)methyl radical have rate constants for bimolecular self-reactions between 1.0×10^8 and 1.5×10^9 l mol^{-1} s^{-1} in the temperature range -60 to $+60°C$. The diffusion-controlled nature of these dimerizations is supported by the Arrhenius activation parameters. Thus, it has to be concluded that there is no kinetic stabilization for captodative-substituted methyl radicals. On the other hand, if captodative-substituted radicals are encountered which are kinetically stabilized (persistent) or which exist in equilibrium with their dimers, then other influences than the captodative substitution pattern alone must be added to account for this phenomenon.

In addition to the stabilization by suitable substituents and the absence of other termination reactions than recombination, it is the strength of the bond formed in the dimerization which is a necessary cofactor for the observation of free radicals by esr spectroscopy. The stability of nitroxides [4] or hydrazyls [5] (Forrester *et al.*, 1968) derives not only from their merostabilized or captodative character but also from a weak N–N bond in the dimer. The same should be the case for captodative-substituted aminyls

[1] (Balaban, 1971). Kosower's radical [3] (Kosower and Poziomek, 1964) and Katritzky's merostabilized radicals [2] (Katritzky and Soti, 1974) are not only stabilized by their substitution pattern but also by an extended π-system. The captodative-substituted radicals [12] which were analysed by esr spectroscopy by Aurich (Aurich *et al.*, 1977; Aurich and Deuschle, 1981) also fall into this category. Similarly, Wurster's radical [13], belonging to the general class of semiquinone radicals generalized by Hünig and Deuchert (Deuchert and Hünig, 1978), and de Vries' *N,N*-dimethylaminodicyano-methyl radical [14] (de Vries, 1977, 1978) are of this class. Several examples of captodative-substituted radicals which are in equilibrium with their dimers, some dissociating in solution at $-50°C$, some only above $+100°C$, were reported by Viehe's group (Van Hoecke *et al.*, 1986; Nootens *et al.*, 1988; Stella *et al.*, 1980, 1981). Typically, additional stabilization by a vinyl group (allylic radicals [15]) (Van Hoecke *et al.*, 1986) or a phenyl group (benzylic radicals [16]) (Korth *et al.*, 1987) is required in order to obtain a stationary concentration of the radicals in solution.

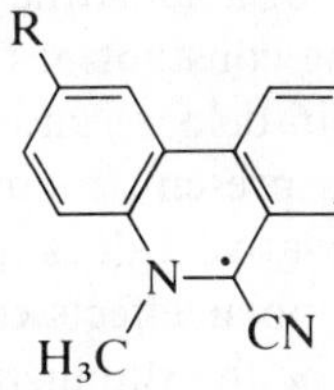

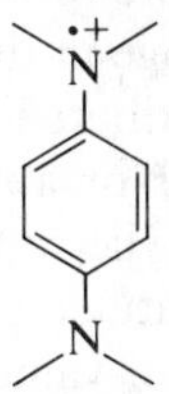

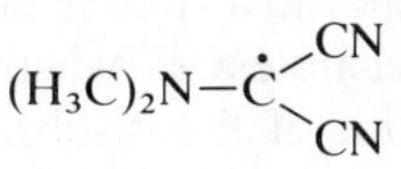

All these examples, and it would be possible to quote more, are a manifestation that captor and donor subtituents stabilize radicals. Judged by the temperature range where dissociation occurs it seems as if captodative substitution stabilizes better than dicaptor substitution (Stella *et al.*, 1981). Mostly, however, these are qualitative or semiquantitative observations which do not allow one to evaluate the magnitude of stabilization in kcal mol^{-1}. In particular, the question of a synergetic action of the captor and the donor substituent cannot be answered satisfactorily. In part, the observed effects might be related to steric interactions of the substituents.

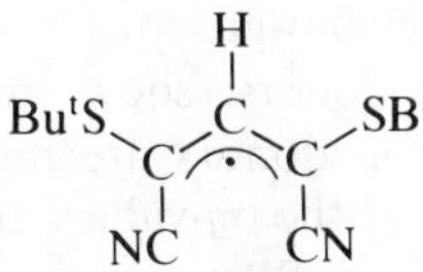

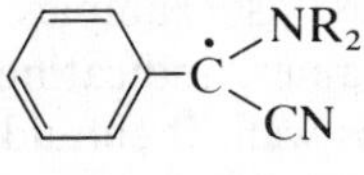

Spin-density distributions obtained from the analysis of esr spectra can be taken to establish a scale for the radical-stabilizing power of substituents.

Arnold's $\sigma_\alpha^\bullet$-scale (Arnold, 1986; Dust and Arnold, 1983; Wayner and Arnold, 1984), which reflects the spin-density distribution in *meta-* and *para-*substituted benzylic radicals and leads to parameters similar to those derived from other sources (Jackson, 1986), provide a good example. Arnold's scheme is based on the hypothesis that hyperfine coupling constants of the α-hydrogen in substituted benzylic radicals give information regarding the effect of substituents on the distribution of spin throughout the radical and that this is connected with the stability of the benzylic radical.

The definition of $\sigma_\alpha^\bullet$ is (9) in which $a(\alpha\text{-H}_x)$ and $a(\alpha\text{-H}_0)$ correspond to the hyperfine coupling constants of the benzylic hydrogen atoms in the

$$\sigma_\alpha^\bullet = 1 - [a(\alpha\text{-H}_x)/a(\alpha\text{-H}_0)] \tag{9}$$

substituted and unsubstituted radicals, respectively. The scale is reproduced in Table 2. Positive values represent a withdrawal of spin density from the α-position and negative $\sigma_\alpha^\bullet$-values indicate higher spin density. These values have been proposed as a measure of the stabilizing power of a substituent. This is justified only if the change in the hyperfine coupling constant of the benzylic α-hydrogen can be attributed to the change in spin delocalization. Regardless of whether a captor or a donor substituent is present, the $\sigma_\alpha^\bullet$-value of the 4-substituted benzylic radical is positive. Negative values are encountered mainly for 3-substituted systems, where only polar effects can operate. The more positive the $\sigma_\alpha^\bullet$-value, the better should be the stabilizing effect of the substituent. Thus, the 4-thiomethyl group heads the list of radical-stabilizing substituents, and the 4-acetyl and 4-methoxycarbonyl groups confer greater stability than the 4-cyano-group. The 4-methoxyl and 4-alkyl groups, as examples of electron donors, show smaller, but also positive, $\sigma_\alpha^\bullet$-values, indicating that electron donors are radical-stabilizing substituents as well. It should be noted that the $\sigma_\alpha^\bullet$-values are not comparable with the resonance parameters of the DSP scale discussed earlier in Section 2.

Arnold's scale is derived for the action of a single substituent on the benzylic π-system. It cannot be used to estimate the influence of several substituents on the system under consideration. In this way it is, therefore, not possible to gain insight into the problem of captodative stabilization of a radical centre. The investigation of the spin-density distribution in benzylic radicals has been extended (Korth *et al.*, 1987) to include multiple substitution patterns. Three types of benzylic radicals were considered: α,p-disubstituted α-methylbenzyl radicals [17], α-substituted p-methylbenzyl radicals [18] and α-substituted benzyl radicals [19]. In [17] and [18] the hyperfine coupling constants of the methyl hydrogens were used to determine the spin-density distribution; in [19] the p-hydrogen hyperfine splitting was taken as the measure.

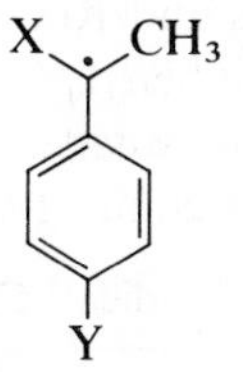

$$X \cdot CH_3 \qquad\qquad Y \cdot X \qquad\qquad Y \cdot X$$

Y   CH_3   H

[17] 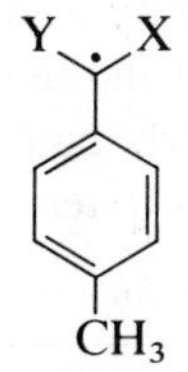   [18] 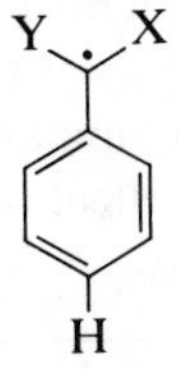   [19]

A model was adopted in which the delocalizing power of the substituent was defined analogously to Fischer's increments (Fischer, 1964) and Arnold's $\sigma_\alpha^{\cdot}$-scale (Arnold, 1986). Experimental delocalization parameters S^{exp} were calculated according to (10). By definition, S^{exp} is zero for the

$$S^{exp}(XY) = 1 - [a(XY)/a(HH)] \qquad\qquad (10)$$

unsubstituted reference radical and the value increases with increasing delocalizing power of the substituent or the substituent couple. Furthermore, for the disubstituted radicals, a quantity S^{calc} was defined (11) which

$$S^{calc}(XY) = 1 - [1 - S^{exp}(XH)][1 - S^{exp}(YH)] \qquad\qquad (11)$$

represents the superimposed effects $S(XH)$ and $S(YH)$ of two independent substituents. The relative difference ΔS between $S^{exp}(XY)$ and $S^{calc}(XY)$, defined by (12), is taken as a measure for the synergetic or antagonistic

$$\Delta S(\%) = 100[S^{exp}(XY) - S^{calc}(XY)]/S^{calc}(XY) \qquad\qquad (12)$$

behaviour of the XY couple on the spin-density distribution. ΔS reveals a synergetic effect of the substituents if it is positive and an antagonistic effect if negative. Experimental and calculated delocalization parameters for α-substituted benzyl radicals are collected in Table 6. The results for the two other types of radicals are similar but less comprehensive.

It is evident from the data in Table 6 that, with only one exception (entry 13), the combination of two captor or two donor substituents does not produce an additive effect, whereas, without exception, the captodative combinations display synergetic behaviour. Thus, the delocalization of the unpaired spin density in captodative radicals is markedly increased in comparison to pure additive superposition of "capto" and "dative" effects. This result is all the more significant since two identical substituents do not lead to an additive influence of the substituents. Comparable results were reported on the basis of the muonium hyperfine splitting constants for suitably substituted cyclohexadienyl radicals (Rhodes and Roduner, 1988). An analysis of the hyperfine splitting in substituted triphenylmethyl radicals leads to similar conclusions, although the effects are smaller due to the dilution effect of the extended π-system (Stewen and Lehnig, 1989).

Table 6 Delocalization parameters for α-substituted benzyl radicals $C_6H_5\dot{C}XY$.[a]

	Radical				
	X	Y	S^{exp}	S^{calc}	$\Delta S(\%)$
1	H	H	0		
2	Me	H	0.034		
3	OMe	H	0.074		
4	COOH	H	0.120		
5	COOEt	H	0.123		
6	COOMe	H	0.126		
7	CN	H	0.150		
8	NH$_2$	H	0.187		
9	SEt	H	0.195		
10	SMe	H	0.199		
11	Me	OMe	−0.010	0.105	−109
12	OMe	OMe	0.026	0.143	−82
13	Me	Me	0.103	0.067	+54
14	COOMe	COOMe	0.115	0.236	−51
15	COOEt	COOEt	0.123	0.231	−47
16	CN	COOMe	0.168	0.257	−35
17	CN	CN	0.187	0.278	−33
18	SMe	SMe	0.323	0.358	−10
19	SEt	SEt	0.344	0.353	−3
20	CN	Me	0.241	0.179	+35
21	COOMe	OMe	0.313	0.191	+64
22	COOMe	SMe	0.32	0.30	+7
23	COOMe	SEt	0.333	0.296	+12
24	CN	OMe	0.339	0.213	+59
25	CN	NH$_2$	0.375	0.309	+21
26	COOMe	NH$_2$	0.378	0.289	+31
27	CN	SEt	0.405	0.316	+28

[a] Korth et al. (1987).

Spin-density distributions are inherent features of free radicals. Esr experiments take place when the radical is in its electronic ground state and the measurement of the spin distribution constitutes only a minute perturbation of the system. This feature and the fact that esr hyperfine splitting can be measured with high precision makes the esr method ideally suited for the study of substituent effects. Therefore, if spin delocalization is accepted as a measure of stabilization, the data in Table 6 provide quantitative information. However, these are percentage values and not energies of stabilization. The relative stabilization, determined by the delocalization of spin density, may differ from results of the more commonly used thermodynamic methods (see below). It should be stressed that spin delocalization as measured by esr spectroscopy can be related to the overall thermodynamic stabilization of a radical (measured, for example, by thermochemical or kinetic methods) if, but only if, the interaction of the substituent with the *radical centre* constitutes the only, or at least the dominant, contribution to the stabilization of the *radical species as a whole*. This means that other (closed-shell) interactions (e.g. hyperconjugative, anomeric, polar or steric) of the substituent with the radical must be unimportant or—within a series of similar radicals—more or less constant. Examples of the latter situation seem to exist (MacInnes and Walton, 1987; Sustmann, 1986; Walton, 1984). An intriguing example that shows that non-radical interactions in a radical molecule may by far override stabilization of the radical centre has been found in the acetoxy migration in carbohydrate radicals (Korth *et al.*, 1988).

C—H BOND-DISSOCIATION ENERGIES

The determination of thermodynamic stability of a radical from C—H bond-dissociation energies (BDE) in suitable precursors has a long tradition. As in other schemes, stabilization has to be determined with respect to a reference system and cannot be given on an absolute basis. The reference BDE used first and still used is that in methane (Szwarc, 1948). Another more refined approach for the evaluation of substituent effects by this procedure uses more than one reference compound. The C—H BDE under study is approximated by a C—H bond in an unsubstituted molecule which resembles most closely the substituted system (Benson, 1965). Thus, distinctions are made between primary, secondary and tertiary C—H bonds. It is important to be aware of the different reference systems if stabilization energies are to be compared.

Some general remarks seem to be appropriate before discussing actual examples of determinations of radical stabilization by C—H BDEs. Nicholas and Arnold (1984) have stressed the point that the formation of the same radical from different precursors will give rise to different stabilization energies. This demonstrates that the leaving group influences the molecule in its ground state. They conclude that C—H and C—C BDEs constitute the best compromise in order to determine stabilization energies. Rüchardt (1970) seems to have been the first to point out the importance of ground-state effects of substituents on bond energies. In his extensive studies on C—C bond dissociations (see below) he takes into account strain energy in the starting molecules but does not admit electronic substituent effects in the ground state. The calculational results of Leroy (Leroy *et al.*, 1986) emphasize, however, that substituents act in the ground state of the radical precursor as well as in the radical produced by C—H bond dissociation. The restriction to an electronic substituent effect in the free radical only seems to be a crude approximation. Conclusions as to the existence of a captodative substituent effect on this basis could be fortuitous. In his review "Bond Dissociation Energies, Isolated Stretching Frequencies, and Radical Stabilization", McKean (1989) has discussed whether the energy of breaking a particular C—H bond is influenced by the conformation of the molecule. Thus, in toluene or propene the individual hydrogen atoms of the methyl groups are not equivalent with respect to the abstraction energetics. In most discussions these hydrogen atoms are, however, treated as equivalent. It is only the analysis of the stretching frequencies which unveils their non-equivalence. The observation confirms the existence of ground-state effects in the precursor which will be particularly important in molecules where atoms with lone pairs interact with the C—H bond to be broken.

In connection with the captodative effect, Rüchardt (Zamkanei et al., 1983) has determined the BDE of the tertiary C—H bond in [20] and compared it with the tertiary bond in isobutane. He concludes that the stabilization of $12.8 \, \text{kcal mol}^{-1}$ which he derives from this comparison falls $4 \, \text{kcal mol}^{-1}$ short of the value of $16.5 \, \text{kcal mol}^{-1}$ which he calculates for the sum of the substituent effects for phenyl ($9 \, \text{kcal mol}^{-1}$), cyano- ($5.5 \, \text{kcal mol}^{-1}$) and methoxyl ($1.5 \, \text{kcal mol}^{-1}$) groups. The latter values were derived from studies on C—C BDEs. Not even additivity of the substituent effects is observed. The existence of a captodative stabilization of radical [21] is denied (see, however, the studies on the thermolysis of [24]).

[20]   [21]

The relative rates of hydrogen abstraction from disubstituted methanes by t-butoxyl radicals were measured by Vertommen et al. (1989). The small rate acceleration found for captodative-substituted systems were of the same order of magnitude as observed for didonor-substituted systems. Recent thermochemical investigations by Rüchardt's group (Beckhaus et al., 1990; Fritzsche et al., 1989; Verevkin et al., 1990) clearly demonstrate that in geminal di- and tri-substituted methanes the interactions of the substituents have a strong effect on the ground state. It must therefore be concluded that the study of hydrogen abstraction in these systems is unsuitable for the evaluation of substituent effects in the radical state.

Bordwell (Bordwell and Bausch, 1986) has developed a method to determine C—H BDEs from a combination of pK_{HA} values and oxidation potentials (E_{ox}) of the corresponding anions in dimethyl sulphoxide solution. These acidity–oxidation potentials (AOP) are taken as measures for BDEs and are related to the stabilization of the radicals formed. This procedure has been recently applied to the subject of captodative stabilization (Bordwell and Lynch, 1989). Values of ΔBDE relative to the C—H BDE in methane are calculated according to (13). These values are set equal to the

$$\Delta BDE = 1.37\Delta pK_{HA} + 23.1\Delta E_{ox}(A^-) \qquad (13)$$

resonance-stabilization energy (RSE) of the corresponding radical. Substituted acetophenonyl radicals [22] and radicals derived from substituted malononitriles [23] provided the model systems. In Table 7 some stabilization energies are listed for substituted acetophenonyl radicals. The unsubstituted parent system [22; X = H] is a captor-substituted radical and becomes a captodative-substituted system if X=—OR, or —NR$_2$. Is there a special captodative effect in these systems, viz., at least additivity or perhaps more?

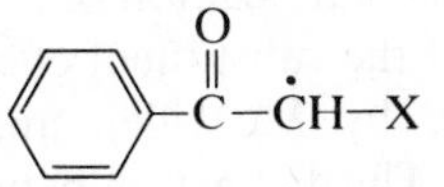

$$X = OMe, OEt, NH_2, NMe_2$$
[22]

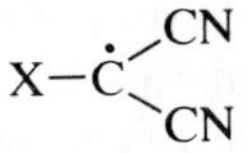

[23]

Table 7 Radical-stabilization energies (RSE) (kcal mol^{-1}) for radicals derived from α-substituted acetophenones, $PhCOCH_2R$.[a]

R	RSE
H	(0.0)
OCH$_3$	13.1
OC$_2$H$_5$	12.9
NH$_2$	~22
N(CH$_3$)$_2$	21

[a] Bordwell and Lynch (1989).

Bordwell compares the alkoxy effect with the C—H BDE for dimethyl ether as reported by McMillen and Golden (1982). The latter is lower by 12 kcal mol^{-1} than that of methane, and from the comparable magnitude of the effect in the α-alkoxy acetophenonyl radical it is concluded that an additive substituent effect exists. Similar arguments hold for the dimethylamino-substituted radical. It is stated that additivity is more than expected for bis-captor-, bis-donor- or other poly-substituted systems, and, therefore, a synergetic action of the captor and donor substituent is postulated for the systems under study. Saturation and steric effects, however, can overshadow electronic substitutent effects in similar systems (Bordwell *et al.*, 1990). The separation and different interpretation of the terms synergetic (synergistic) and captodative by these authors seems rather artificial and does not follow the intention of Viehe (Viehe *et al.*, 1985).

McKean's work, where BDEs are predicted from stretching frequencies and where radical-stabilization energies for molecules exhibiting conformational multiplicity can be derived, should be mentioned in this context. From an analysis of the stretching frequencies he predicts two different C—H BDEs for dimethyl ether, 102.9 and 93.9 kcal mol^{-1}. The value of 93.9 kcal mol^{-1} is allocated to a C—H bond *trans* to a lone pair of oxygen. The ground-state destabilization of an adjacent C—H bond is one of a number of phenomena induced by lone pairs (McKean, 1978; Schäfer *et al.*, 1981; Williams *et al.*, 1981). The BDE determined by thermochemical measurements, is 93.3 kcal mol^{-1} (McMillen and Golden, 1982), i.e. it is similar to the weaker C—H bond as deduced from the stretching frequencies. According to McKean there is, therefore, almost no stabilization of a radical centre by a methoxyl group. On the other hand, if the substituent effect is ascribed to the radical, as is done, for instance, by McMillen and Golden, a stabilization of 12 kcal mol^{-1} is calculated. The distinction between different types of C—H bonds in these systems by McKean corresponds to what is expected on the basis of theoretical considerations, and it provides a more realistic picture in cases where conformational ambiguities are present. However, it also leads to a different interpretation: the lowering of the C—H BDE is due to a ground-state destabilization of the parent molecule and does not reflect a stabilization of the radical. It seems to be completely unrealistic to incorporate McKean's interpretation into Bordwell's work and to conclude that the value of 12.9 kcal mol^{-1}, reported in the latter work, constitutes an increase over additive stabilization of the α-methoxy acetophenonyl radical. This value could reflect a destabilization of the ground state as well, either of the α-methoxyacetophenone or the corresponding anion.

The discussion shows that it is difficult and sometimes ambiguous to interpret C—H BDEs in terms of radical stabilization only. Consequently, their usefulness in the context of captodative substitution appears to be questionable.

C—C BOND-DISSOCIATION ENERGIES

The formation and the breaking of C—C bonds is one of the basic reactions in organic chemistry. Experimental studies on C—C BDEs have been of the utmost importance in the development of the theory of organic chemistry (Rüchardt, 1984). In recent years, Rüchardt and his group (Birkhofer *et al.*, 1986, 1989; Rüchardt and Beckhaus, 1986) have systematically studied various factors that influence the strength of the C—C bond. They established a linear relationship between BDEs and strain energies in many sterically congested, substituted ethanes (Rüchardt and Beckhaus, 1980). Through exchange of alkyl groups in these molecules by substituents like phenyl, methoxyl, cyano- and other groups they obtained measures for the change in BDE upon substitution. These values were interpreted in terms of the stabilization of the free radicals formed. The stabilization energies in Table 8 are defined for the exchange of CH_3 ($SE = 0$ kcal mol^{-1}) by X. The magnitude of the effect is similar to the substituent influence determined by other methods (for a discussion see Merényi *et al.*, 1986) but is different from what was derived from C—H BDEs (Table 7). What is the reason for this difference? As pointed out above, BDEs measure the difference between the influence of substituents in the ground state and in the radical. Obviously, the action of substituents on C—H and C—C bond energies differs, thus leading to different stabilization scales.

Table 8 Selected stabilization energies (SE/kcal mol^{-1}) of alkyl radicals.[a]

Radical	SE
$H_3C\dot{C}R_2$	$\equiv 0$
$H_5C_6\dot{C}R_2$	8.4
$RCO\dot{C}R_2$	6.5
$NC\dot{C}R_2$	5.5
$H_3COCO\dot{C}R_2$	3.5
$H_3CO\dot{C}R_2$	1.3

[a] Birkhofer *et al.* (1989).

    Rüchardt originally made the assumption that substituents do not exert a noticeable electronic effect on the ground state of the radical precursor. He attributed the full electronic substituent power to the radical. On the basis of this presumption, the BDEs in [24] and [25] were interpreted in terms of radical stability. The original value for the BDE in [24] (Zamkanei *et al.*, 1983) was later slightly modified (Birkhofer *et al.*, 1987). The BDEs were compared with those where only one cyano- or one methoxyl group was incorporated. From Table 9 (Birkhofer *et al.*, 1989) it can be derived that phenyl, cyano-, and methoxy-groups exert an additive substituent effect on

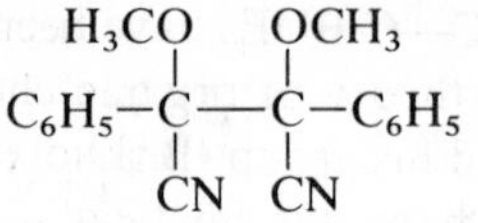

$$\underset{\underset{CN \quad CN}{|}}{\overset{\overset{H_3CO \quad OCH_3}{|}}{C_6H_5-C-C-C_6H_5}}$$

[24]

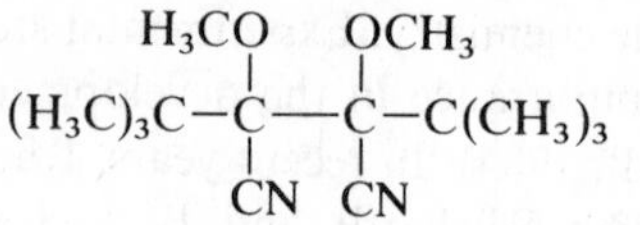

$$\underset{\underset{CN \quad CN}{|}}{\overset{\overset{H_3CO \quad OCH_3}{|}}{(H_3C)_3C-C-C-C(CH_3)_3}}$$

[25]

$$\underset{\underset{EtO_2C \quad CO_2Et}{|}}{\overset{\overset{NR_2 \quad NR_2}{|}}{H-C-C-H}}$$

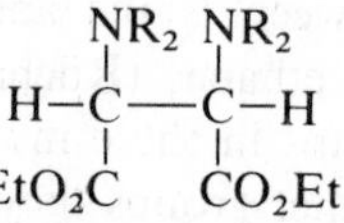

R = CH$_3$, C$_2$H$_5$, i-C$_3$H$_7$

[26]

Table 9 Stabilization energies (SE/kcal mol^{-1}) for substituted methyl radicals.[a]

Radical	SE_{exp}	SE_{calcd}
$(H_3C)_3\dot{C}$	$\equiv 0$	$\equiv 0$
$C_6H_5-\dot{C}(CH_3)CN$	15	14
$C_6H_5-\dot{C}(CH_3)OCH_3$	9.4	9.7
$C_6H_5-\dot{C}(CN)OCH_3$	14	15.2

[a] Birkhofer et al. (1989).

the BDEs. Rüchardt interprets the results as evidence for the non-existence of a special captodative effect. For this system, the conclusion seems to be correct if one adopts the notion that "captodative" corresponds to an *increase* above the additive influence. In terms of the above discussion that an additive effect is normally more than the stabilization observed for systems with two like substituents (see also Merényi et al., 1988), it can also be concluded that the additive behaviour is indeed the manifestation of a special substituent effect. The systems with two like substituents, *gem*-substituted dicyano- and dimethoxy-1,2-diphenylethanes, were also analysed. A conclusive answer was not given (Birkhofer et al., 1989) due to difficulties in dissecting ground-state effects in the radical precursors from the substituent effect in the radical. It might, however, also be that in the former cases it is not easy to separate ground-state effects properly from the influence of the substituents in the free radical. Recently, the thermolysis of diethyl 1,2-di-(*N,N*-dialkylamino)succinates [26] was investigated (Schulze et al., 1990). In this case, C—C bond cleavage leads to a radical which is stabilized by a dialkylamino- and an alkoxycarbonyl group [27]. The quantitative interpretation analogous to the above examples leads to *ca.*

4 kcal mol^{-1} more stabilization than is expected from the sum of the stabilization by the individual substituents. Even though ground-state electronic effects of the substituents in [26] are neglected, a captodative substituent effect is found for this system. The comparison of this result with the earlier ones on the thermolysis of [24] and [25] again reveals that the nature of the captor and the donor substituents plays an important role. It seems that only the strongest donor and strongest acceptor substituents lead to an increase above the additive substituent effect in these reactions. According to the σ_R-values, the dialkylamino-group is the strongest donor.

Katritzky (Katritzky et al., 1986) has recently advanced the idea that captodative-substituted radicals should be stabilized significantly by polar solvents. This hypothesis, which is qualitatively derived from the polar resonance structures for these radicals, was supported by semiempirical molecular orbital calculations. An experimental test was carried out by Beckhaus and Rüchardt (1987). For the dissociation of [24] and [25] into the radicals [21] and [28], they were unable to confirm Katritzky's hypothesis. The rate of thermolysis of [24] and [25] is not affected by a change in solvent polarity. If the stabilization were of the order of Katritzky's prediction, it should, however, have become evident in the rate measurements. The experiments thus suggest that the contribution of polar resonance structures to the ground state of the radicals is not appreciable. See, however, the results obtained by Koch (1986) on the dl–meso isomerization of [47].

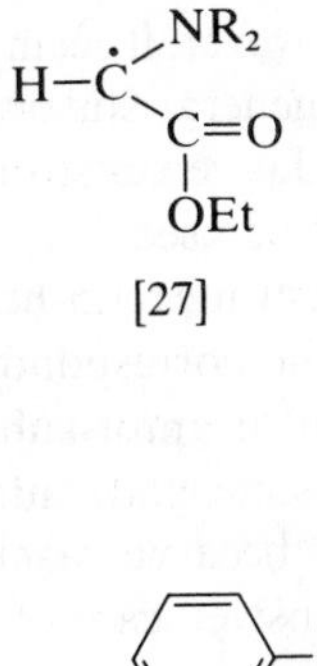

[27] 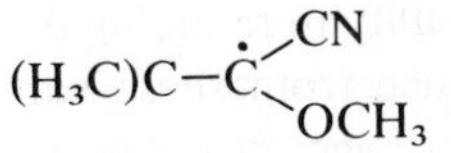   [28] 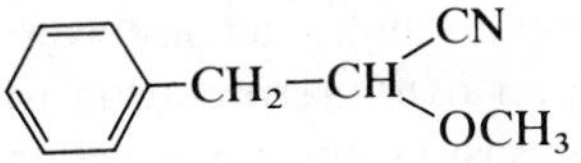   [29]

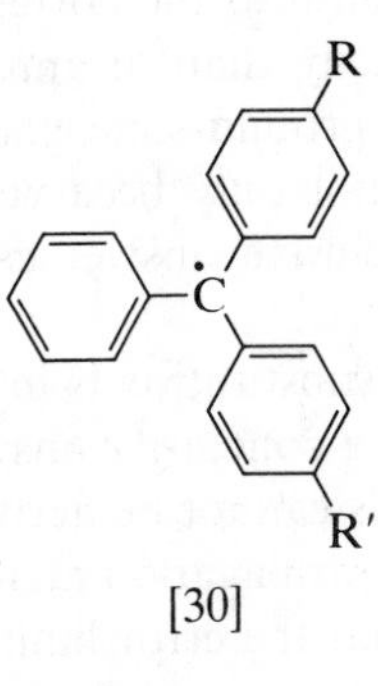

[30] 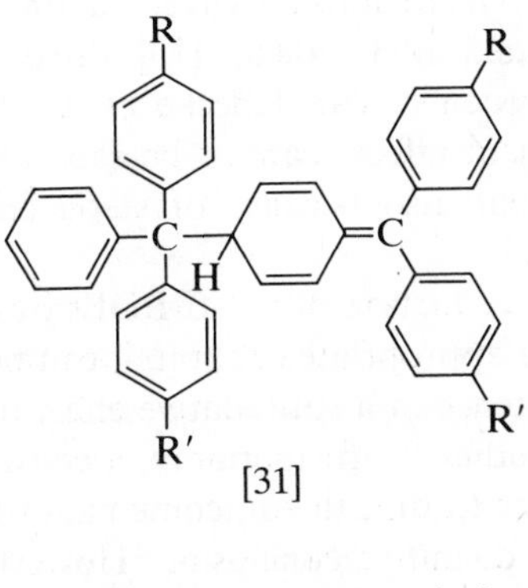   [31]

Louw and Bunk (1983) have studied the C—C bond dissociation in [29] and compared it with that in bibenzyl. They arrive at the conclusion that there is a small extra captodative stabilization of *ca.* 4 kcal mol^{-1} for the cyanomethoxymethyl radical.

Neumann (Neumann *et al.*, 1986, 1989) studied the equilibrium between 4,4'-disubstituted triphenylmethyl radicals [30] and their corresponding quinoid dimers [31] by esr spectroscopy, and determined the change in free energy for the equilibrium as a function of the substituents. It is expected that substituents will influence the degree of dissociation of the weak C—C bond in the Gomberg dimer ($\Delta G^0 = 4.75$ kcal mol^{-1} for the unsubstituted system [30; R = R' = H]). The analysis is carried out on a system which displays a small ΔG^0 for the reference compound, a fact from which one can delineate that the effect of individual substituents will be small. This consequence of "dilution" by extended conjugation in the triphenylmethyl radicals becomes evident in the experiments. The variation of the ΔG^0-values is small, in the best case about 2 kcal mol^{-1}. An expectation value ΔG^0_{calcd} is calculated according to (14) for unsymmetrically 4,4'-disubsti-

$$\Delta G^0_{calcd}(R/R') = 0.5\,[\Delta G^0_{found}(R/R) + \Delta G^0_{found}(R'/R')] \tag{14}$$

tuted triphenylmethyl radicals from the corresponding symmetrically substituted radicals. In Table 10 the ΔG^0-values for the unsymmetrically substituted systems are listed. The expected small variation of the ΔG^0-values is confirmed. The captodative cyano- and methoxy-substituted system shows a slight increase over additivity in reducing ΔG^0. However, it seems to be an overinterpretation to deduce from these values a general statement concerning the existence/non-existence of a special captodative substituent effect. Also, for a rigorous evaluation, ΔH^0 values should be used.

Similarly to the triphenylmethyl system, captodative-substituted 1,5-hexadienes, which can be cleaved thermally in solution into the corresponding substituted allyl radicals [15], dissociate more easily than dicaptor-substituted systems (Van Hoecke *et al.*, 1986). Since ground-state and radical substituent effects cannot be separated cleanly, not only because of electronic but also because of steric effects, a conclusive answer cannot be provided.

If the reduction of C—C BDEs by captodative substitution is interpreted with the appropriate caution, it can be stated that a conclusive answer as to the existence of a captodative effect in free radicals cannot be derived from these studies. If, furthermore, a consequent error-propagation analysis had been carried out, the outcome might have been that the error limits do not allow a definitive conclusion. However, the results convey a feeling that— regardless of the pros and cons for the different determination procedures— a possible captodative effect will not be great.

Table 10 Comparison of experimental and calculated ΔG^0 values (kcal mol^{-1}) for substituent combinations R/R' in dissociation reactions of dimers of substituted triphenylmethanes [30].[a]

R	R'	ΔG^0_{found}	ΔG^0_{calcd}	$\Delta\Delta G^0$
Bu^t	CF$_3$	3.31	3.56	−0.25
Bu^t	CN	2.84	3.01	−0.17
Bu^t	OMe	3.63	3.42	+0.21
Bu^t	Ph	3.53	2.70	+0.83
Bu^t	OPh	3.58	3.53	+0.05
Bu^t	SMe	3.22	2.81	+0.41
CF$_3$	OMe	3.79	3.72	+0.07
CF$_3$	Ph	3.58	3.00	+0.58
CF$_3$	SMe	3.18	3.10	+0.08
CN	OMe	2.91	3.17	−0.26
CN	Ph	3.02	2.45	+0.57
COPh	OPh	3.22	3.30	−0.08
OMe	Ph	2.73	2.86	−0.13
OMe	SMe	3.74	2.97	+0.77
Ph	SMe	2.84	2.25	+0.59

[a] Neumann et al. (1989).

ROTATIONAL BARRIERS

Several attempts have been made to analyse the captodative effect through rotational barriers in free radicals. This approach seems to be well suited as it is concerned directly with the radical, i.e. peculiarities associated with bond-breaking processes do not apply. However, in these cases also one has to be aware that any influence of a substituent on the barrier height for rotation is the result of its action in the ground state of the molecule and in the transition structure for rotation. Stabilization as well as destabilization of the two states could be involved. Each case has to be looked at individually and it is clear that this will provide a trend analysis rather than an absolute determination of the magnitude of substituent effects. In this respect the analysis of rotational barriers bears similar drawbacks to all of the other methods.

The study of substituted allyl radicals (Sustmann and Brandes, 1976; Sustmann and Trill, 1974; Sustmann et al., 1972, 1977), where pronounced substituent effects were found as compared to the barrier in the parent system (Korth et al., 1981), initiated a study of the rotational barrier in a captodative-substituted allyl radical [32]/[33] (Korth et al., 1984). The concept behind these studies is derived from the stabilization of free radicals by delocalization of the unpaired spin (see, for instance, Walton, 1984). The stabilized allyl radical will be stabilized further if substituents are introduced. This stabilization occurs to different degrees in the ground state and the transition structure for rotation. In the ground state the substituent acts on a delocalized radical. Its influence on this state should be smaller than in the transition structure, where it acts on a localized radical. In the transition state the double bond and the atom with the unpaired electron are decoupled, i.e. in the simple Hückel molecular orbital picture, the electron is localized in an orbital perpendicular to the π_{C-C}-bond.

Table 11 Rotational barriers (kcal mol^{-1}) in allylic radicals.[a]

Radical	E_a
syn-CH_2=CH—$\overset{\cdot}{C}HD$ $\rightleftharpoons$ anti-CH_2=CH—$\overset{\cdot}{C}HD$	15.7 ± 1.0
anti-CH_2=CH—$\overset{\cdot}{C}HCN$ $\rightleftharpoons$ syn-CH_2=CH—$\overset{\cdot}{C}HCN$	9.8 ± 1.1
syn-CH_2=CH—$\overset{\cdot}{C}HCN$ $\rightleftharpoons$ anti-CH_2=CH—$\overset{\cdot}{C}HCN$	10.6 ± 1.3
anti-CH_2=CH—$\overset{\cdot}{C}H(OCH_3)$ $\rightleftharpoons$ syn-CH_2=CH—$\overset{\cdot}{C}H(OCH_3)$	14.3 ± 1.4
syn-CH_2=CH—$\overset{\cdot}{C}H(OCH_3)$ $\rightleftharpoons$ anti-CH_2=CH—$\overset{\cdot}{C}H(OCH_3)$	14.7 ± 1.5
CH_2=C(C(CH$_3$)$_3$)—$\overset{\cdot}{C}(CN)_2$	9.6 ± 0.3
CH_2=C(CH$_3$)—$\overset{\cdot}{C}(CN)_2$	>12
CH_2=CH—$\overset{\cdot}{C}(OCH_3)_2$	>12
CH_2=CH—$\overset{\cdot}{C}(CN)OCH_3$	6.0 ± 0.4^b
	6.1 ± 0.6^c

[a] Sustmann (1986).
[b] syn-Cyano-conformation of primary radical.
[c] anti-Cyano-conformation of primary radical.

The experimental result seems to support this model. Table 11 lists values for rotational barriers in some allyl radicals (Sustmann, 1986). It includes the rotational barrier in the isomeric 1-cyano-1-methoxyallyl radicals [32]/ [33] (Korth et al., 1984). In order to see whether the magnitude of the rotational barriers discloses a special captodative effect it is necessary to compare the monocaptor and donor-substituted radicals with disubstituted analogues. As is expected on the basis of the general influence of substituents on radical centres, both captor *and* donor substituents lower the rotational barrier, the captor substituent to a greater extent. Disubstitution by the same substituent, i.e. dicaptor- and didonor-substituted systems, do not even show additivity in the reduction of the rotational barrier. This phenomenon appears to be a general one and has led to the conclusion that additivity of substituent effects is already a manifestation of a special behaviour, viz., of a captodative effect. The barrier in the 1-cyano-1-methoxyallyl radicals [32]/

[33] is lower by 2.9 kcal mol^{-1} than the expectation value from additivity. An error-propagation analysis allows the conclusion that the captodative substituent effect on the rotational barrier in this allyl radical is at least additive and perhaps slightly greater.

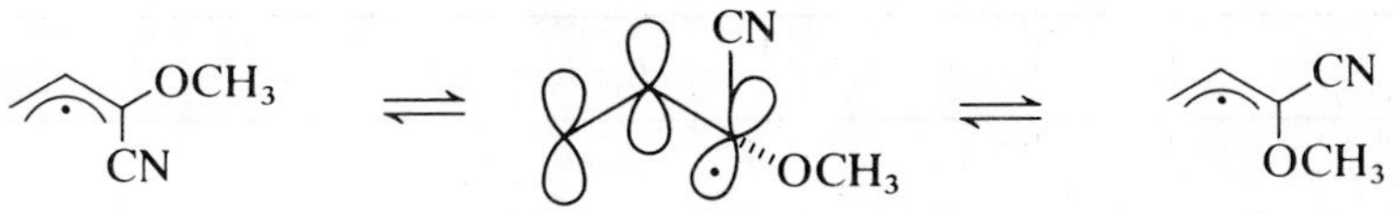

[32]   [33]

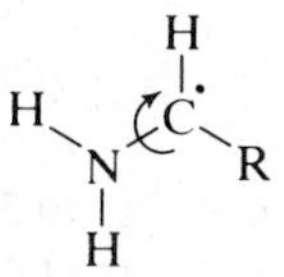

[34]

Benzylic radicals offer themselves to a similar analysis. Some barriers to rotation have been determined (Conradi et al., 1979). The barrier to rotation of 9.8 ± 0.8 kcal mol^{-1} for the α-cyano-α-methoxybenzyl radical [21] (Korth et al., 1985) could not be interpreted rigorously in terms of a captodative effect because estimates had to be made for the effect of a single captor or donor substituent on the rotational barrier. Within these limitations the barrier does not reflect more than an additive substituent effect.

An analogous approach was followed in the analysis of rotational barriers in substituted α-aminoalkyl radicals (MacInnes et al., 1985; MacInnes and Walton, 1987). The rotational barrier about the C—N bond in [34] is a measure of the ability of substituents to stabilize a radical centre by delocalization. Values for rotational barriers in these radicals are given in Table 12. Two captodative-substituted radicals are recognized (R = Bu^{t}OCO, CN). They display the highest barriers in the series. An increase of the rotational barrier with increasing delocalization is expected in these radicals because it is the ground state which is affected more by the substituent. The amino-group is, to a first approximation, uncoupled from the π-system in the transition structure for rotation. Apart from the conclusion that the barriers for rotation in the captodative-substituted radicals are significantly greater than those for the other aminoalkyl radicals, it is difficult to derive quantitative information about the captodative effect. Necessary quantitative stabilization energies for disubstituted radicals are not available (see Walton, 1984) and, in the case of the captodative-substituted radical, the influence of each individual substituent alone is not

known accurately enough. For [34] with R = Bu^{t}OCO a complicating factor
is the possibility of ground-state stabilization by hydrogen bonding.

Table 12 C—N Rotational barriers (kcal mol^{-1}) of substituted aminoalkyl radicals
H$_2$NĊHR.[a]

R	E_a
CH$_3$	7.6 ± 0.4
CH$_2$CH$_3$	∼7.5
CH(CH$_3$)$_2$	7.3 ± 0.4
C(CH$_3$)$_3$	7.5 ± 0.2
CF$_3$	10.9 ± 0.6
COOBu^t	14.9 ± 1.2
C≡CH	10.5 ± 1.2
C≡N	11.0 ± 2

[a] MacInnes and Walton (1987).

The rotational barrier about the C—O bond in the cyanomethoxymethyl
radical, [35]/[36], constitutes a similar case, although the situation is somewhat more complicated (Beckwith and Brumby, 1987). As oxygen carries
two lone pairs of electrons, the transition structure for rotation about the
C—O bond can still be stabilized by conjugation. Compared to the methoxy-
methyl radical, the barrier in the captodative-substituted radical is 1–2 kcal
mol^{-1} higher.

A last example concerns the rotational barrier in phenoxyl radicals
(Gilbert *et al.*, 1988). Compared to the parent phenols [37] and [39] the
rotational barrier in [38] is increased by a factor of seven, whereas, with a
captor substituent [40], the barrier increases only by a factor of 1.2. This
could be interpreted in terms of a captodative stabilization in [38]. The
captodative character of the radical [38] is represented by a resonance
structure [41].

The study of the rotational barriers in captodative-substituted radicals
leads to the following conclusions: the barriers are noticeably lower or
higher than in cases of dicaptor or didonor substitution. This can be
interpreted as the consequence of a captodative effect in these systems.
However, the amount of special influence on the barrier height in energetic
terms is small and may sometimes not exceed the numerical uncertainties. A
derivation of absolute values for stabilization energies of captodative-
substituted radicals by this procedure is not possible, since both ground and
transition states are affected by substitution. The lowering of the barriers

represents a decrease of the energetic separation of the two states. This problem has been discussed in detail by Leroy (1986).

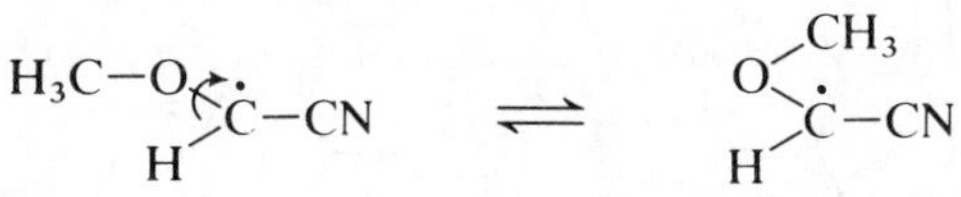

[35]   [36]

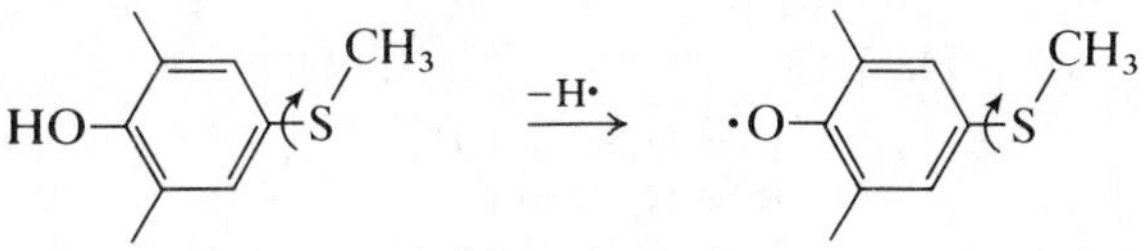

[37]   [38]

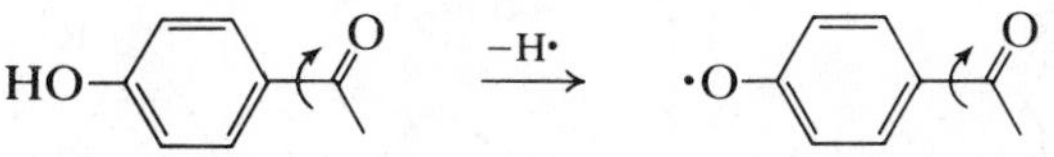

[39]   [40]

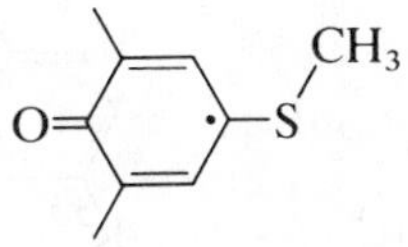

[41]

ISOMERIZATION REACTIONS

Isomerizations in which C—C bonds are cleaved homolytically have been chosen several times as probes for the study of substituent effects on radical stabilization. The nature of the intermediates—in some cases there may not even be intermediates but only biradical-like transition states—is often not known in detail. It may thus be uncertain whether the radicals include fully evolved radical centres, especially in the case of intramolecular isomerizations where biradicaloids might be involved. On this basis it is not expected that stabilization energies which derive from rate measurements for isomerizations will be identical to those obtained by other procedures.

Leigh and Arnold (1981) investigated *cis–trans* isomerizations about C—C bonds in substituted tetraphenylethylenes [42]. This isomerization involves a biradical or biradical-like transition structure which is attained by a 90° rotation about the double bond. The predominant action of the substituent is assumed to become evident mainly at the orthogonal stage of

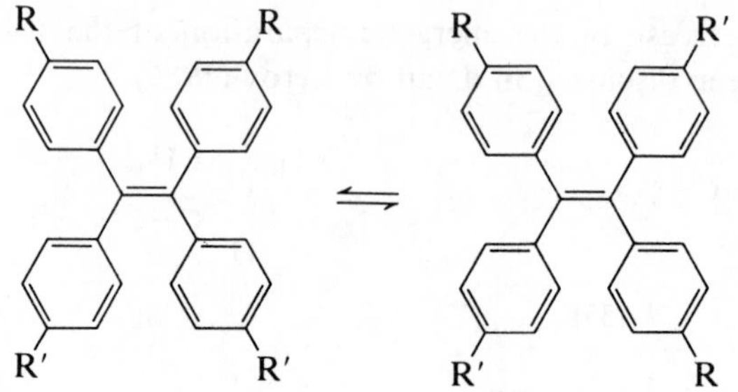

[42a]   [42b]

$$R = H, R' = CH_3$$
$$R = H, R' = CN$$
$$R = H, R' = OCH_3$$
$$R = CN, R' = OCH_3$$

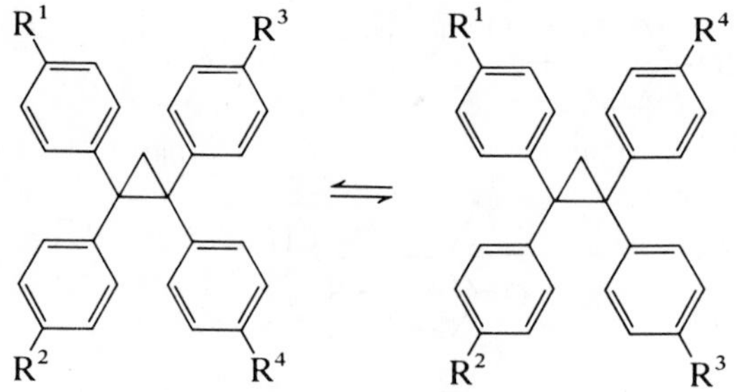

[43a]   [43b]

$$R^1 = R^3 = H, R^2 = R^4 = OCH_3$$
$$R^1 = R^3 = H, R^2 = R^4 = CN$$
$$R^1 = R^3 = H, R^2 = OCH_3, R^4 = CN$$
$$R^1 = R^3 = OCH_3, R^2 = R^4 = CN$$

the reaction. Any influence of the substituent in the ground state is neglected. Table 13 displays the Arrhenius activation parameters for isomerization, evaluated according to a non-linear least-squares analysis of the Arrhenius equation. The monosubstituted systems show very similar activation energies, whereas the barrier for the captodative-substituted molecule is about 2 kcal mol^{-1} lower. Thus, a slight acceleration is encountered. A plot of the rate data against the sum of the $\sigma_\alpha^{\cdot}$-values (Nicholas and Arnold, 1981) for the *para*-substituents gives a linear correlation in the case of the monosubstituted examples. The captodative system reacts faster by a factor of two than expected from the correlation line. If a significant captodative effect is postulated, it has to be concluded that the small observed effect is due to cancellation of ground and transition-state effects, or it has to be admitted that, indeed, the captodative effect consists only of a small perturbation. Whether the observed effect should be termed captodative seems to be disputable.

Table 13 Arrhenius activation energies and pre-exponential factors for thermal isomerization of tetraphenylethenes [42] and the standard enthalpy differences in benzene solution.[a]

R	R'	E_a/kcal mol^{-1}	10^{-12} A/s^{-1}	ΔH^0/kcal mol^{-1}
H	CH$_3$	35.5 ± 0.6	1.8 ± 1.2	0.1 ± 0.1
H	CN	35.3 ± 0.6	2.3 ± 1.8	0.1 ± 0.2
H	OCH$_3$	35.1 ± 0.3	2.2 ± 0.8	0.0 ± 0.2
CN	OCH$_3$	32.9 ± 1.2	0.6 ± 0.8	0.5 ± 0.2

[a] Leigh and Arnold (1981).

Cis–trans isomerization of analogously substituted tetraphenylcyclopropanes [43] gave similar results (Arnold and Yoshida, 1981). The isomerization via an assumed intermediate 1,3-biradical requires again 2 kcal mol^{-1} less activation energy for the captodative-substituted molecule. Whereas it is likely that steric and hyperconjugative effects of the substituents play a negligible role in systems [42] and [43], the "dilution" of the influence of the substituents by the extended π-system diminishes the predictability of this approach.

Table 14 Arrhenius activation energies and free energies of activation (kcal mol^{-1}) for isomerization of substituted cyclopropanes in CDCl$_3$.[a]

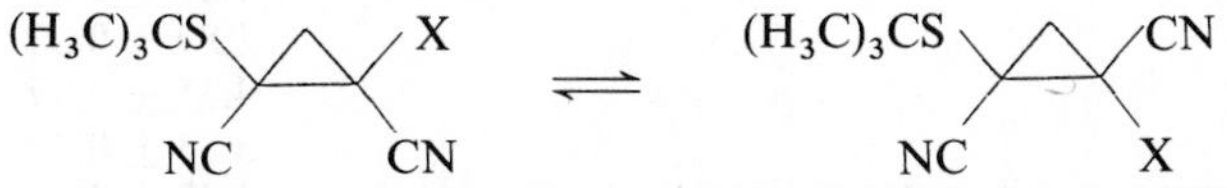

$$(H_3C)_3CS \quad \diagdown \quad X \quad \rightleftharpoons \quad (H_3C)_3CS \quad \diagdown \quad CN$$
$$NC \qquad CN \qquad\qquad\qquad NC \qquad X$$

Substituent	$E_a(cis \to trans)$	$E_a(trans \to cis)$	$\Delta G^{\ddagger}_{373}$ [b]
CO$_2$CH$_3$	31.7	31.7	31.0
Cl	28.2	29.4	29.3
C$_6$H$_5$	27.0	26.9	25.5
C$_6$H$_5$S	26.9	26.9	27.4
H$_3$CO	24.1	24.2	28.4

[a] Merényi *et al.* (1983, 1986).
[b] Average value for *cis → trans* and *trans → cis* isomerization.

Viehe's group has studied *cis–trans* isomerizations of captodative-substituted cyclopropanes in more detail (Table 14) (Merényi *et al.*, 1983; De Mesmaeker *et al.*, 1982). The lowest activation energy is observed for X = OCH$_3$ and the highest for X = CO$_2$CH$_3$. Thus, a donor instead of a captor substituent, giving the assumed biradical intermediate a disubstituted capto-dative character, lowers the activation energy by about 7 kcal mol^{-1}. Obviously, this can be interpreted in terms of an energetically favourable captodative intermediate. However, it is impossible to decide whether the lowering of the barrier is due to ground-state destabilization, to transition-state stabilization or to both. As in the other approaches where C—C bonds are being broken (p. 154), hyperconjugative interactions or polar effects (dipole interactions) might strongly influence the strength of the central C—C bond. It is difficult to extract more conclusive evidence from these data, in particular about the magnitude of a possible captodative effect. If, for the moment, we attribute the total lowering of the activation energy to the stabilization of a biradical intermediate, it is still required to compare this with the effect of other substituents in cyclopropane isomerization. For instance, two phenyl groups in the 1,2-positions of a cyclopropane (Rodewald and DePuy, 1964) give an activation energy for isomerization which is not far off the value for the captodative-substituted cyclopropane. This suggests either that the amount of captodative stabilization is notice-able, or that, indeed, ground-state destabilization contributes significantly to the lowering of the barrier for isomerization. A clear, quantitative analysis of these data in terms of a captodative effect is thus impossible.

Table 15   Rearrangement rates of [44] in isooctane at 50°C.^{a}

Substituent	k_{rel}
4-H	1.00
4-CO$_2$C$_2$H$_5$	1.63 ± 0.04
4-CH$_3$	1.89 ± 0.03
4-OCH$_3$	4.35 ± 0.09
4-SCH$_3$	3.83 ± 0.33
4-F	1.42 ± 0.04
4-SOCH$_3$	1.22 ± 0.04
4-SO$_2$CH$_3$	1.17 ± 0.03

^{a} Creary et al. (1988).

The methylenecyclopropane rearrangement [44] → [45], which is suggested to proceed by thermal fragmentation of the cyclopropane bond, with the transition state resembling the singlet perpendicular biradical [46], served Creary as a probe for his studies on the stabilization of radicals by substituents (Creary, 1986; Creary and Mehrsheikh-Mohammadi, 1986; Creary et al., 1987, 1988). Relative rates for the rearrangement with various groups in the para-position of the phenyl ring are given in Table 15. The influence of substituents in this position is rather small, being at most within

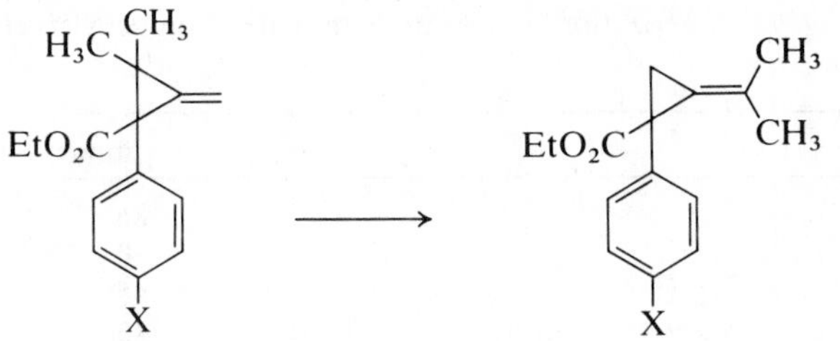

[44]   [45]

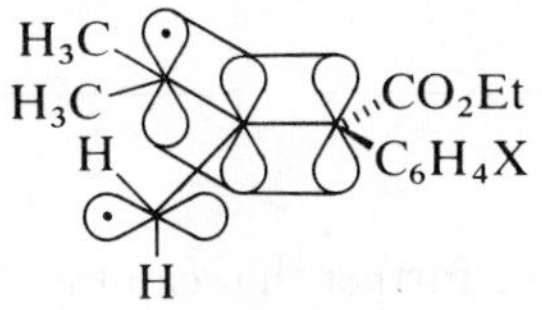

[46]

a factor of four (4-OCH$_3$ and 4-SCH$_3$). The latter molecules will form a captodative-substituted biradical as an intermediate. One might be tempted to see in the rate enhancement the manifestation of a captodative effect. However, even the 4-CO$_2$C$_2$H$_5$-substituted system—giving a dicaptor-substituted biradical—reacts about twice as fast as the unsubstituted system. On the energy scale the observed rate enhancements are very low. This might be related to a diradicaloid transition state that is not fully evolved and/or the "dilution" of the substituent effect by the aromatic system. Thus, the system is extremely sensitive to influences other than radical-stabilizing effects. Apart from the statement of the detection of a small rate increase in the case of captodative substitution, no further deductions can be made from the results of Table 15. They certainly cannot serve as a basis for the quantitative evaluation of a captodative radical stabilization.

Meso–dl isomerizations of molecules carrying both captor and donor substituents provide a further means to gain insight into the captodative effect. Viehe has studied the isomerization of 2,3-dimethyl-2,3-diphenyl succinonitriles (Merényi *et al.*, 1982). As shown in Table 16, the shortest half-life at 160°C is observed for dimethylamino-substitution in the *para*-position. Here, a captodative-substituted benzyl radical should be formed in the course of the reaction. A factor of 43 in half-life between the slowest and fastest isomerizations seems to be appreciable. However, in energetic terms it amounts only to about 2 kcal mol^{-1}. Even though one is inclined to ascribe this stabilization to the stability of the intermediate, such a small energetic change might easily be related to ground-state destabilization effects. The

Table 16 *meso–dl* Isomerization of 4,4′-disubstituted 2,3-diphenylsuccinonitriles at 160°C.[a]

Substituent	$\tau_{1/2}/\text{min}$
H	86
Cl	50
CN	45
CO_2CH_3	45
OCH_3	18
SCH_3	12
NH_2	4
$N(CH_3)_2$	2

[a] Merényi *et al.* (1982).

data in Table 16 demonstrate further that captor groups in the 4-position of the phenyl ring also decrease the half-life, i.e. both dicaptor and captodative substitution lower the isomerization barrier, even though the latter slightly more. It could be argued that the observed rate acceleration would be higher if both of the substituents were attached to the central carbon atoms, thus eliminating the dilution effect of the phenyl group. Such systems have been studied (Birkhofer *et al.*, 1986; Eiching *et al.*, 1983; Lin *et al.*, 1986). Activation enthalpies and entropies for the *meso* → *dl* and *dl* → *meso* conversion are shown in Table 17. The compound with $X = CH_3$ serves as the reference system. It can be seen that the captodative-substituted systems have barriers to isomerization which are lowered by 5–10 kcal mol^{-1}. Complications due to the presence of steric and electronic effects in the ground state again do not allow a final answer as to the existence or magnitude of a captodative effect.

Table 17 Activation parameters for *meso–dl* isomerization of 2,3-disubstituted 2,3-diphenylsuccinonitriles.[a]

$$H_5C_6 \overset{\overset{\displaystyle X}{|}}{\underset{\underset{\displaystyle CN}{|}}{C}} \text{——} \overset{\overset{\displaystyle X}{|}}{\underset{\underset{\displaystyle CN}{|}}{C}} C_6H_5$$

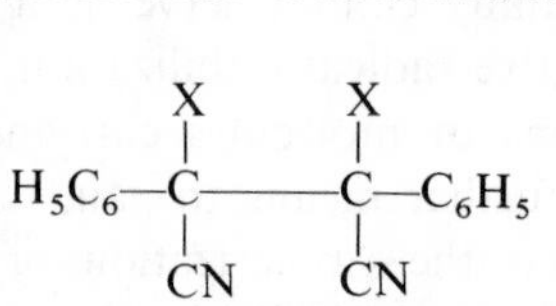

X	$\Delta H^{+}/\text{kcal mol}^{-1}$	$\Delta S^{+}/\text{cal (mol K)}^{-1}$
CH_3	42.2	17.0
OCH_3 *meso* ⇌ *dl*	36.6	15.1
$\quad$ *dl* ⇌ *meso*	36.0	12.2
SCH_3 *meso* ⇌ *dl*	30.7	19.7
$\quad$ *dl* ⇌ *meso*	31.7	22.0

[a] Birkhofer *et al.* (1986); Eiching *et al.* (1983).

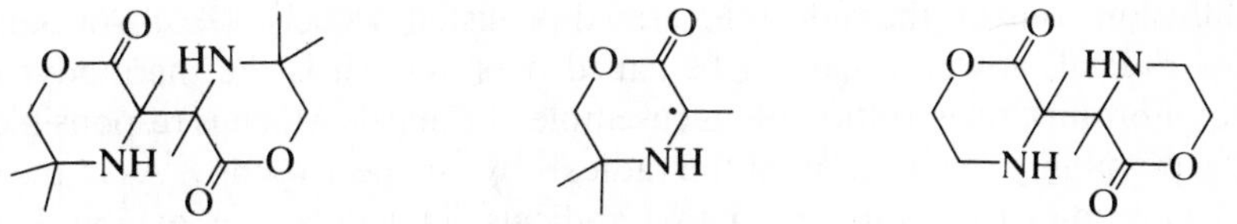

[47]   [48]   [49]

Meso–dl isomerization of [47] was described by Koch (Koch *et al.*, 1975; Koch, 1986; Olson and Koch, 1986. The intermediate radical [48] is in equilibrium with the dimer and can be easily recognized by esr spectroscopy. The thermodynamic parameters for bond homolysis, as a function of medium, are reported in Table 18. A strong solvent effect is observed, in contrast to Rüchardt's example ([24], [25]) reported above. This is interpreted as a manifestation of the polar character of the intermediate radical. The easy detection of [48] by esr spectroscopy is traced back, at least in part, to its captodative character. However, the strong solvent effect on homolysis of [47] need not necessarily be related to the captodative character of radicals [48].

Table 18   Thermodynamic parameters for bond homolysis of [47] as a function of medium.[b]

Medium	K/mol l^{-1}	ΔG^0/kcal mol^{-1}	ΔH^0/kcal mol^{-1}	ΔS^0/cal (mol K)$^{-1}$
EtOH/Mg^{2+c}	1.3×10^{-9}	12	23	36
EtOH/Na^{+c}	7.0×10^{-11}	14	22	28
EtOH	3.2×10^{-11}	14	22	24
Glyme	6.8×10^{-14}	18	30	40
Benzene	5.8×10^{-16}	21	36	51

[a] Koch (1986).
[b] Values of K and ΔG^0 are reported at 25°C.
[c] Ionic strength = 0.3 μ.

As can be deduced from the structures of [47] and [48], hydrogen bonding is likely to have a strong influence on the dissociation equilibrium. This view is supported by the fact that, in particular, alcoholic solvents display a pronounced decrease in the ΔG^0-values. Other factors must also contribute to the easy homolysis of the central C—C bond, since replacement of the two *gem*-dimethyl groups by hydrogen leads to a system [49] which undergoes bond homolysis reluctantly (Himmelsbach *et al.*, 1983; Kleyer *et al.*, 1983). It is interesting to note that the recombination of [48] shows negative activation enthalpies (-8 kcal mol^{-1} in benzene and -1 kcal mol^{-1} in ethanol). The rate constant for combination of [48] in benzene at 25°C is 6×10^8 l mol^{-1} s^{-1}. Even though this rate constant should be close to the limit of diffusion control, the radical is termed persistent (Koch, 1986). An individual radical, however, cannot be called persistent; it is the thermodynamic equilibrium concentration of an ensemble of radicals which is responsible for the possibility of detection of the radicals by esr spectroscopy. As in the case of captodative-substituted methyl radicals, [48] does not exhibit kinetic stability, and the question is whether the radical should be characterized as persistent.

ADDITION REACTIONS

Radical additions to double bonds have been investigated mechanistically and, recently, also for synthetic applications (Giese, 1989). It has been established that the reactivity trends can be described properly in qualitative and quantitative terms by an FMO interpretation (Fischer, 1986; Fischer and Paul, 1987; Giese, 1983; Münger and Fischer, 1985).

   Already, at an early stage of the studies on the captodative effect, Viehe's group (Lahousse et al., 1984) measured relative rates for the addition of t-butoxyl radicals to 4,4'-disubstituted 1,1-diphenylethylenes and to substituted styrenes. This study did not reveal a special character of captodative-substituted olefins in such reactions. It might be that the stability of the radical to be formed does not influence the early transition state of the addition step. The rationalization of the kinetic studies mentioned above in terms of the FMO model indicates, indeed, an early transition state for these reactions, with the consequence that product properties should not influence the reactivity noticeably.

   In a systematic study of the addition of cyclohexyl radicals to α-substituted methyl acrylates, Giese (1983) has shown that the captodative-substituted example fits the linear correlation line of log k_{rel} with σ-values as perfectly as the other cases studied. Thus, no special character of the captodative-substituted olefin is displayed. More recently, arylthiyl radicals have been added to disubstituted olefins in order to uncover a captodative effect in the rate data (Ito et al., 1988). Even though α-N,N-dimethyl-aminoacrylonitrile reacts fastest in these additions, this observation cannot per se be interpreted as the manifestation of a captodative effect. Owing to the lack of rate data for the corresponding dicaptor- and didonor-substituted olefins, it is not possible to postulate a special captodative effect. The result confirms only that the N,N-dimethylamino-group, as expected from its $\sigma_\alpha^{\cdot}$-value, enhances the addition rate. In the sequence α-alkoxy-, α-chloro-, α-acetoxy- and α-methyl-substituted acrylonitriles, it reacts fastest.

   Newcomb and his group (Park et al., 1986) have carried out a study on the rate of cyclization of substituted ω-hexenyl radicals. They discovered a small rate enhancement if the olefin was substituted terminally by a methoxyl and a cyano-group compared to the monosubstituted cases. If accepted as a demonstration of a special captodative effect, it shows that the effect is rather small. In conclusion, it seems that addition reactions in general are not good models in this context because they are characterized by early transition states.

AZOALKANE DECOMPOSITION

The thermal decomposition of symmetrically and unsymmetrically substituted azoalkanes has been used to study substituent effects in radical chemistry. Even though the mechanism of the reactions is still under debate and no unanimous opinion is found in the literature (for a review, see Engel, 1980; Schmittel and Rüchardt, 1987) one may adopt an operational standpoint: as long as the mechanism remains the same in a series of compounds, the results can be interpreted in a uniform way. In particular, Timberlake (1986) has taken this position and has studied extensively the decomposition of symmetrically substituted azo-compounds. From the influence of various substituents on the activation parameters for azoalkane decomposition [50] → [51] he derived a substituent scale which shows many similarities with other scales of radical stabilization energies. Recently, the thermolysis was investigated of some unsymmetrical azocompounds [52], which were designed to make a contribution to the problem of captodative substitution (Janousek et al., 1988). The free energies of activation are: X = CO_2CH_3

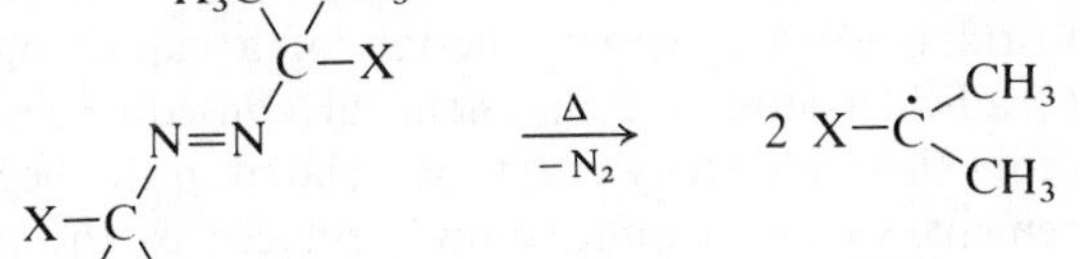

[50] $\xrightarrow[-N_2]{\Delta}$ 2 X–Ċ(CH$_3$)(CH$_3$)

[51]

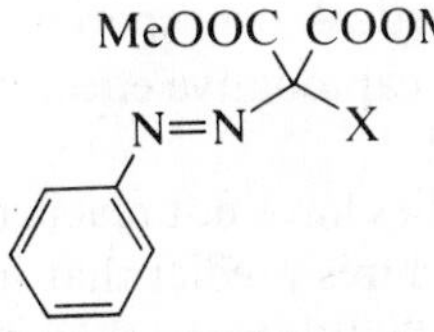

X = COOMe, OMe, NHCOOMe, SMe

[52]

(33.6), $X = OCH_3$ (30.8), $X = NHCOCH_3$ (30.1), and $X = SCH_3$ (27.3 kcal mol^{-1}). The decrease in free energy of activation in the case of donor substituents, which renders the radicals formed captodative, is in line with the concept of captodative stabilization. Thus, operationally, this is a confirmation of the expectation. The difficulties become more obvious if the system is analysed more thoroughly. As reference compounds, the molecules with $X = H$ or CH_3 would have been desirable in order to see the donor effect better. Further, could it be that the ground state of the azoalkanes is destabilized by introduction of substituents with lone pairs adjacent to the azo-group? Is there perhaps a destabilizing anomeric effect? These contributions might be only of the order of a few kcal mol^{-1} but this is the magnitude one is concerned with in the change in free energy of activation if captodative effects are analysed.

6 Conclusion

Although the concept of captodative stabilization of free radicals has been suggested several times in the past, either without coining a special name or using terms like "merostabilization" or "push–pull stabilization", it was only Viehe's formulation of the subject which aroused interest among many experimental chemists. As a consequence, numerous articles have appeared in the last decade, applying this substitution pattern in synthesis or trying to establish a quantitative basis for the amount of stabilization in this type of radical. In this article we have tried to analyse whether the claim of the proponents of the captodative effect has found experimental or theoretical support. The postulate of a synergetic action of a captor and a donor substituent at a radical centre and its chemical consequences have been discussed. The more than additivity effect, postulated in the beginning as a necessary requirement, was later omitted and replaced by the view that an additive substituent effect is already more than is normally found for two like substituents. Therefore, such an observation should be considered as a manifestation of a synergetic interaction, i.e. a captodative effect.

Have the numerous theoretical or experimental investigations shown evidence for the existence of a captodative effect in free radical chemistry? If yes, what then is its magnitude?

The quantum chemical studies have not reached a unanimous conclusion. The more sophisticated procedures predict that in some captodative substituted systems an additive or a slightly more than additive substituent effect is possible. The calculations, particularly those of Leroy, have also contributed to the belief that the study of substituent effects requires the consideration of their influence in the ground and final states of the model system.

All physical organic studies which were undertaken to establish a captodative effect quantitatively suffer from the drawback that an experiment can provide only the energetic difference between two states. A trivial consequence of this fact is that any manifestation of a captodative effect depends entirely on the chemical system under consideration. All "positive" investigations, theoretical or experimental, have shown, so far, that in energetic terms any captodative effect is not very large, in most cases only a few kcal mol^{-1}. Stabilizing or destabilizing influences of this magnitude may easily be achieved in a particular system by other interactions (e.g. dipolar, anomeric, steric) which are not related to the radical nature of the intermediates. It is thus difficult to decide by experiment whether a seeming verification of the postulate is solely due to the stabilization of the radical/radicaloid transition state, to a destabilization of the ground state or to the sum of both. It can be confirmed only that the energetic separation of the two states has been reduced. Of course, one may adopt a practical standpoint in the sense that *any* synergetic, i.e. additive or more than additive, influence of two substituents may be termed captodative, regardless of whether this is the result of the action on the radical centre or a non-radical state, or on other parts of the radical species. Then, several of the experimental studies suggest a captodative effect (isomerization of cyclopropanes, *meso–dl* isomerization, rotational barriers, azo decomposition). To relate the observed effects to a captodative effect in its original sense, i.e. to a dominating interaction of the substituents with the radical centre, requires a careful analysis of possible closed-shell interactions. This requirement leads to some dilemmas in the selection/usefulness of model systems. In systems where the substituents are bonded to the atom which is or will be the radical centre, the strongest captodative effect is expected. However, closed-shell interactions are also likely to be the strongest in these cases and might dominate the net energetic change in the course of the reaction. On the other hand, in systems where the substituents are far from the reaction centre, in order to minimize direct non-radical interactions, the "dilution" effect may reduce the interaction to an extent which brings its value close to the systematic error limits of the applied method.

The only unambiguous confirmation of a synergetic, i.e. more than additive, substituent effect derives from esr measurements of spin delocalization. This is due to the fact that the measurement is done on the radical and constitutes only a small perturbation of the ground state. But, to stress it again, spin delocalization provides no values for stabilization energies. Thus, it is not possible to make a statement about the magnitude of the stabilization or destabilization from these measurements. Furthermore, even a significantly reduced spin density at the radical centre as a consequence of delocalization might not be decisive for the reaction pathway of a particular class of radical.

Another claim of captodative substitution was that chemical consequences should be connected with this substitution pattern. For synthetic
purposes, a captodative effect of a few kcal mol^{-1} might be helpful for the
selection of a lower energy pathway in cases where a radical can react by
several reaction paths, discriminated by activation barriers of closely similar
energy. Synthetic applications have shown that captodative-substituted
radicals generally behave like other stabilized, short-lived radicals. Left
alone these radicals usually give dimers in high yield, as, for instance, allylic
or benzylic radicals also do. (2 + 2)-Cycloadditions occur in many cases via
1,4-biradicals; the better the reaction the greater is the stabilization of the
intermediate radical by substituents. As captodative substitution provides
stability to a radical centre it is no surprise that captodative olefins are good
partners in (2 + 2)-cycloadditions (Coppe-Motte *et al.*, 1986). However, as
far as one can tell, they do not convert the normally concerted (4 + 2)-
cycloaddition, 1,3-dipolar cycloadditions (Döpp and Henseleit, 1982; Döpp
and Libera, 1983; Döpp and Walter, 1986a,b; Döpp *et al.*, 1986) or Diels–
Alder reactions (Boucher and Stella 1984, 1988a,b, 1989) into stepwise
processes.

The merits of the formulation of the captodative effect by Viehe lies in the
fact that the numerous attempts to prove or disprove its existence have led to
a better general understanding of substituent effects in free radical chemistry
during the past decade.

Acknowledgements

Critical comments and helpful discussions from K. U. Ingold, R. Merényi,
C. Rüchardt, L. Stella and H. G. Viehe are acknowledged.

References

Arnold, D. R. (1986). *In* "Substituent Effects in Radical Chemistry" (eds H. G.
   Viehe, Z. Janousek and R. Merényi), p. 171. Reidel, Dordrecht
Arnold, D. R. and Yoshida, M. (1981). *J. Chem. Soc., Chem. Commun.* 1203
Aurich, H. G. and Deuschle, E. (1981). *Liebigs Ann. Chem.* 719
Aurich, H. G., Deuschle, E. and Weiss, W. (1977). *J. Chem. Res. (S)* 301; *(M)* 3457
Balaban, A. T. (1971). *Rev. Roum. Chim.* **16**, 725
Balaban, A. T., Caprois, M. T., Negoita, N. and Baican, R. (1977). *Tetrahedron* **33**,
   2249
Baldock, R. W., Hudson, P. and Katritzky, A. R. (1973). *Heterocycles* **1**, 67
Baldock, R. W., Hudson, P. and Katritzky, A. R. (1974). *J. Chem. Soc., Perkin
   Trans. I* 1422
Beckhaus, H.-D. and Rüchardt, C. (1987). *Angew. Chem.* **99**, 807; *Angew. Chem.,
   Int. Ed. Engl.* **26**, 770

Beckhaus, H.-D., Dogan, B., Verevkin, S., Hädrich, J. and Rüchardt, C. (1990). *Angew. Chem.* **102**, 313; *Angew. Chem., Int. Ed. Engl.* **29**, 320

Beckwith, A. L. J. and Brumby, S. (1987). *J. Magn. Res.* **73**, 252

Benson, S. W. (1965). *J. Chem. Educ.* **42**, 502

Birkhofer, H., Beckhaus, H.-D. and Rüchardt C. (1986). *In* "Substituent Effects in Radical Chemistry" (eds H. G. Viehe, Z. Janousek and R. Merényi), p. 199. Reidel, Dordrecht

Birkhofer, H., Hädrich, J., Beckhaus, H.-D. and Rüchardt, C. (1987). *Angew. Chem.* **99**, 592; *Angew. Chem., Int. Ed. Engl.* **26**, 573

Birkhofer, H., Hädrich, J., Pakusch, J., Beckhaus, H.-D., Rüchardt, C., Peters, K. and von Schnering, H.-G. (1989). *In* "Free Radicals in Synthesis and Biology" (ed. F. Minisci), p. 27. Kluwer, Dordrecht

Bordwell, F. G. and Bausch, M. J. (1986). *J. Am. Chem. Soc.* **108**, 1979

Bordwell, F. G. and Lynch, T.-Y. (1989). *J. Am. Chem. Soc.* **111**, 7558

Bordwell, F. G., Bausch, M. J., Cheng, J.-P., Lynch, T.-Y. and Mueller, M. E. (1990). *J. Org. Chem.* **55**, 58

Boucher, J. L. and Stella, L. (1985). *Tetrahedron* **41**, 875

Boucher, J. L. and Stella, L. (1988a). *Tetrahedron* **44**, 3607

Boucher, J. L. and Stella, L. (1988b). *Tetrahedron* **44**, 3595

Boucher, J. L. and Stella, L. (1989). *J. Chem. Soc., Chem. Commun.* 187

Clark, T. (1988). Private communication

Conradi, M. S., Zeldes, H. and Livingston, R. (1979). *J. Phys. Chem.* **83**, 2160

Coppe-Motte, G., Borghese, A., Janousek, Z., Merényi, R. and Viehe, H. G. (1986). *In* "Substituent Effects in Radical Chemistry" (eds H. G. Viehe, Z. Janousek and R. Merényi), p. 371. Reidel, Dordrecht

Cox, J. D. and Pilcher, G. (1970). "Thermochemistry of Organic and Organometallic Compounds", Ch. 7. Academic Press, New York

Crans, D., Clark, T. and Schleyer, P. V. R. (1980). *Tetrahedron Lett.* **21**, 3681

Creary, X. (1986). *In* "Substituent Effects in Radical Chemistry" (eds H. G. Viehe, Z. Janousek and R. Merényi), p. 245. Reidel, Dordrecht

Creary, X. and Mehrsheikh-Mohammadi, M. E. (1986). *J. Org. Chem.* **51**, 2664

Creary, X., Mehrsheikh-Mohammadi, M. E. and McDonald, S. (1987). *J. Org. Chem.* **52**, 3254

Creary, X., Sky, A. F. and Mehrsheikh-Mohammadi, M. E. (1988). *Tetrahedron Lett.* **29**, 6839

De Mesmaeker, A., Vertommen, L., Merényi, R. and Viehe, H. G. (1982). *Tetrahedron Lett.* **23**, 69

Deuchert, K. and Hünig, S. (1978). *Angew. Chem.* **90**, 927; *Angew. Chem., Int. Ed. Engl.* **17**, 875

Devaquet, A. and Salem, L. (1969). *J. Am. Chem. Soc.* **91**, 3793

de Vries, L. (1977). *J. Am. Chem. Soc.* **99**, 1982

de Vries, L. (1978). *J. Am. Chem. Soc.* **100**, 926

Dewar, M. J. S. (1952) *J. Am. Chem. Soc.* **74**, 3353

Döpp, D. and Henseleit, M. (1982). *Chem. Ber.* **115**, 798

Döpp, D. and Libera, H. (1983). *Tetrahedron Lett.* **24**, 885

Döpp, D. and Walter, J. (1986a). *Heterocycles* **20**, 1055

Döpp, D. and Walter, J. (1986b). *In* "Substituent Effects in Radical Chemistry" (eds H. G. Viehe, Z. Janousek and R. Merényi), pp. 375 and 379. Reidel, Dordrecht

Döpp, D., Walter, J. and Holz, S. (1986). *In* "Substituent Effects in Radical Chemistry" (eds H. G. Viehe, Z. Janousek and R. Merényi), p. 379. Reidel, Dordrecht

Duerden, M. F. and Godfrey, M. (1980). *J. Chem. Soc., Perkin Trans. 2* 330

Dust, J. M. and Arnold, D. R. (1983). *J. Am. Chem. Soc.* **105**, 1221, 6531

Eiching, K. H., Beckhaus, H.-D., Hellmann, S., Fritz, H., Peters, E. M., Peters, K., von Schnering, H. G. and Rüchardt, C. (1983). *Chem. Ber.* **116**, 1787

Engel, P. S. (1980). *Chem. Rev.* **80**, 99

Fischer, H. (1964). *Z. Naturforsch. Teil A* **19**, 866

Fischer, H. (1986). *In* "Substituent Effects in Radical Chemistry" (eds H. G. Viehe, Z. Janousek and R. Merényi), p. 123. Reidel, Dordrecht

Fischer, H. and Paul, H. (1987). *Acc. Chem. Res.* **20**, 220

Forrester, A. R., Hay, J. M. and Thomson, R. H. (1968). "Organic Chemistry of Stable Free Radicals", chs 4 and 5. Academic Press, London

Fritzsche, K., Dogan, B., Beckhaus, H. D. and Rüchardt, C. (1990). *Thermochimica Acta*, **160**, 147

Gallo, R. (1983). *Progr. Phys. Org. Chem.* **14**, 115

Giese, B. (1983). *Angew. Chem.* **95**, 771; *Angew. Chem., Int. Ed. Engl.* **22**, 753

Giese, B. (1989). *Angew. Chem.* **101**, 993; *Angew. Chem., Int. Ed. Engl.* **28**, 969

Gilbert, B. C., Hanson, P., Isham, W. J. and Whitwood, A. C. (1988). *J. Chem. Soc., Perkin Trans. 2* 2077

Goldschmidt, S. (1920). *Ber. Dtsch. Chem. Ges.* **53**, 44

Goldschmidt, S. (1929). *Liebigs. Ann. Chem.* **473**, 137

Griller, D. and Ingold, K. U. (1976). *Acc. Chem. Res.* **9**, 13

Himmelsbach, R. J., Barone, A. D. and Koch, T. H. (1983). *J. Org. Chem.* **48**, 2989

Ito, O., Arito, Y. and Matsuda, M. (1988). *J. Chem. Soc., Perkin Trans. 2* 869

Jackson, R. A. (1986). *In* "Substituent Effects in Radical Chemistry" (eds H. G. Viehe, Z. Janousek and R. Merényi), p. 325. Reidel, Dordrecht

Janousek, Z., Bougeois, J.-L., Merényi, R., Viehe, H. G., Luedtke, A. E., Gardner, S. and Timberlake, J. W. (1988). *Tetrahedron Lett.* **27**, 3379

Katritzky, A. R. and Soti, F. (1974). *J. Chem. Soc., Perkin Trans. 1* 1427

Katritzky, A. R., Zerner, M. C. and Karelson, M. M. (1986). *J. Am. Chem. Soc.* **108**, 7213

Klessinger, M. (1980). *Angew. Chem.* **92**, 937; *Angew. Chem., Int. Ed. Engl.* **19**, 908

Kleyer, D. L., Haltiwanger, R. C. and Koch, T. H. (1983). *J. Org. Chem.* **48**, 147

Klopman, G. (1968). *J. Am. Chem. Soc.* **90**, 223

Koch, T. H. (1986). *In* "Substituent Effects in Radical Chemistry" (eds H. G. Viehe, Z. Janousek and R. Merényi), p. 263. Reidel, Dordrecht

Koch, T. H., Oleson, J. A. and DeNiro, J. (1975). *J. Am. Chem. Soc.* **97**, 7285

Korth, H.-G., Trill, H. and Sustmann, R. (1981). *J. Am. Chem. Soc.* **103**, 4483

Korth, H.-G., Sustmann, R., Viehe, H. G. and Merényi, R. (1983). *J. Chem. Soc., Perkin Trans. 2* 67

Korth, H.-G., Lommes, P. and Sustmann, R. (1984). *J. Am. Chem. Soc.* **106**, 663

Korth, H.-G., Lommes, P., Sicking, W. and Sustmann, R. (1985). *Chem. Ber.* **118**, 4627

Korth, H.-G., Lommes, P., Sustmann, R., Sylvander, L. and Stella, L. (1987). *Nouv. J. Chim.* **11**, 365

Korth, H. G., Sustmann, R., Gröninger, K. S., Leising, M. and Giese, B. (1988). *J. Org. Chem.* **53**, 4364

Kosower, E. M. and Poziomek, E. J. (1964). *J. Am. Chem. Soc.* **86**, 5515

Lahousse, F. Merényi, R., Desmurs, J. R., Allaime, H., Borghese, A. and Viehe, H. G. (1984). *Tetrahedron Lett.* **25**, 3823

Leigh, W. J. and Arnold, D. R. (1981). *Can. J. Chem.* **59**, 609

Leroy, G. (1983). *Int. J. Quant. Chem.* **23**, 271

Leroy, G. (1985a). *Bull. Soc. Chim. Belg.* **94**, 945

Leroy, G. (1985b). *Adv. Quant. Chem.* **17**, 1

Leroy, G. (1988). *J. Mol. Struct. (Theochem)* **168**, 77

Leroy, G., Peeters, D. and Wilante, C. (1982). *J. Mol. Struct. (Theochem)* **88**, 217

Leroy, G., Wilante, C., Peeters, D. and Ueyawa, M. M. (1985). *J. Mol. Struct. (Theochem)* **124**, 107

Leroy, G., Peeters, D., Sana, M. and Wilante, C. (1986). *In* "Substituent Effects in Radical Chemistry" (eds H. G. Viehe, Z. Janousek, and R. Merényi), p. 1. Reidel, Dordrecht

Leroy, G., Sana, M., Wilante, C., Peeters, D. and Dogimont, C. (1987). *J. Mol. Struct. (Theochem)* **153**, 249

Leroy, G., Sana, M., Wilante, C. and Nemba, R. M. (1989). *J. Mol. Struct. (Theochem)* **198**, 159

Lin, Y. C., Wu, L. M., Wang, P. M. and Liu, P. Q. (1986). *Acta Chim. Sinica* **44**, 664

Louw, R. and Bunk, J. J. (1983). *Recl. Trav. Chim. Pays-Bas* **102**, 119

MacInnes, I. and Walton, J. C. (1987). *J. Chem. Soc., Perkin Trans. 2* 1789

MacInnes, I., Walton, J. C. and Nonhebel, D. C. (1985). *J. Chem. Soc., Chem. Commun.* 712

McKean, D. C. (1978). *Chem. Soc. Rev.* **7**, 399

McKean, D. C. (1989). *Int. J. Chem. Kinet.* **21**, 445

McMillen, D. F. and Golden, D. M. (1982). *Ann. Rev. Phys. Chem.* **33**, 493

Merényi, R., Daffe, V., Klein, J., Masamba, W. and Viehe, H. G. (1982). *Bull. Soc. Chim. Belg.* **91**, 456

Merényi, R., De Mesmaeker, A. and Viehe, H. G. (1983). *Tetrahedron Lett.* **24**, 2765

Merényi, R., Janousek, Z. and Viehe, H. G. (1986). *In* "Substituent Effects in Radical Chemistry" (eds H. G. Viehe, Z. Janousek and R. Merényi), p. 301. Reidel, Dordrecht

Merényi, R., Janousek, Z. and Viehe, H. G. (1988). *In* "Organic Free Radicals" (eds H. Fischer and H. Heimgartner), Springer-Verlag, Berlin-Heidelberg, p. 123.

Münger, K. and Fischer, H. (1985). *Int. J. Chem. Kinet.* **17**, 809

Neumann, W. P., Uzick, W. and Zarkadis, A. K. (1986). *J. Am. Chem. Soc.* **108**, 3762

Neumann, W. P., Penenory, A., Stewen, U. and Lehnig, M. (1989). *J. Am. Chem. Soc.* **111**, 5845

Nicholas, A. M. de P. and Arnold, D. R. (1984). *Can. J. Chem.* **62**, 1850

Nootens, C., Merényi, R., Janousek, Z. and Viehe, H. G. (1988). *Bull. Soc. Chim. Belg.* **97**, 1045

Olson, J. B. and Koch T. H. (1986). *J. Am. Chem. Soc.* **108**, 756

Park, S.-U., Chung, S.-K. and Newcomb, M. (1986). *J. Am. Chem. Soc.* **108**, 240

Pasto, D. J. (1988). *J. Am. Chem. Soc.* **110**, 8164

Pasto, D. J., Krasnansky, R. and Zercher, C. (1987). *J. Org. Chem.* **52**, 3062

Reynolds, W. F. (1983). *Progr. Phys. Org. Chem.* **14**, 165

Rhodes, C. J. and Roduner, E. (1988). *Tetrahedron Lett.* **29**, 1437

Rodewald, L. B. and DePuy, C. H. (1964). *Tetrahedron Lett.* 2951

Rüchardt, C. (1970). *Angew. Chem.* **82**, 845; *Angew. Chem., Int. Ed. Engl.* **9**, 830

Rüchardt, C. (1984). "Sitzungsberichte der Heidelberger Akademie der Wissenschaften, Math.-Nat. Klasse, 3. Abhandlung". Springer-Verlag, Berlin

Rüchardt, C. and Beckhaus, H.-D. (1980). *Angew. Chem.* **92**, 417; *Angew. Chem., Int. Ed. Engl.* **19**, 429

Rüchardt, C. and Beckhaus, H.-D. (1986). *Top. Curr. Chem.* **130**, 1

Salem, L. (1968). *J. Am. Chem. Soc.* **90**, 543, 553
Schäfer, L., van Alsenoy, C., Williams, J. O. and Scarsdale, J. N. (1981). *J. Mol. Struct.* **76**, 349
Schleyer, P. v. R. (1987). *Pure Appl. Chem.* **59**, 1647
Schmittel, M. and Rüchardt, C. (1987). *J. Am. Chem. Soc.* **109**, 2750
Schulze, R., Beckhaus, H.-D. and Rüchardt, C. (1990). *Z. Chem.*, in press
Stella, L., Janousek, Z., Merényi, R. and Viehe, H. G. (1978). *Angew. Chem.* **90**, 741; *Angew. Chem., Int. Ed. Engl.* **17**, 691
Stella, L., Merényi, R., Janousek, Z., Viehe, H. G., Tordo, P. and Munoz, A. (1980). *J. Phys. Chem.* **84**, 304
Stella, L., Pochat, F and Merényi, R. (1981). *Nouv. J. Chim.* **5**, 55
Stewen, U. and Lehnig, M. (1989). *Chem. Ber.* **122**, 2319
Sustmann, R. (1986). In "Substituent Effects in Radical Chemistry" (eds H. G. Viehe, Z. Janousek and R. Merényi), p. 143. Reidel, Dordrecht
Sustmann, R. and Binsch, G. (1971). *Mol. Phys.* **20**, 1, 9
Sustmann, R. and Brandes, D. (1976). *Chem. Ber.* **109**, 354
Sustmann, R. and Trill, H. (1974). *J. Am. Chem. Soc.* **96**, 4343
Sustmann, R. and Vahrenholt, F. (1973). *Theor. Chim. Acta (Berl.)* **29**, 305
Sustmann, R., Ansmann, A. and Vahrenholt, F. (1972). *J. Am. Chem. Soc.* **94**, 8099
Sustmann, R., Trill, H., Vahrenholt, F. and Brandes, D. (1977). *Chem. Ber.* **110**, 255
Szwarc, M. (1948). *J. Chem. Phys.* **16**, 128
Timberlake, J. W. (1986). In "Substituent Effects in Radical Chemistry" (eds H. G. Viehe, Z. Janousek and R. Merényi), p. 271. Reidel, Dordrecht
Topsom, R. D. (1976). *Progr. Phys. Org. Chem.* **12**, 1
Van Hoecke, M., Borghese, A., Penelle, J., Merényi, R. and Viehe, H. G. (1986). *Tetrahedron Lett.* **27**, 4569
Verevkin, S., Dogan, B., Beckhaus, H.-D. and Rüchardt, C. (1990) *Angew. Chem.*, **102**, 693; *Angew. Chem., Int. Ed. Engl.* **29**, 674
Vertommen, L., Beaujean, M., Merényi, R., Janousek, Z. and Viehe, H. G. (1989). In "Free Radicals in Synthesis and Biology" (ed. F. Minisci), p. 329. Kluwer, Dordrecht
Viehe, H. G., Merényi, R., Stella, L. and Janousek, Z. (1979) *Angew. Chem.* **91**, 982; *Angew. Chem., Int. Ed. Engl.* **18**, 917
Viehe, H. G., Janousek, Z., Merényi, R. and Stella, L. (1985). *Acc. Chem. Res.* **18**, 148
Viehe, H. G., Merényi, R. and Janousek, Z. (1988). *Pure Appl. Chem.* **60**, 1635
Viehe, H. G., Janousek, Z. and Merényi, R. (1989). In "Free Radicals in Synthesis and Biology" (ed. F. Minisci), p. 1. Kluwer, Dordrecht
Walton, J. C. (1984). *Rev. Chem. Intermed.* **5**, 249
Wayner, D. D. M. and Arnold, D. R. (1984). *Can. J. Chem.* **62**, 1164
Williams, J. O., Scarsdale, J. N. and Schäfer, L. (1981). *J. Mol. Struct.* **76**, 11
Zamkanei, M., Kaiser, J. H., Birkhofer, H., Beckhaus, H.-D. and Rüchardt, C. (1983). *Chem. Ber.* **116**, 3216

High-spin Organic Molecules and Spin Alignment in Organic Molecular Assemblies

HIIZU IWAMURA

Department of Chemistry, Faculty of Science, The University of Tokyo, Japan

ADVANCES IN PHYSICAL ORGANIC CHEMISTRY
VOLUME 26   ISBN 0-12-033526-3

1   Introduction

The design and synthesis of molecular magnets are subjects of increasing current interest. There were at least four international conferences on these topics held in the recent past (Okazaki, Japan, September 1986; Tashkent, USSR, November 1987; Dallas, USA, April 1989; Boston, USA, November 1989). The idea is to establish unprecedented macroscopic spins of long-range order in molecular systems. Why is this field of study interesting? This is because strong magnetic properties constitute the antithesis to molecular systems in general and organic molecules in particular. The large majority of organic molecules have closed-shell electronic structures, i.e. they have a singlet ground state with equal numbers of electrons having α- and β-spins. Most organic compounds are therefore good electric insulators and are magnetically inactive. Strictly speaking, they are diamagnetic, i.e. they are weakly repelled out of an external magnetic field. There are some organic molecules that have open-shell structures where not all electrons are paired. Many of these molecules have one unpaired electron and, therefore, a doublet ground state. These are called free radicals and are known to show paramagnetic properties. The alignment of spins in molecular systems becomes an issue when there is an interaction between two doublet centres, as in diradicals and radical pairs. The Coulombic repulsion between electrons lifts the zeroth-order degeneracy for these chemical entities and gives rise to singlet and triplet states of different total energy. Which state lies lower in energy is highly important in the chemistry of diradicals (Borden, 1982a) and radical pairs (Bethell and Brinkman, 1973; Lepley and Closs, 1973; McBride, 1983). The conditions for having ground triplet states and the importance of the singlet/triplet radical pairs in free radical reactions have been amply discussed.

While diradicals and radical pairs, as well as excited triplets, are accepted as important intermediates, relatively little is known about molecular entities that have more than two unpaired electrons, i.e. entities for which the quantum number (S) of the spin angular momentum is equal to or greater than 3/2. This is due firstly to the high bonding energy of the K- and L-shell valence electrons in typical organic molecules that leads to their closed-shell electronic structure. There is a large energy gap between the valence and conduction bands making most organic compounds good electric insulators. Secondly, the geometrical symmetry of organic molecules is not necessarily high and, therefore, there are not very many degenerate orbitals. In contrast to these situations in organic molecules, high-spin states are the rule rather than the exception in transition- and lanthanoid-metal salts and complexes. The 3d and 4f atomic orbitals in these metal atoms are originally five- and seven-fold degenerate, respectively. Even when the degeneracy is lifted by a set of ligands of octahedral symmetry, for example, to give doubly degenerate t_{2g}-orbitals and triply degenerate e_g-orbitals, they are still high spin as long as the ligand field is not very strong (Fig. 1).

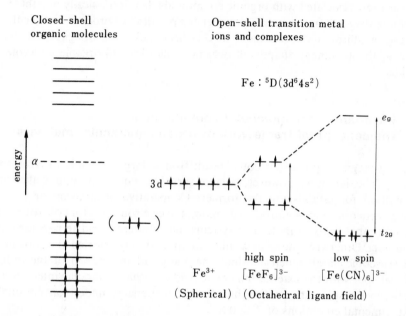

Fig. 1 Schematic orbital diagrams of organic molecules compared with transition metal ions represented by Fe(III).

The same is true for spin alignment in organic molecular assemblies. In crystals, liquid crystals, membranes and other organized systems of persistent organic radicals, the electron spins have a strong tendency to behave as randomly oriented spins or to align themselves antiparallel to one another. After all, the Heitler–London theory of chemical bonds dictates the stabilization of the antiparallel alignment of two spins in two approaching orbitals with increasing overlap. Very few examples are known to date in which the electron spins of neighbouring molecules are aligned in parallel (see Section 6).

In this review, we focus our discussion on organic molecules with more than two coupled spins ($S \geq 3/2$). They are arbitrarily called high-spin molecules (Itoh, 1978; Weltner, 1983). Molecular solids in which more than two unpaired electrons are aligned in parallel among neighbouring molecules will also be discussed.

The chemistry of high-spin organic molecules and highly ordered spin alignment in organic molecular assemblies is expected to open up a new field in science and is considered also to be the best area in which to look for interesting magnetic properties from organic materials. Ferromagnetism is found in some minerals, inorganic compounds, metals and alloys, but has never been associated with organic compounds. Is it intrinsically possible to prepare magnets from organic polymers or plastics? In order to answer this question affirmatively, we have to establish molecular designs that will permit the alignment of spins in organic molecules and organic molecular solids.

2   Intra- and inter-molecular spin alignment: the conceptual framework of organic molecular magnets

Strong magnetic properties could result from a large assembly of electron spins. The latter is, however, a necessary but not in itself a sufficient condition for establishing the former. Cooperative phenomena or long-range order have to be introduced among the electron spins for constructing macroscopic spins in molecular systems, namely ferro- and ferri-magnetic coupling. Otherwise, the spins would end up randomly oriented and showing macroscopic paramagnetism, since the energy of interaction of the individual spin with the external magnetic field and that of dipolar interaction between the spins are smaller than the thermal energy under conventional experimental conditions of $T > 0$ K.

There appear to be at least four approaches to the realization of strong molecular magnets (Buchachenko, 1989; Iwamura, 1989). The first step would be the construction of high-spin molecules, namely, molecules having a large magnetic moment. Since the magnetization (I, the energy of the interaction of magnetic moment M with the external magnetic field H; see Section 4) is proportional to the size of the magnetic moment, it is desirable to have the magnetic moment as high as possible. Since the electron spins have a tendency to align antiparallel to one another (this is called antiferromagnetic coupling) to make lower-spin states more stable in molecular systems, rigorous molecular design against Nature is needed. Even when this is established, however, the result would lead at best to paramagnetism of high spins, i.e. unordered spins being aligned only in an external magnetic field and at a rather low temperature. Interaction leading to the ferromagnetic ordering of the high-spin molecules is needed next. Again, we are against "Mother Nature" who favours very little interaction or, if present, antiferromagnetic spin alignment between molecules carrying unpaired spins. Special molecular design is needed to align molecular magnetic moments parallel to one another. Once the latter design, i.e. how to align spins intermolecularly, is rigorously established, we shall be in a position to have real organic ferromagnets. We shall also be able to establish macroscopic spins even from an assembly of doublet molecules (Fig. 2). Ferromagnetism in α-iron and chromium dioxide would serve as original models for this approach.

High-spin Molecules

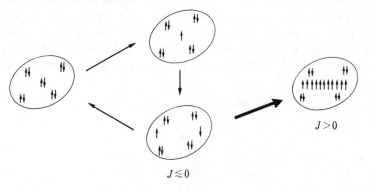

$J \lesssim 0$

$J > 0$

Intermolecular Spin Alignment

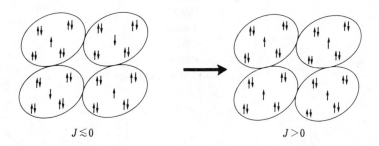

$J \lesssim 0$ $J > 0$

Molecular Ferromagnets

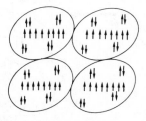

Fig. 2 A conceptual sketch for constructing molecular ferromagnets. Arrows and ellipsoids represent the electron spins and molecular boundaries, respectively. Bold arrows show where careful molecular design is necessary.

Another approach starts by accepting the natural trend and uses the strong tendency of two electrons in two overlapping orbitals to align antiparallel to one another. When two magnetic moments of equal strength are taken into account, this leads of course to antiferromagnetic interaction and consequent closed-shell molecules. But what if two spins of unequal size are aligned alternately? The neighbouring two spins cancel each other out, but only partly. The residual moments are in parallel and give macroscopic spins. These can be arranged within a molecule as well as among molecules. Strong magnetism in ferrites is a prototype of this category, which is called ferrimagnetism (Fig. 3). The following discussion is classified according to the above framework.

Ferrimagnetic Interaction in Molecules

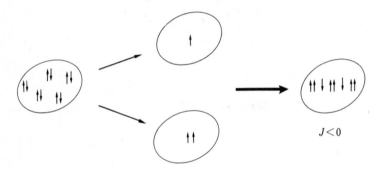

Intermolecular Ferrimagnetic Coupling

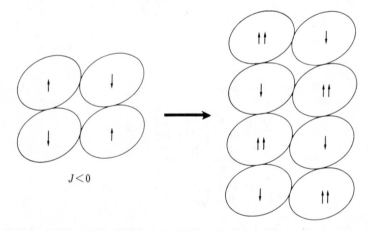

Fig. 3 A conceptual sketch showing approaches to molecular ferrimagnets (paired spins in the open-shell molecules have been left out).

3 Design of high-spin organic molecules

High-spin organic molecules are not without precedents. The Leo triradical [1], a classical representative (Leo, 1937), was later modified to give triradical [1'] in a slightly more stabilized quartet state (Schmauss *et al.*, 1965; Kothe *et al.*, 1971; Brickmann and Kothe, 1973; Wilker *et al.*, 1975). Tetraradical [2] has been reported more recently to be in a ground quintet state (Seeger and Berson, 1983; Seeger *et al.*, 1986). Dicarbene [3] had been established by Itoh (Itoh, 1967) and Wasserman (Wasserman *et al.*, 1967) to be a quintet species. A straightforward extension would be the construction of polyradicals like [4] (Braun *et al.*, 1981). However, the electron spins in the last polyradicals are either randomly oriented or have a strong tendency to align antiparallel to one another, since the exchange interaction between them is either negligibly small or negative in sign when non-vanishing. Careful design is necessary to align many electron spins in parallel within a molecule.

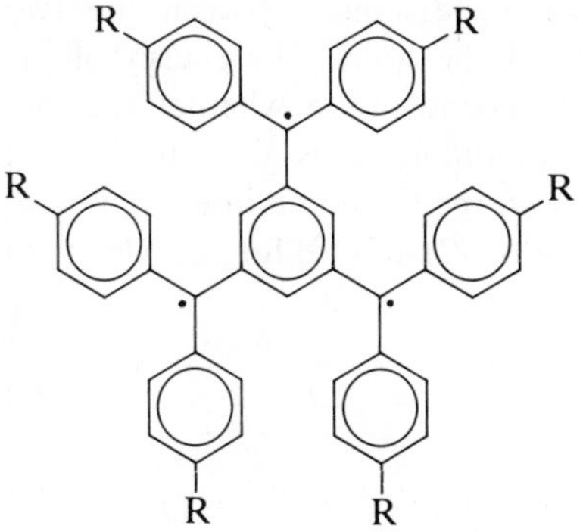

[1] R = H
[1'] R = Ph

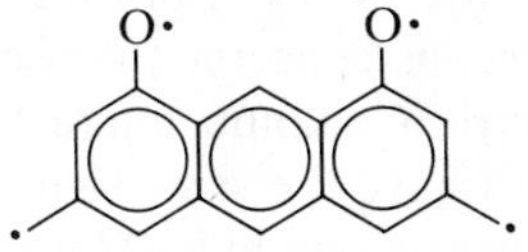

[2]

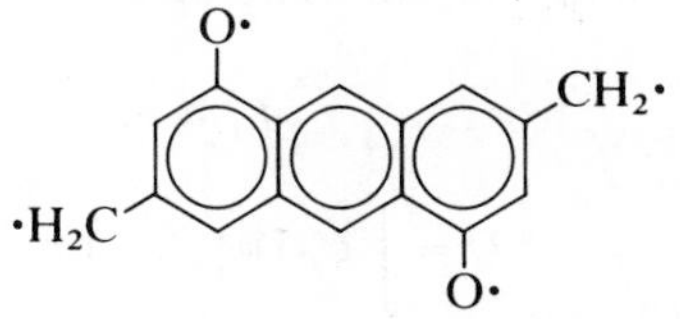

[2']

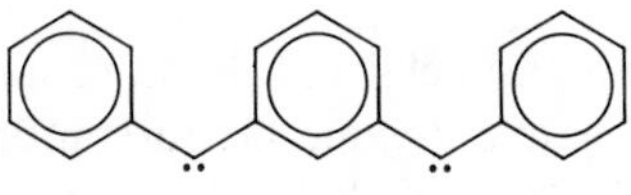

[3]

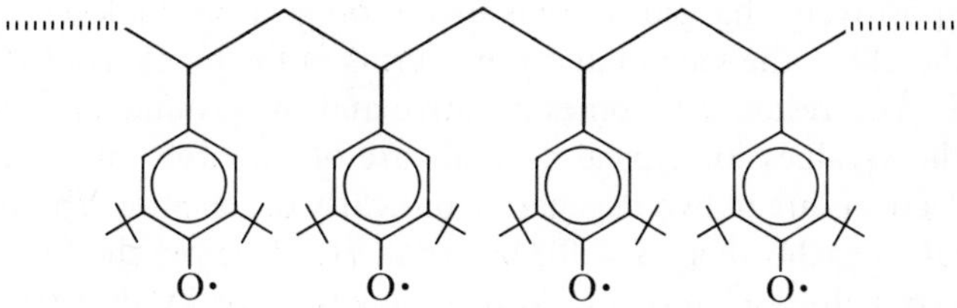

[4]

Furthermore, if we are able to construct very high-spin molecules that have thousands of parallel spins, the molecules themselves may be able to function as single-grain magnetic domains. In α-iron, for example, a microcrystalline particle of radius 200 pm contains *ca.* 5000 spins and is considered to form a single domain structure. The magnetic properties of such superhigh-spin organic molecules will be of special interest.

EXCHANGE INTERACTION

The Coulombic interaction expressed by the form e^2/r_{12} between two electrons 1 and 2 at a distance of r_{12} depends on the sign of the electron spins concerned (Fig. 4) (Salem, 1982). This is because Pauli's principle excludes the presence of two spins of the same sign from occupying the same space simultaneously. The exclusion does not apply when the two spins have opposite signs. For this reason, the average distance between the two electrons concerned depends on the signs of the spins. The energy of the electric Coulomb interaction is therefore dependent on whether the two spins have the same sign or not. The energy difference is given by (1) and approximated by $-2J$ when $S^2 \ll 1$. Here, Q, the Coulomb integral, and J, the exchange integral, are given respectively by (2) and (3) for two electrons 1 and 2 at atoms a and b (Fig. 4).

$$E_T - E_S = -2(J - S^2 Q)/(1 - S^4) = -2J \qquad (1)$$

$$Q = \int \varphi_a^*(1)\varphi_b^*(2)e^2[1/r_{12} - 1/r_{b1} - 1/r_{a2} + 1/R_{ab}]\varphi_a(1)\varphi_b(2) \, dv_1 \, dv_2 \qquad (2)$$

$$J = \int \varphi_a^*(1)\varphi_b^*(2)e^2[1/r_{12} - 1/r_{b1} - 1/r_{a2} + 1/R_{ab}]\varphi_a(2)\varphi_b(1) \, dv_1 \, dv_2 \qquad (3)$$

When J is positive, two parallel spins are more stable than the antiparallel spins [Fig. 4(b)]. The negative J-value favours antiparallel spins relative to parallel spins [Fig. 4(c)]. Between molecular orbitals that have an appreciable overlap integral S, the exchange of electrons is feasible by attraction from the positively charged nucleus of the other atom making r_{b1} and r_{a2} smaller; therefore, the second and third terms in the brackets of (3) become dominant. As a result, J becomes negative and the ground state becomes a singlet. The J-values for typical π-bonds are of the order of -1 eV. When the two electrons are in two degenerate or nearly degenerate orbitals that are orthogonal to each other ($S = 0$), but spatially close by, the first term will dominate and the exchange of electrons takes place by direct Coulombic repulsion between the two electrons concerned; J can then become positive.

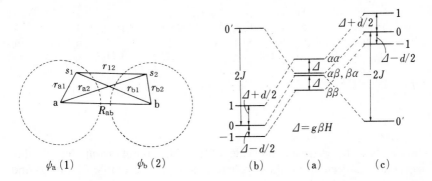

Fig. 4 The Coulombic repulsion of the electron spins 1 and 2 and singlet and triplet states. $\psi_T = [\psi_a(1)\psi_b(2) - \psi_b(1)\psi_b(2)]/\sqrt{2}$; $\psi_S = [\psi_a(1)\psi_b(2) + \psi_b(1)\psi_a(2)]/\sqrt{2}$.

In organic molecules, it is typically the one-centre exchange integral that contributes significantly to the positive J-values, although those between two adjacent positions in perpendicular ethylene (Borden, 1982b) and biphenyl derivatives (Veciana *et al.*, 1989b) should not be neglected. In contrast, the two electrons of the two transition-metal ions in binuclear metal complexes often exhibit an exchange interaction in spite of the large distance (300–500 pm) involved. This kind of exchange coupling occurs via polarization of the spins in the p-orbital on the ligand atoms and is called superexchange.

Equation (1) is equivalent to the Heisenberg Hamiltonian (4); where s_1 and s_2 are spin operators. Since $2s_1 \cdot s_2 = S^2 - s_1^2 - s_2^2 = S(S+1) - 3/2$, the difference in the eigenvalues $<S|H|S>$ for the triplet and singlet states becomes $<1|H|1> - <0|H|0> = -2J'$ and, therefore, $J' = (J - S^2 Q)/(1 - S^4)$.

$$H = -2J's_1 \cdot s_2 \tag{4}$$

The question of how to design high-spin organic molecules now reduces to the problem of how to arrange within a molecule many singly occupied degenerate orbitals that are orthogonal to one another.

ORGANIC MOLECULES HAVING MANY DEGENERATE ORBITALS

A straightforward answer to the above question would be a consideration of molecular symmetry. According to group theory, doubly degenerate molecular orbitals, denoted by the symbol e, can arise if a given molecule has one three-fold or higher axis of rotation, or if it has D_{2d} symmetry. It is well

[5] 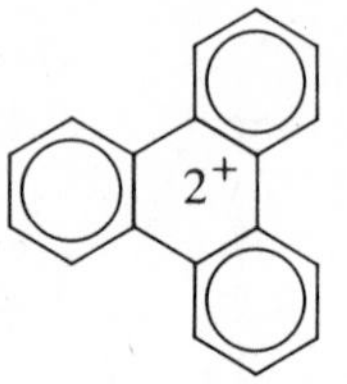   [6] 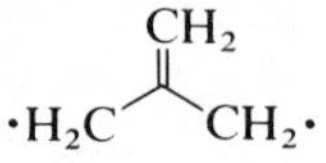   [7]

known that the benzene molecule of the D_{6h} point group has doubly degenerate molecular orbitals. Allene is a typical example of the second criterion. If there is more than one rotation axis that is of three-fold or higher symmetry, as in methane of T_d symmetry and cubane of O_h symmetry, triply degenerate wave functions appear; these are generally denoted by the symbol t or f. However, as is usually the case, all the hydrocarbons mentioned above have too many electrons to form open-shell structures; each degenerate orbital has a pair of electrons with antiparallel spins. If we are able to reduce the number of electrons to form the dication radicals [5] and [6] of benzene and triphenylene, respectively, we will have the last two electrons occupying doubly degenerate HOMOs, and triplet species will result, as dictated by Hund's rule. This has been shown to be the case by construction of benzene and triphenylene derivatives stabilized with electron-donating substituents (Breslow, 1982; Breslow et al., 1982). These compounds will be mentioned later when we discuss potential intermolecular spin alignment in Section 6.

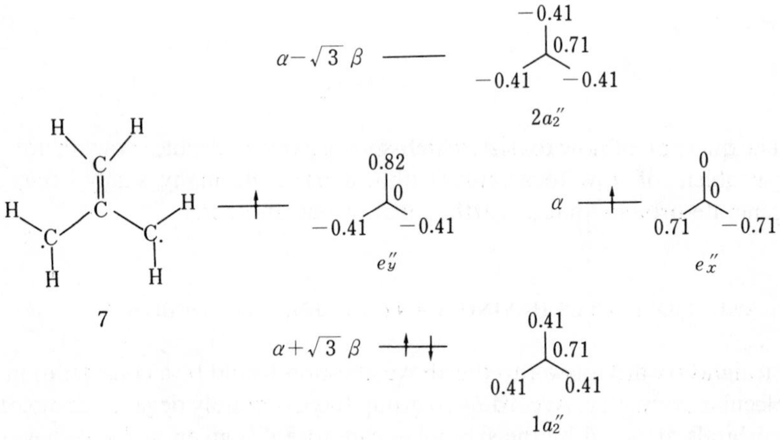

Fig. 5 Trimethylenemethane [7] and its Hückel π-MOs.

Triplet trimethylenemethane [7] in the D_{3h} point group satisfies the above condition as shown in Fig. 5. We note that this molecule constitutes, at the same time, the simplest example of another series of molecules that can have degenerate orbitals for another reason (Berson, 1978, 1982; Dowd, 1966, 1972). There is a series of alternant hydrocarbons called non-Kekulé molecules (Coulson and Longuet-Higgins, 1947; Coulson and Rushbrooke, 1940). Longuet-Higgins proposed that an alternant hydrocarbon (AH) has at least $(N-2T)$ singly occupied non-bonding molecular orbitals (NBMOs) molecules (Coulson and Rushbrooke, 1940; Coulson and Longuet-Higgins, 1947). Longuet-Higgins proposed that an alternant hydrocarbon (AH) has at least $(N-2T)$ singly occupied non-bonding molecular orbitals (NBMOs) where N is the number of carbon atoms in the AH and T is the maximum number of double bonds occurring in any resonance structure (Longuet-Higgins, 1950). Let us take for example three isomeric benzoquino-dimethanes (xylylenes) [8]. There are eight carbon atoms providing 2p π-orbitals ($N = 8$). The maximum number of double bonds are four, three and four ($T = 4$, 3 and 4) for the o-, m and p-isomers, respectively. Therefore, the numbers of NBMOs ($N-2T$) are zero, two and zero, respectively. The o-[8] and p-benzoquinodimethanes p-[8] are predicted to have closed-shell electronic structures with singlet ground states (see Fig. 6), in good agreement with our chemical intuition that classical Kekulé structures can be written for these two isomers. Actually, ^{1}H-nmr spectra have been obtained unambiguously for p-[8] and o-[8] by means of low-temperature trapping (Williams et al., 1970) and by a flow method (Trahanovsky et al., 1988), respectively.

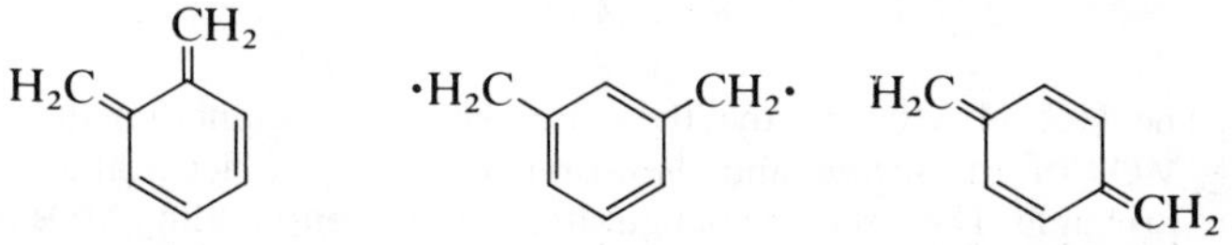

(o)   (m)   (p)

[8]

The m-isomer m-[8] is predicted to have two degenerate orbitals of zero energy ($N - 2T = 2$) for which the seventh and eighth electrons are supplied after three bonding MOs are filled with six electrons in accordance with Pauli's exclusion principle. According to Hund's rule, the last two electrons should be accommodated one each to the two degenerate NBMOs (Fig. 6). Although in the C_{2v} point group and without any axis higher than two-fold symmetry, m-[8] is therefore predicted to be in a triplet ground state as suggested by its non-Kekulé structure. A more quantitative evaluation was put forward by Baudet who estimated by SCF–CI calculations that the triplet state is more stable than the singlet by 0.34 eV (7.8 kcal mol^{-1}) (Baudet, 1971).These predictions were confirmed more recently by experiment (Migirdicyan and Baudet, 1975; Wright and Platz, 1983) and by more sophisticated calculations (Kato *et al.*, 1983). Whereas the original preparative method employed extended 254 nm photolysis of an alkane matrix containing *m*-xylene at 77 K, an exceptionally simple method for preparing *m*-[8] by photolysis of *m*-xylylene dichloride in ethanol glass has been proposed (Haider *et al.*, 1988).

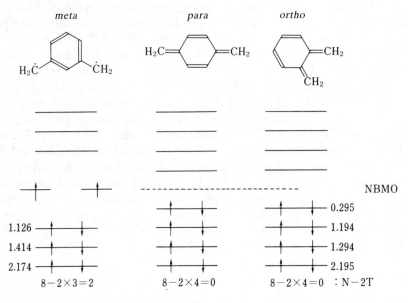

Fig. 6 The Hückel MOs of the three isomeric benzoquinodimethanes [8]. The bonding MOs of the *ortho*- and *para*-isomers are filled according to the Pauli exclusion principle. The electron configuration of the non-bonding MOs of the *meta*-isomer is dictated by Hund's rule.

The double degeneracy of NBMOs in *m*-[8] has nothing to do with the geometrical symmetry of the molecule, but, rather, with the connectivity of the two radical centres, or the phase relationship of the atomic orbitals in the conjugated system. Therefore, the term "topological symmetry" has been proposed to describe the connectivity of the carbon atoms carrying the π-electrons and the periodicity of the π-orbitals in this class of non-Kekulé hydrocarbons.

The Longuet-Higgins theory based on Hückel MO theory and Hund's rule was pioneering work in the area of molecular design of high-spin alternant hydrocarbons and has been appreciated for its predictability.

Historically, it took quite some time before the "diradical paradox" associated with the Chichibabin [9], Thiele [10] and Schlenk [11] hydrocarbons was resolved by epr experiments and by the reasoning based on these topological considerations (Platz, 1982); only the last hydrocarbon was concluded to have a ground triplet state (Kothe *et al.*, 1970; Luckhurst *et al.*, 1971).

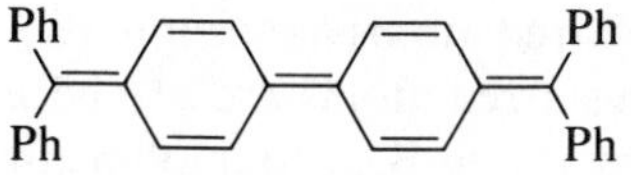

[9] 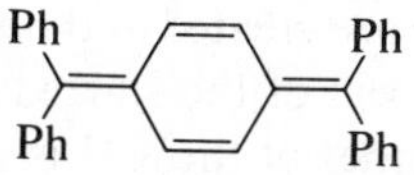   [10] 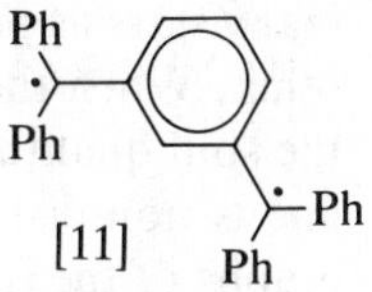   [11]

More recently, however, the ground spin states of some alternant hydrocarbons have been found to be contradictory to the Longuet-Higgins prediction. For example, whereas the ground state is predicted to be a triplet for tetramethyleneethane [12], and triplet signals have been observed for the parent and the derivatives (Dowd, 1979; Dowd *et al.*, 1896, 1989; Roth *et al.*, 1987), it is difficult to establish triplet ground states experimentally (Du and Borden, 1987). *m,m'*-Biphenyldicarbene [13] is predicted to have a ground

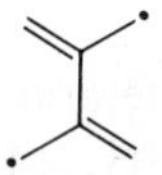

[12]

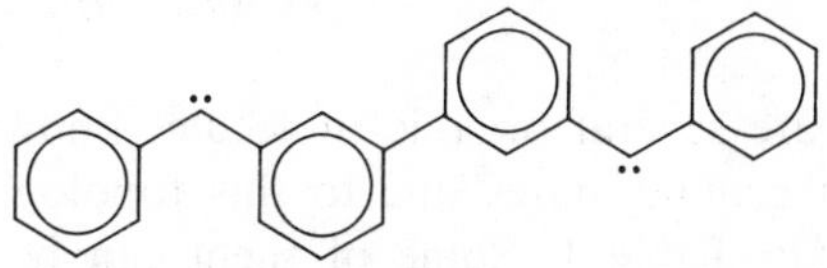

[13]

quintet state in contrast to the observed singlet state (Itoh, 1978). This inconsistency is ascribed to the use of the simple Hückel MO theory in the Longuet-Higgins treatment. More recently, the MO theory has been elaborated by taking into account configuration interaction (Borden and Davidson, 1977). The results for non-Kekulé alternant hydrocarbons are summarized as follows. Non-Kekulé diradicals are classified into two groups. When one or more constituent atomic orbitals are shared between the two NBMOs (the system is called non-disjoint), the one-centre Coulombic repulsion between the two electrons favours the parallel orientation of the two spins according to Hund's rule, making the triplet the ground state. On the other hand, when a set of atomic orbitals constructing an NBMO are not shared by another set of atomic orbitals constructing the other NBMO, the system is called disjoint. A typical example is [12] in which two NBMOs consist of two isolated allyl radicals joined at the nodal points. Hund's rule cannot be applied straightforwardly; in these situations, the difference in the Coulombic repulsion energy between the singlet and triplet states becomes insignificant. The singlet could become the ground state by taking into account higher order effects.

Valence bond theories have been developed by Ovchinnikov (Ovchinnikov, 1978) and Klein (Klein, 1982, 1983; Klein *et al.*, 1982) using the expansion of the Heisenberg Hamiltonian. According to these theories, the carbon atoms in alternant hydrocarbons are starred in such a way that as many stars as possible are created and no two starred atoms are next to each other. When the numbers of the starred and unstarred atoms are n^* and n, the spin quantum number of the AH is given by (5). *m*-Benzoquinodimethane is non-disjoint, and $n^* = 5$, $n = 3$ and $S = 1$. In other words, the Γ-graphs of the isomeric quinodimethanes [8] are as shown in Fig. 7. Once an up-spin is placed at one benzylic radical centre (starred carbon atom), spin polarization places down-spins at the unstarred carbon atoms, ending in the other radical centres with up-spin in the *m*-isomer and down-spins in the *o*- and *p*-isomers, respectively. Organic chemists will recall precedents for such topological considerations in the discussion of the signs and sizes of long-range nuclear spin–spin couplings and contact shifts in nmr spectroscopy (Eaton and Phillips, 1965).

$$S = (n^* - n)/2 \tag{5}$$

There are several organic molecules known in the literature that have high-spin ground states due to this topological symmetry, and these are collected in Table 1. Some of them can be used as building blocks for constructing higher-spin systems, but others may be structurally dead-ends and difficult to extend further systematically. In this sense, the poly(*m*-phenylenecarbene) system has special significance for its structural uniqueness and versatility, and it will therefore be discussed in some detail.

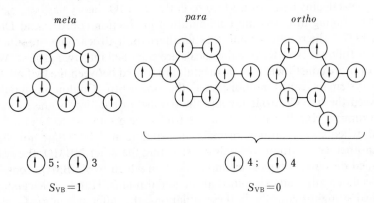

Fig. 7 Γ-Graphs of the isomeric benzoquinodimethanes [8].

Table 1 High-spin organic molecules.

	D/cm^{-1}	E/cm^{-1}	g	Reference
$S = 3/2$				
[1']	0.0041	0		Brickmann and Kothe (1973)
[29]	0.075	0	2.03	Koga and Iwamura[a]
[Cu(ii), monocarbene]				
$S = 2$				
[2]	0.01174	0.00315	2.0054	Seeger et al. (1986)
[3]	0.07131	0.01902	2.0023	Itoh (1967), Wasserman et al. (1967)
b	0.0983	0.0170	2.003	Izuoka et al. [a]
[25]	0.0944	−0.0024	2.0023	Huber and Schwoerer (1980)
[27]	0.0207	0.0047		Novak et al. (1989),
m,p'-[20]				Murata et al. (1987)
c	0.1575	0.020		Iwamura and Murata (1989)
[28]	0.124	0.002		Tukada et al. (1987)
[29]				Koga and Iwamura (1989)
(syn-dicarbene)				
$S = 3$				
d	0.0548	0		Wasserman et al. (1968)
[15; $m = 3$]	0.04874	0.00889		Teki et al. (1985)
[23; $m = 3$]	0.04158	0.01026	2.0038	Takui and Itoh (1973)
$S = 4$				
[24; $m = 4$]	0.01400	0.00050	2.003	Iwamura et al.[a]
[15; $m = 4$]	0.03161	0.00394	2.002	Teki et al. (1986)
$S = 5$				
[15; $m = 5$]	−0.0168	0.0036	2.003	Fujita et al. (1990)

[a] Unpublished results from these sources.

b

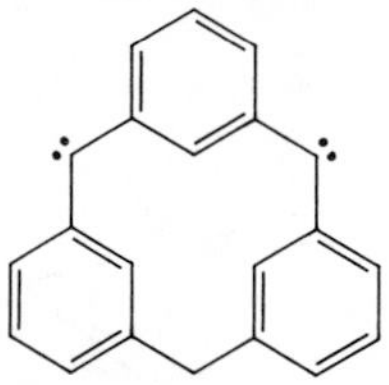

c

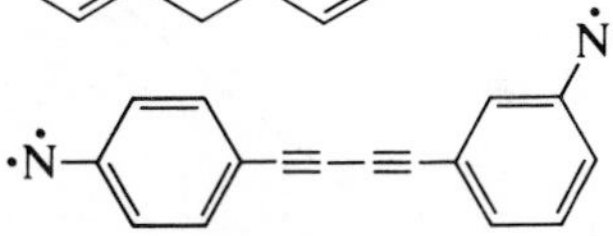

d

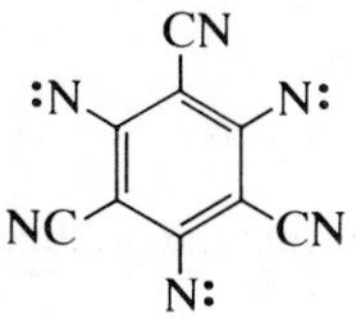

POLY(m-PHENYLENECARBENES)

All the theories predict that polyradicals [14] consisting of m-diphenylmethyl units connected at m-positions should have $S = m/2$ and spin multiplicities of $m + 1$. The exchange interactions between the radical centres in [14] are estimated to be 0.2055 and 0.0012 eV for the first and second neighbours, respectively (Tyutyulkov and Karabunarliev, 1986).

There is a third class of orbital degeneracy which does not require any symmetry in apparent geometrical shape of the molecule. When two orthogonal orbitals are nearly degenerate, more than one electronic configuration is possible for the two electron system. Both electrons may reside in the orbital of lower energy with the spins antiparallel to each other to form a singlet state as dictated by Pauli's exclusion principle. The electron correlation does not necessarily favour this configuration. As long as the energy gap of the two orbitals is not considerably larger than the Coulomb repulsion between the two electrons, Hund's rule prevails and the triplet configuration with the electrons residing one in each orbital with parallel spins will be favoured. A typical example is found in triplet carbenes and nitrenes where two orthogonal orbitals belong to the same carbon and nitrogen atoms, respectively. In a canonical expression of diphenylcarbene [15; $m = 1$], for example, the divalent carbon atom assumes a nearly sp^2-hybridized planar structure with the remaining $2p_z$-orbital perpendicular to the plane [Fig. 8]. The last two electrons are assigned by the operation of Hund's rule one each to these nearly degenerate orbitals to give a triplet ground state (Brandon *et al.*, 1965; Higuchi, 1963a,b; Hutchison and Kohler, 1969).

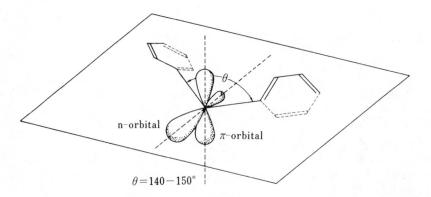

Fig. 8   Molecular structure of diphenylcarbene [15; $m = 1$] showing the orthogonal orbitals at the divalent carbon atom.

[14]   [15]

The last two factors for orbital degeneracy have been combined by replacing each doublet radical centre in [14] by a triplet carbene to form poly(carbenes) [15]. The system was originally conceived and proposed by Mataga as early as 1968 as a model for one-dimensional organic ferromagnets (Mataga, 1968). Diphenylcarbene had been established to have a ground triplet state due to a strong exchange interaction ($J = +0.20\,\text{eV}$) between two spins, one each on the orthogonal σ- and π-orbitals at the carbenic centre (Higuchi, 1963a,b). Higuchi noted the presence of two kinds of spins; one is more or less localized in the σ-orbital at the divalent carbon and the other is more or less delocalized over the conjugated π-framework. The duality is very much reminiscent of the sd-interaction of the electron spins in magnetic metals. In the latter, the localized spins of high 3d-character at the lattice centres are aligned via strong exchange coupling with the metal conduction electrons of high 4s-character. Itoh and Iwamura have recently found computationally as well as experimentally that the MO picture (Fig. 9) is very typical of cross-conjugated systems (Alexander and Klein, 1988; Itoh, 1971; Iwamura, 1986; Tyutyulkov et al., 1988). As the chain is elongated, the highest occupied and lowest unoccupied molecular orbitals (HOMO and LUMO, respectively) remain unperturbed at the energy levels of β and $-\beta$, respectively. The energy gap between the valence band and the conduction band of the polymer remains constant at 2β, with $2m$ parallel spins in the impurity band ($=$ non-bonding levels). The same is true for the similar two-dimensional network [14′] (Hughbanks and Kertesz, 1989; Tyutyulkov et al., 1985). We have also learned from experiment that [15] has additional merits over [14]. First, whereas it is rather difficult to generate [14] systematically, the series [15] has the corresponding diazo-compounds as suitable precursors from which members can be produced by photolysis at cryogenic temperatures. Secondly, the epr fine structures of [15] are usually an order of magnitude more widespread than those of [14], making the analysis of the exchange interaction easier.

Tyutyulkov analysed cross-conjugated systems like [14] theoretically, using the Coulson–Rushbrooke theorem and Wannier transformation of the Bloch MO. His results hold multiple significance. First, it was suggested, by taking into account the interaction between the one-dimensional chains of the quasi-one-dimensional polymer, that the critical temperature T_c may reach 10^2–10^3 K (see p. 226). Secondly, the radical centres of hydrocarbons

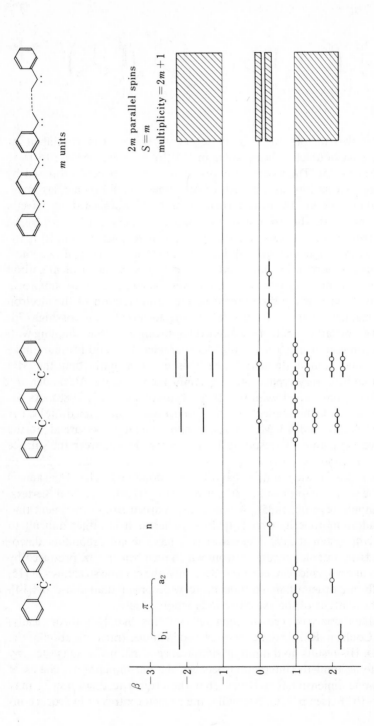

Fig. 9 Hückel MO diagrams of diphenylcarbene [15; $m = 1$], m-phenylenebis(phenylcarbene) [15; $m = 2$] and poly(m-phenylene-carbenes) [15].

[14] can be replaced by heteroatom-centred stable radicals (Tyutyulkov and Polansky, 1987; Tyutyulkov *et al.*, 1988, 1989) such as aminoxyls (Forrester *et al.*, 1968). Their band structure is again characterized by a wide energy gap, in the middle of which there is a band of nearly degenerate quantum states (Tyutyulkov *et al.*, 1983, 1988).

POLY(ACETYLENES) AND OTHER CONJUGATED POLYMERS

As will be described in Section 5, a series of [15] gave rise to novel high-spin organic molecules and served as good models for one-dimensional organic ferromagnets. In order to extend the system further and construct macroscopic spins of practical use, namely, to achieve organic superparamagnets or ferromagnets, it became clear that [15] is far from ideal. First, only when the carbenic centres are generated from the diazo-groups without fail and survive without any further chemical reaction will the expected spin multiplicity be attained in these systems. Otherwise, the cross-conjugated system will be severed at the reaction site. Such a probability should increase statistically as the chain length is increased. Secondly, there are no straightforward polymerization or condensation reactions leading to this skeleton. The construction of the backbone appears to require stepwise synthesis. The hexaketone precursor has already shown another constraint of limited solubility in typical organic solvents. Thirdly, the carbenic centre is a reactive intermediate and thus survives at temperature only up to 100–150 K in host crystals and solid solutions. Even by introduction of stabilizing substituents, as realized to some extent by fluorine substitution in triplet phenylnitrene (Dunkin and Thomson, 1982), it would be extremely difficult to make carbenes that would survive at 300 K. Substituents carrying a lone pair of electrons are known to stabilize neighbouring carbene centres, but they stabilize the singlet state relative to the triplet (Schuster, 1986). Based on these considerations, it is deemed more practical and promising to introduce new structures in which the radical or carbene centres are attached as pendents to fully conjugated main chains (Iwamura and Izuoka, 1987). In this way, one could be able to bypass one or even two open-shell centres that might fail to be generated and yet keep the cross-conjugation between the remote radical centres in the side chains effective.

Poly(acetylenes) [16], poly(diacetylenes) [17], poly(phenylenes) [18] and poly(phenylenevinylenes) [19] have been proposed as main chains. The new molecular design that has to be developed now concerns the question of where to place the open-shell centres on the pendents in the side chains of the polymers (Iwamura, 1987, 1988; Iwamura and Murata, 1989; Murata *et al.*, 1987; Ovchinnikov, 1978; Tyutyulkov *et al.*, 1985; Yamaguchi *et al.*, 1987).

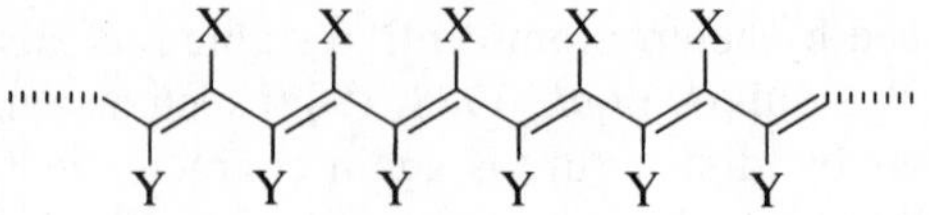

[16]

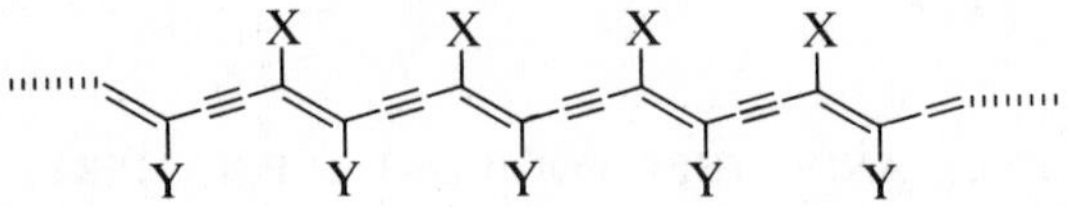

[17]

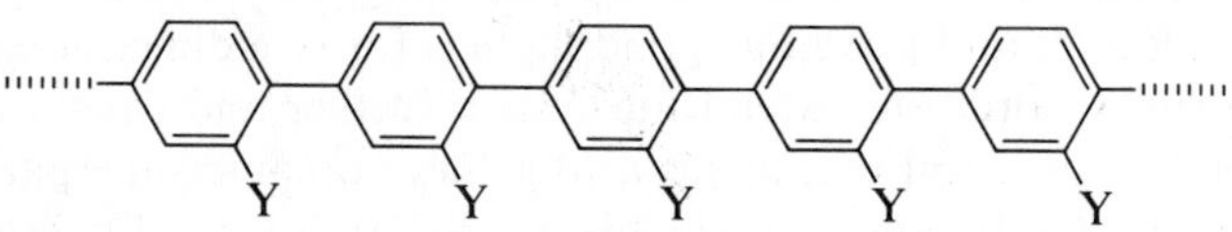

[18]

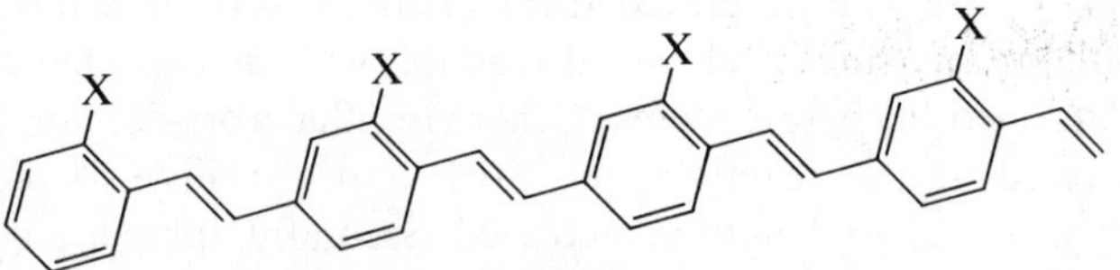

[19]

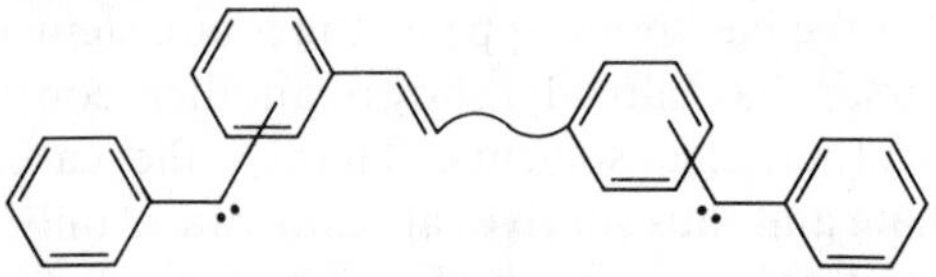

[20]

As a model for the unit of poly(diphenylacetylenes) [16b][1] or [19], isomeric stilbene derivatives [20] carrying the phenylmethynyl groups are considered. According to the Longuet-Higgins theory on π-NBMOs in alternant hydrocarbons (Longuet-Higgins, 1950), the p,p'-isomer (p,p'-[20]) is predicted to have no NBMO, a result consistent with the intuition that a stable quinonoid structure can be drawn for this isomer. The number of NBMOs ($N - 2T$) is calculated to be two, both for m,p'-[20] and m,m'-[20]. However, according to the modern MO theory of Borden and Davidson described on p. 191, the last two isomers are differentiated as disjoint and non-disjoint, respectively (Borden, 1982; Borden and Davidson, 1977). Therefore, m,m'-[20] is predicted to have a singlet ground state. The interaction of the two radical centres is expected to be ferromagnetic in m,p'-[20] and to lead to a high-spin ground state, a quintet state rather than a triplet since the σ-spins at the carbenic sites are added. The valence bond

[1] See the Appendix on p. 245 for designations of lower case letters used as suffixes in structural formula numbers.

theory [see (5)] reveals that $n = n^*$ in m,m'-[20] and they differ by two in m,p'-[20]. Therefore, the same theoretical prediction is obtained either by the modern MO or valence bond theories.

Table 2(a) Ground electronic states (GS) of oligomeric poly(phenoxyl) radicals.

	n	S_{GS}	ΔDE^a
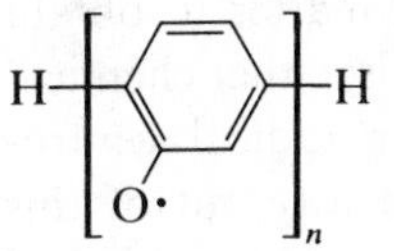	2	1	7.3
	3	3/2	4.7
	4	2	4.3
	5	5/2	3.9
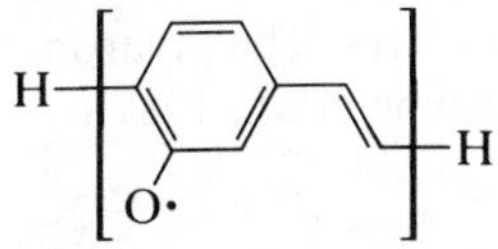	2	1	3.5
	3	3/2	2.7
	4	2	2.7
	5	5/2	2.2

[a] Energy gap in kcal mol^{-1} between ground states and next higher excited state of spin $S_{GS} - 1$.

Table 2(b) Coupling of phenoxyl radicals by various coupling spacers (X).

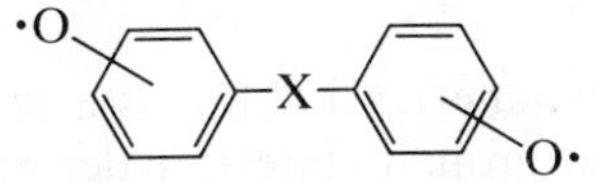

	S–T gap^a		
X	m,p'-	m,m'-	p,p'-
None	12.5	−0.6	≪0^b
CH=CH	6.5	1.4	≪0
CH=CH—CH=CH	6.0		≪0
C≡C—C≡C	2.7	0.8	≪0
p-C_6H_4	1.8	1.8	≪0
m-C_6H_4	−0.2	2.3	2.3
C=O	1.9	0.7	0.8
O	2.8	−0.1	−4.5
NH	4.2	0.3	−10.3

[a] AM1-CI triplet–singlet energy gaps in kcal mol^{-1}. The triplet is the GS where the gap is positive.
[b] Kekulé structures can be drawn.

When the monophenyl-substituted polymers, e.g. [16c], are taken into account, radical centres are predicted to be placed at the same position of every phenyl ring: all p- or all m-positions. This is a very fortunate message for synthetic chemists in that regiostereoregular homopolymerization of phenylacetylenes, phenyldiacetylenes and so on can be employed for the construction of the desired polymers. A polymer carrying the p-phenoxyl radical [16d] is already on a "short list" of high-spin polymers suggested theoretically by Ovchinnikov (Ovchinnikov, 1978). Lahti performed a computational modelling of a pair of phenoxyls using the AM-1 semiempirical MO method with configurational interaction (Lahti and Ichimura, 1989). His results are collected in Table 2.

As we will see in Section 5, it is not a straightforward matter to obtain conjugated polymers with a high content of pendent radicals; either chemical introduction of radicals into polymers is inefficient or radical centres cannot be kept intact during polymerization. A possible way out of this problem has been proposed by Fukutome, namely, doping of closed-shell cross-conjugated polymers to generate the open-shell centres. The polaronic interaction in these systems has been discussed (Fukutome $et\ al.$, 1987).

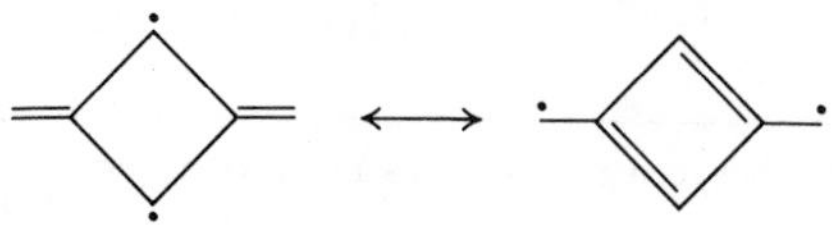

[21]

2,4-Dimethylenecyclobutane-1,3-diyl [21] is a typical non-Kekulé hydrocarbon that has a triplet ground state (Snyder and Dougherty, 1989). A theoretical study has been carried out on the chemical- vs band-structure relationship for related oligomers and polymers, poly(cyclobuta-1,3-diene-1,3-diyl), to show that there are various spin states including high-spin ones in degeneracy. The magnetic properties of such a polymer would be very interesting (Pranata $et\ al.$, 1989).

Once the radical centres are placed in non-alternant systems, it is difficult to tell $a\ priori$ if Hund's rule can be applied to the unpaired electrons in the NBMOs (Berson, 1987, 1989). The band structures of the polymers consisting of such non-alternant hydrocarbons have been discussed (Tyutyulkov $et\ al.$, 1985, 1989; Tyutyulkov and Karabunarliev, 1985). Poly(fulvene-3,6-diyl), for example, has a degenerate valence band corresponding to the doubly degenerate sets of orbitals for a single ring at $(\sqrt{5} + 1)\beta/2$. If this system could be heavily doped with an appropriate acceptor, a cationic polymer with a half-filled degenerate band might be realized (Hughbanks and Kertesz, 1989).

In the above discussion, we have taken into account only the through-bond exchange interaction among the pendent radical centres. In these rather congested polymers, the exchange interaction among neighbouring radical centres through space should not be neglected. Kamachi and his coworkers reported such interactions among verdazyls and transition-metal porphyrins [22] attached as pendents to poly(methacrylate) and other non-conjugated polymer main chains (Kamachi, 1987; Kamachi *et al.*, 1984; 1987). The interactions are usually weakly antiferromagnetic, i.e. detectable at temperatures below 100 K. Should the through-bond ferromagnetic interaction via the π-framework be operative as designed, the latter effect is expected to surpass the through-space interaction by an order of magnitude.

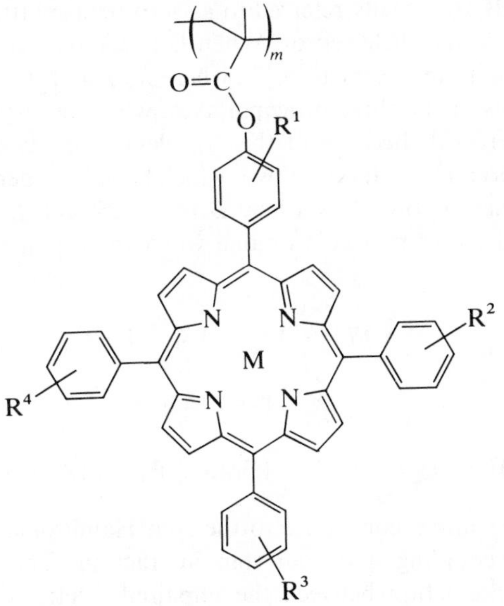

[22]

4 Analytical methods and characterization

There are several independent measurements available for confirming that high-spin molecules are really obtained.

Determinations of epr fine structure and paramagnetic susceptibilities are most often used. For characterization of higher spin orders, neutron diffraction and other physical methods may be useful. On the other hand, a successful measurement of normal high-resolution nmr spectra would serve as good evidence for singlet ground states of the chemical entities at issue.

EPR FINE STRUCTURE

When a molecule contains more than one unpaired electron ($S > 1/2$), another term of the form SDS, representing the spin–spin interaction, must be added to the spin Hamiltonian of $S = 1/2$ molecules (which consists mainly of the Zeeman term) as shown in (6). The new term leads to further

$$\mathbf{H} = \beta HgS + SDS \tag{6}$$

splittings of the energy levels, and the additional spectral transitions due to these splittings are called the fine structure. Since the spin–spin interactions are present regardless of whether a magnetic field is applied, their effects on the energy levels are usually referred to as zero-field splitting (zfs), and $\mathbf{D}$ is referred to as the zero-field tensor. When $\mathbf{D}$ is diagonalized, it is customary to express the spin–spin term as in (7), where X, Y and Z are the eigenvalues in the directions of the three principal axes, with the further property that their sum is zero (8). Because of this traceless property of $\mathbf{D}$, these three principal values can be reduced to two, which have been denoted as D and E. Equation (6) then becomes (9), where $D = -3/2Z$ and $E = -1/2(X - Y)$. E is zero for a linear molecule or one when there is a three-fold axis of symmetry.

$$SDS = -XS_X^2 - YS_Y^2 - ZS_Z^2 \tag{7}$$

$$X + Y + Z = 0 \tag{8}$$

$$\mathbf{H} = \beta HgS + D[S_Z^2 - 1/3S(S + 1)] + E(S_X^2 - S_Y^2) \tag{9}$$

Two distinct sources contribute to the spin Hamiltonian of the form of (6), spin–orbit coupling and spin–spin interaction. The latter, magnetic dipole–dipole interaction between the unpaired electrons, is dominant in organic molecules that do not contain heavy elements.

The spin-basis functions for the m unpaired electrons for calculating the energy levels, for example, become (10), and $S_2^2 = (S_{z1} + S_{z2} + S_{z3})^2$, etc.

$$|+m/2> = |\alpha_1\alpha_2\alpha_3 \ldots \alpha_m>$$

$$|+(m-2)/2> = (1/\sqrt{m})|\alpha_1\alpha_2\alpha_3 \ldots \beta_m + \ldots + \beta_1\alpha_2\alpha_3 \ldots \alpha_m> \tag{10}$$

$$\ldots \qquad \ldots \qquad \ldots$$

$$|-(m-2)/2> = (1/\sqrt{m})|\alpha_1\beta_2\beta_3 \ldots \beta_m + \ldots + \beta_1\beta_2\beta_3 \ldots \alpha_m>$$

$$|-m/2> = |\beta_1\beta_2\beta_3 \ldots \beta_m>$$

Then the $(m + 1) \times (m + 1)$ eigenvalue matrix can be constructed from (9) and can be solved exactly by direct diagonalization or with the aid of perturbation methods for the eigenvalues with the external magnetic field along any principal axis. $(m + 1)$ Energy values and therefore $m \, \Delta m_s = \pm 1$

transitions are obtained in principle for each principal axis. For epr spectra of randomly oriented multiplets, consult publications by Wasserman *et al.* (1964) and Weltner (1983). The theory of the magnetic dipolar interaction of two coupled carbene triplets has been advanced by Itoh (Itoh, 1978) and Benk and Sixl (Benk and Sixl, 1981). When the electronic coupling is small, the dipolar coupling tensor of the resulting quintet ($\mathbf{D_Q}$) is given by the sum of those of the triplets (11). In higher-spin systems, anomalous lines are sometimes produced (Weltner, 1983).

$$\mathbf{D_Q} = (\mathbf{D_T^1} + \mathbf{D_T^1})/6 \tag{11}$$

Since the tensor $\mathbf{D}$ is strictly proportional to the inverse cube of the distance between the unpaired electrons for two spin systems and is regarded as a measure of the inverse volume of the delocalized spins in higher-spin systems, the absolute D-values become smaller as S increases. Since the epr spectral width is approximated by $2D$, it is customarily the case that the $g = 2$ region becomes crowded with spectral peaks with increasing S.

The second term in (6) is due to dipolar interaction; the energy levels are therefore dependent on $3\cos^2\varphi - 1$, where φ is the angle between the external magnetic field and any principal axis. The experimental data of such an angular dependence will be of use in determining the zfs values.

Unlike typical optical spectroscopy, the energy gap between spin sublevels and consequently the epr transition energies are rather small (*ca.* 9300 MHz = 6.2×10^{-24} J). The population of the upper spin sublevels is not negligible at $T > 0$. It is the rule rather than the exception that the transition probability and, therefore, signal intensity is temperature dependent. Since the Boltzmann factor $\exp(-\Delta W_m/kT)$, reduces to $-\Delta W_m/kT$, the signal intensity (I_n) is linearly proportional to the inverse of the absolute temperature. This relation is called the Curie law in magnetic resonance and usually holds true for any transition of any spin multiplicity, unless there are other states populated in equilibrium. In the latter case, the population of the magnetic state under consideration becomes subject to another Boltzmann distribution. Let us take for example the equilibration of a triplet with a singlet state. Since the latter is epr silent, the signal intensities of the triplet are the observables. Since I_n is inversely proportional to temperature and the population of the triplet may be represented by a Boltzmann factor, I_n is represented by a product of the two factors (12), where C is a proportionality constant and ΔE is the potential energy difference between the singlet and triplet states. Some theoretical curves for representative ΔE-values are drawn in Fig. 10.

$$I_n = \frac{C}{T} \frac{\exp(-\Delta E/RT)}{1 + 3\exp(-\Delta E/RT)} \tag{12}$$

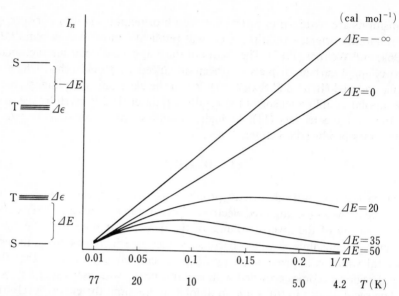

Fig. 10 Theoretical curves [equation (12)] representing the temperature dependence of the signal intensities due to a triplet species that has a singlet manifold in equilibrium. The ΔE-values represent the energy gap of the two states.

In the case of a quintet state consisting of two weakly coupled triplets and populating with the triplet and singlet manifolds, the signal intensity of the quintet will be given by (13).

$$I_n = \frac{C}{T}\ \frac{5\exp(-3\Delta E/RT)}{1 + 3\exp(-\Delta E/RT) + 5\exp(-3\Delta E/RT)} \tag{13}$$

MAGNETIZATION AND MAGNETIC SUSCEPTIBILITY

Let us consider a *paramagnetic* substance that contains, in a unit volume, N magnetic atoms, ions or molecules each of which has a magnetic moment M. At a sufficiently low temperature the spins may order and, therefore, the magnetization $I = NgJM_B$, where M_B is the Bohr magneton. At a finite temperature, since the electron spins undergo thermal vibration, its direction also fluctuates. The energy of this vibration at T K is in the order of kT, which is about 4.1×10^{-21} J at room temperature. The potential energy of the magnetic moment $M = 1\ M_B$ in the external magnetic field of 1 MA m^{-1} ($= 1.26$ T) amounts to 1.2×10^{-23} J. This value is two orders of magnitude smaller than kT at room temperature. Therefore, the alignment of an assembly of electron spins with respect to the direction of the external magnetic field is dictated by the Boltzmann distribution. Since the direction of the spin is quantized, the z-component of the magnetic moment M is given by (14), where $J_z = J, J - 1, \ldots -(J - 1), -J$; here J is the quantum number of total angular momentum, and the z-direction is parallel to the external magnetic field H.

$$M_z = g M_B J_z \tag{14}$$

$$I = NgJM_B B_J(\alpha) \tag{15}$$

$$\alpha = gJM_B H/kT \tag{16}$$

$$\chi = I/H = Ng^2 J(J + 1)M_B^2/3kT = NM_{\mathrm{eff}}^2/3kT = C/T \tag{17}$$

The average magnetization I is given by (15), where $B_J(\alpha)$ is called a Brillouin function, and α by (16). When $\alpha \ll 1$, $B_J(\alpha)$ may be expanded, and, if we take only the first term, then (17) results. The paramagnetic susceptibility χ is inversely proportional to the absolute temperature T. This relation is called the Curie law, and the proportionality constant C is the Curie constant.

The above statements apply to an assembly of independent spins. Deviation from proportionality, if any is observed, suggests the presence of cooperative magnetic phenomena, i.e. ferro-, antiferro-, ferri-, meta-, micto-magnetism, and so on. The magnetic susceptibility at above the spin-ordering temperature (T_C) can be usually fitted by the Curie–Weiss expression (18) with the Weiss temperature $\theta > 0$ for the sample with dominant

$$\chi = (T - \theta)^{-1} \tag{18}$$

ferromagnetic interactions and $\theta < 0$ for those with dominant antiferromagnetic interactions (Fig. 11). Data on transition-metal complexes are often presented in the form of plots of the effective moment $M_{\mathrm{eff}}\{= (8\chi T)^{1/2}\}$ vs temperature as shown in Fig. 12. Independent spins should show a horizontal line, while the ferro- and antiferro-magnetically coupled spins are expected to show curves deviating from the line upward and downward, respectively, as the temperature is lowered.

At temperatures lower than T_C required for long-range spin ordering, a macroscopic spontaneous magnetization at zero applied magnetic field appears when the interaction is such that spins are aligned parallel (*ferromagnetism*) (Fig. 13). The M vs H plots show spontaneous magnetization at $H = 0$ and hysteresis will be observed. A characteristic saturation moment M_s will also appear. No spontaneous macroscopic moment occurs when the spins are antiparallel to one another (*antiferromagnetism*). It goes without saying that, when different local moments are aligned antiferromagnetically and incomplete cancellation of the spins occurs, reduced but finite macroscopic spontaneous magnetization again appears and is called *ferrimagnetism*. *Metamagnetism* results when spin ordering is induced by a first-order transition under a strong applied magnetic field. Organic polymers are often amorphous and therefore *J*-values can fluctuate and even change their sign. There would be a chance of spin ordering of relatively short range appearing at a certain temperature T_g as the temperature is lowered. Plots of magnetic susceptibility *vs* temperature would show a maximum at T_g. These systems are called *spin glasses* (Binder and Young, 1986).

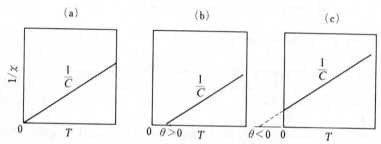

Fig. 11 Schematic drawings of inverse χ *vs* temperature relations for paramagnetic samples (a) without interspin coupling, (b) with ferro- and (c) antiferro-magnetic interspin interactions. The slope of the lines is equal to the inverse of the Curie constants.

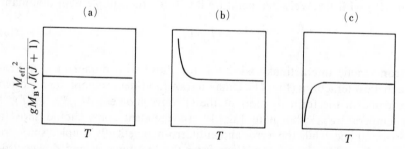

Fig. 12 Schematic drawings of M_{eff} *vs* temperature relations for (a) unordered spins with (b) weak ferro- and (c) antiferro-magnetic interactions.

As suggested earlier, high-spin molecules that have thousands of parallel spins would correspond to single domain particles of ferromagnetic substances. In α-iron, for example, a microcrystalline particle of radius 200 pm contains *ca.* 5000 spins and is considered to form a single domain structure.

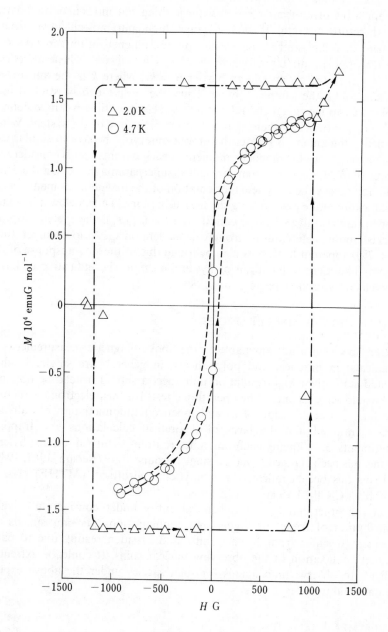

Fig. 13 A typical hysteresis curve for a sample with extended ferromagnetic coupling (see page 239).

The analogy breaks down when domain walls are taken into account; the domain walls of molecular systems are nothing but molecular boundaries and therefore discrete. If such a super high-spin molecule experiences magnetocrystalline and shape anisotropy, the magnetization can take two possible opposite directions when anisotropy is uniaxial. These directions may be separated by an energy barrier $w = kv$, where k is the anisotropy constant and v the sample volume. Thus, the system can be studied in a similar way to paramagnetic relaxation. A relaxation time τ can be defined as $\tau = \tau_0 \exp(w/kT)$, where τ_0 is a characteristic time of the system. When the measuring time τ_m is greater than τ, one observes a behaviour similar to a paramagnet. Since the magnetic moment of each paramagnet is considerably greater than M_{eff}, such systems are called superparamagnetic. On the other hand, for $\tau_m < \tau$ the complete reorientation of the magnetic moments of the cluster cannot take place during the measuring time. Thus, below a blocking temperature T_B given by the condition $\tau_m = (\tau_{av}(T_B))$, the system appears blocked and its behaviour is strongly dependent on τ_m. At low temperatures $(T < T_B)$, apparent hysteresis may be found due to the slow response of the system. The mere observation of hysteresis cannot be indicative of ferromagnetism by itself (Palacio *et al.*, 1989a,b).

NMR SPECTRA OF SINGLET SPECIES

Singlet species are not amenable to the above magnetic measurements. If diradicals, in practice, and polyradicals, in general, are singlet in their ground state, their high-resolution nmr spectra should instead be obtained and would serve as one of the operational tests for their electronic structure. The nmr characterization of *o*- and *p*-benzoquinodimethanes has already been mentioned. On the basis of theoretical calculations and trapping experiments, 3,4-dimethylenefuran, a heteroatom-perturbed tetramethyleneethane, has been claimed to have a singlet ground state (Berson, 1987, 1989), and this has been confirmed in methyltetrahydrofuran (MTHF) glass at 77 K by ^{13}C CP/MAS spectra (Zilm *et al.*, 1987).

On the other hand, if the chemical entity under consideration has a contribution of triplet or higher-spin configurations, its nmr signals are expected to suffer from contact shifts and broaden readily due to paramagnetic relaxation of the observing nuclear spins. It could be extremely difficult to observe nmr signals for such species under the above experimental conditions.

OTHER MEASUREMENTS

Since the neutron has a nuclear magnetic moment of $-1.913\,M_N$, a beam of thermal neutrons undergoes scattering by interaction with magnetic moments as well as by atomic nuclei that are present in its path. There are several methods established for separating the observed scattering into the two factors. Dependence of the amplitude of the magnetic scattering on the diffraction angle can be used for the analysis of the arrangements of the magnetic moments in the solid samples, just as X-ray diffraction reveals the arrangement of atoms in a crystal lattice. In the case of a sample that has a magnetic transition, the variation of the scattering intensity as a function of scattering angle is determined at temperatures above and below the transition. The enhancement of the scattering below the transition temperature can be regarded as magnetic in origin. An increase in enhancement of the Bragg scattering shows ferromagnetic ordering and a doubling of the unit cell is expected for antiferromagnetic order. Such a study has been performed (Epstein and Miller, 1989) for a charge-transfer salt (see Section 6). In order to reduce the incoherent background scattering which would arise due to the hydrogen content of the samples, it is often the case that perdeuterated samples have to be employed.

The magnetic entropy can be obtained from the specific heat. Most of the entropy in the disordering of the spins with increasing temperature occurs at T_C or a somewhat higher temperature.

Whereas stereoselective addition and insertion reactions are often used to distinguish between singlet and triplet carbenes, there are no good operational tests available for multicentre high-spin reactants. Stereochemical studies on the reaction of the Schlenk hydrocarbon with 2,4-hexadienes show a two-step cycloaddition mechanism involving a long-lived adduct biradical, but the control experiment showing the concerted reaction of [11] when generated in the singlet state is missing (Goodman and Berson, 1985). It is not clear, in the earlier studies on the photochemical reactions of [15a; $m = 2$] (Murahashi et al., 1972), if the dicarbene was ever generated under the experimental conditions before the intermediate monodiazomonocarbene reacted with the substrates.

THEORETICAL APPROACHES

Various predictive methods based on molecular graphs of π-systems as described in Section 3 have been critically compared by Klein (Klein et al., 1989) and can be extended to more quantitative treatments. In principle, the effective exchange integrals J_{ab} in the Heisenberg Hamiltonian (4) for the interaction of localized electron spins at sites a and b are calculated as the difference in energies of the high-spin and low-spin states. It was Hoffmann who first tried to calculate the dependence of J_{ab} on the M—L—M bond angle in metal coordination compounds using the extended Hückel method (Hay *et al.*, 1975). Yamaguchi's *ab initio* approaches are called spin-projected unrestricted Hartree–Fock (PUHF) and Møller–Plesset perturbation (PUMP) methods (Yamaguchi *et al.*, 1988). Instead of describing the singlet pair as ↑↓ in a classical Heisenberg model and the ordinary UHF method, a quantum mechanically more accurate function $2(\uparrow\downarrow - \downarrow\uparrow)$ is used in PUHF. J_{ab} is given by (19), where X = PUHF or PUMP, and $E(X)$ and $s^2(X)$ are total UHF energies and total angular momenta of the high-spin (HS) and low-spin (LS) states, respectively.

$$J_{ab} = [{}^{LS}E(X) - {}^{HS}E(X)]/[{}^{HS}s^2(X) - {}^{LS}s^2(X)] \qquad (19)$$

The basis-set dependence and correlation correction for the J_{ab}-values were first studied by using the idealized face-to-face interaction between two methyl radicals. The interaction was found to be always antiferromagnetic, and the APUMP2 4-31G procedure was found to be best for semiquantitative evaluation; the APUHF STO-3G method was still acceptable for qualitative discussion of larger systems (Yamaguchi *et al.*, 1989a).

Lahti and Ichimura (Lahti *et al.*, 1985, 1989) take empirical approaches to similar MO treatments of various molecular coupling units (see Table 2).

The variation of the signs of intermolecular J_{ab}, the absolute values of which decrease with distance R in an exponential manner, can be explained by the extended McConnell model (20).

$$J_{ab} = J_{ab}(OO) + J_{ab}(SDP) \qquad (20)$$

Here, the first and second terms denote the orbital overlap (OO) and spin-density product (SDP), respectively.

5   Generation and characterization of high-spin organic molecules

CROSS-CONJUGATED POLY(CARBENES)[15]

Encouraged by the experimental finding (Itoh, 1967; Wasserman *et al.*, 1967) that the dicarbene [15; $m = 2$] (=[3]) had a ground quintet state, i.e. all four spins were ferromagnetically coupled, Iwamura and Itoh have been engaged in a project directed towards the construction of the higher series of poly(carbenes) [15], [23] and [24] (Iwamura *et al.*, 1985; Teki *et al.*, 1983, 1985, 1986).

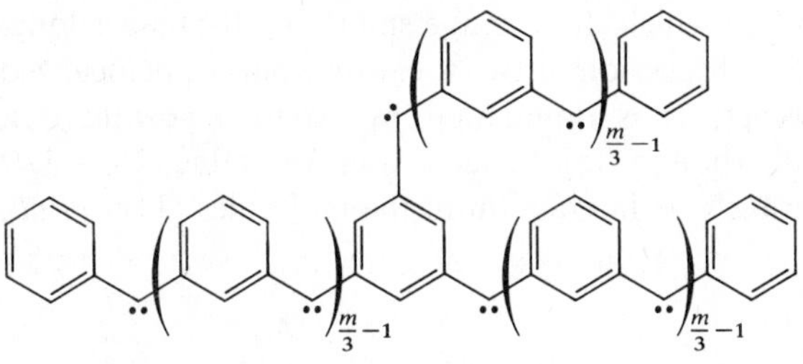

[23] 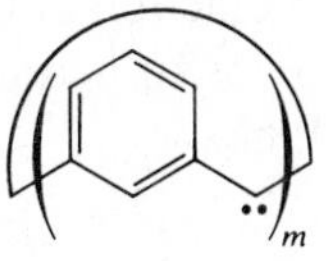   [24]

All the poly(carbenes) were generated by photolysis at cryogenic tempera-tures of the corresponding polydiazo-compounds, e.g. [15a][1], [23a], [24a], which in turn were obtained through a series of standard synthetic reactions. The tetradiazo-compound [15a; $m = 4$] was doped (5×10^{-4} mol dm^{-3}) in a single crystal of benzophenone of known crystal structure (space group $P2_12_12_1$ with $Z = 4$) and photolysed with the 405 nm mercury line in an epr cavity at 4.2 K to give [15; $m = 4$] oriented with respect to the direction of an external magnetic field. When the principal z-axis is oriented parallel to the external magnetic field, the system with $S = 4$ should generate nine spin sublevels. Under a high-field approximation, eight allowed transitions ($\Delta M_S = \pm 1$) are expected to appear in ratios of 4:7:9:10:10:9:7:4, and that is exactly what was observed (Fig. 14). The resonance fields and signal intensities observed at the K-band (25 GHz) were well-reproduced by a third-order perturbation calculation based on the spin Hamiltonian (9) with $g = 2.002$, $D = +0.03161$ and $E = -0.00394$ cm^{-1}, and $S = 4$, confirming the tetracarbene to be in the nonet state. The temperature dependence of the total signal intensity in the range 1.8–56 K showed that the observed nonet state is the ground state, while the other states are located at least 300 cm^{-1} above the ground state. The effective exchange energy between the carbenic centres are estimated by a theoretical study based on the periodic Kondo–Hubbard model to be *ca.* 0.2 eV (Nasu, 1986).

The spin distribution in the lower homologue [15; $m = 2$] was most elegantly studied by Takui *et al.* (1989) by means of a combination of ENDOR experiments and theoretical calculations within the framework of a generalized UHF Hubbard model (Teki *et al.*, 1987a) and a Heisenberg model (Teki *et al.*, 1987b). The spin distribution obtained is just as expected qualitatively in Fig. 7.

Higher homologues [15a; $m = 5$ and 6] have been prepared (Fujita *et al.*, 1990) and some of the epr data are collected together with those of the lower homologues in Table 3. In these homologous series, D decreases with increasing S. This might be regarded as a result of the larger size of the

[1] See the Appendix on p. 245 of suffixes to structural formula numbers.

molecules with the larger S in which the average spin–spin distance is long. This seems to be true for the discussion of two localized spins or delocalized π–π^* triplet states. However, D is dominated by large one-centre σ–π interaction at the divalent carbon in [15]. In such cases, the value $(2S - 1)D$ seems to measure the average σ–π interaction in the molecules (Teki et al., 1986).

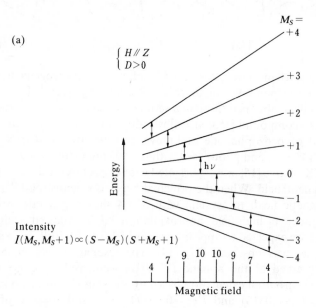

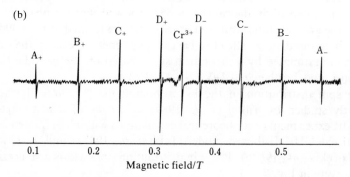

Fig. 14 (a) Epr fine structure expected of $S = 4$ at a high field. (b) An X-band epr spectrum observed after photolysis of a mixed crystal of benzophenone with [15a; $m = 4$] at 4.2 K in an esr cavity. The external magnetic field is along the direction 26° from the a axis in the ab plane of the benzophenone crystal lattice. The microwave frequency is 9550.6 MHz. The central line is due to Cr(III) in the MgO powder used as a reference.

Table 3 Comparison of D, E, and $(2S - 1)D$ (in cm^{-1}) of the known high-spin hydrocarbons with similar electronic structures.

Hydrocarbons	S	D	$|E|$	$(2S - 1)D$
Diphenylmethylene [15; $m = 1$]	1	0.40505	0.01918	0.4050
m-Phenylenebis(phenylmethylene) [15; $m = 2$]	2	0.07131	0.01902	0.2139
Benzene-1,3,5-tris(phenylmethylene)	3	0.04158	0.01026	0.2079
3,3′-Diphenylmethylenebis(phenylmethylene) [15; $m = 3$]	3	0.04874	0.00889	0.2437
m-Phenylenebis[(diphenylmethylen-3-yl)methylene] isomer I	4	0.03161	0.00394	0.2213
[15; $m = 4$] isomer II	4	0.03227	0.00225	0.2260
[15; $m = 4$] isomer III	4	0.03347	0.00216	0.2343
[15; $m = 4$] isomer IV	4	0.03241	0.00403	0.2269
[15; $m = 5$]	5	−0.0168	0.0036	−0.1512

Tetracarbene [24; $m = 4$] has been generated in a 2-methyltetrahydro-
furan (2-MTHF) matrix and analysed based on (9) to give preliminary data:
$g = 2.0030$, $D = 0.01400$ and $E = 0.00050\,\mathrm{cm}^{-1}$ with $S = 4$. Note that the
E-value of the last compound is almost vanishing due to the cylindrical
symmetry of the molecule.

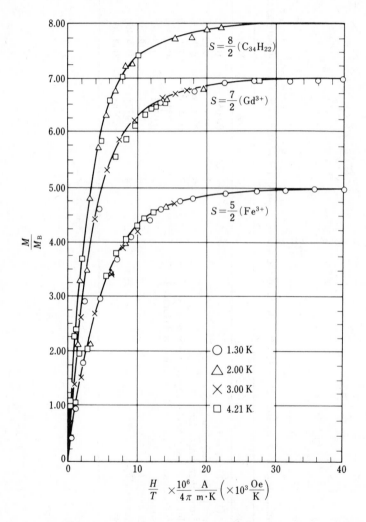

Fig. 15 Magnetization curves for paramagnetic species with various $J = S$-values.
These data represent a landmark in that the magnetization of a hydrocarbon
molecule [15; $m = 4$] surpassed for the first time those of transition-metal and
lanthanide ions, represented here by $NH_4Fe(SO_4)_2 \cdot 12H_2O$ and $Gd_2(SO_4)_3 \cdot 8H_2O$,
respectively.

Magnetization of a sample of [15; $m = 4$] dispersed in a benzophenone host was determined on a Faraday balance (Kimura and Bandow, 1984a,b) in the range 0–5 T at 2.1 and 4.2 K (Sugawara et $al.$, 1984, 1986a). The data are plotted in Fig. 15 together with those of typical paramagnetic transition-metal salts (Henry, 1952). The observed values follow the curves given by (15). For iron alum, there are five singly occupied 3d atomic orbitals and therefore $J = S = 5/2$. In good agreement with this, the saturation value of the magnetization (M_S) is $5M_B$. For gadolinium sulphate octahydrate, there are seven singly occupied 4f-orbitals, $J = S = 7/2$, and $M_S = 7M_B$. These two M_S-values are maxima for 3d transition metals and 4f lanthanoids, respectively. The data for [15; $m = 4$] agree nicely with the theoretical Brillouin function in which $J = S = 4$. The M_S-value of a hydrocarbon thus surpassed for the first time those of transition metal ions (Iwamura et $al.$, 1985; Sugawara et $al.$, 1984, 1986a). It should be noted that the magnetization value of [15] can still be increased in principle as well as in practice by increasing the number of the repeating units m. The higher homologues of the series are expected to behave at least as organic molecular superparamagnets (see Section 4).

On the other hand, a sample of [15; $m = 4$] in MTHF glass (8.7×10^{-7} M) gave an inverse χ vs T plot that showed a negative Weiss temperature of -22 K. The nonet species was concluded to experience an antiferromagnetic molecular field at temperatures lower than 65 K. This observation suggests that molecular clusters of [15; $m = 4$] must have been formed during the formation of the glassy sample in spite of the rather low concentration of the diazo-precursor, and that magnetic coupling in the clusters is antiferromagnetic, as is usually the case in molecular solids (see Section 6).

OTHER HIGH-SPIN MOLECULES

As pointed out at the beginning of Section 3, the first tailor-made, high-spin hydrocarbon is the 1,3,5-benzenetriyltris(bis(p-biphenylyl)methyl) radical [1′] (Schmauss et $al.$, 1965), the quartet ground state of which was demonstrated by susceptibility measurements (Kothe et $al.$, 1971). The epr fine structure was determined to have a strong centre line at 3274 G and a weak $\Delta M_S = 2$ transition at 1637 G: $D = 0.0041$ cm^{-1} with an isotropic g tensor (Brickmann and Kothe, 1973). 2,4,6-Tricyano-1,3,5-trinitrenobenzene had been for some time the highest-spin species of this class (Wasserman et $al.$, 1968). It is noteworthy that [23; $m = 3$] is a septet with $D = +0.04158$ and $E = 0.01026$ cm^{-1} (Takui and Itoh, 1973). The non-vanishing E-value clearly demonstrates the absence of a three-fold axis; the high-spin state is not due to geometrical symmetry but ascribed to topological symmetry. The stability of the orbital degeneracy and, therefore, the high-spin multiplicity due to topological symmetry demonstrated here should be contrasted with the vulnerability of those of geometrical symmetry to Jahn–Teller distortion.

Epr signals due to the triplet and quintet states were obtained when single crystals of diacetylene were photolysed (Huber and Schwoerer, 1980; Schwoerer et al., 1981; Sixl et al., 1986). The quintet states are considered to arise from substructures such as [25].

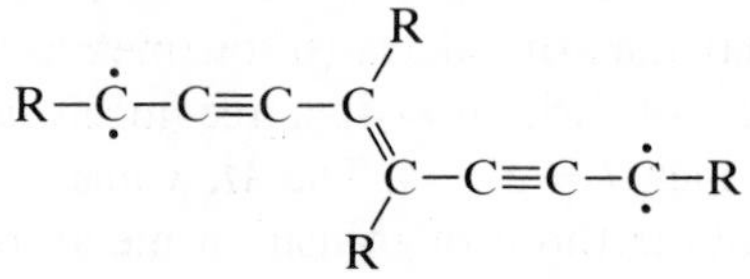

$$R-\overset{\cdot}{\underset{\cdot}{C}}-C\equiv C-C\overset{R}{\underset{\diagdown}{\diagup}}$$
$$\phantom{R-\overset{\cdot}{C}-C\equiv C-C}C-C\equiv C-\overset{\cdot}{\underset{\cdot}{C}}-R$$
$$\phantom{R-\overset{\cdot}{C}-C\equiv C-C}R$$

[25]

More recently, Dougherty advanced the idea of regarding high-spin molecules as consisting of two or more radical centres connected one to another through ferromagnetic coupling units. According to this expression, [14] and [15] are described as having m doublet and triplet centres, respectively, connected together through the m-phenylenediyl moiety, one of the most dependable ferromagnetic coupling units ever studied. Two 2-methylenecyclopenta-1,3-diyls and other new triplet units have been attached to the m-phenylenediyl unit to give quintets (Dougherty, 1989). He has also found that the cyclobutanediyl [26] serves as another good ferromagnetic coupling unit by demonstrating that [27] generated by photolysis

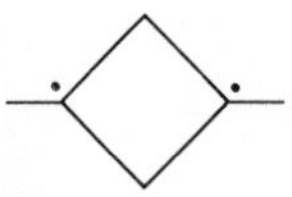

[26] 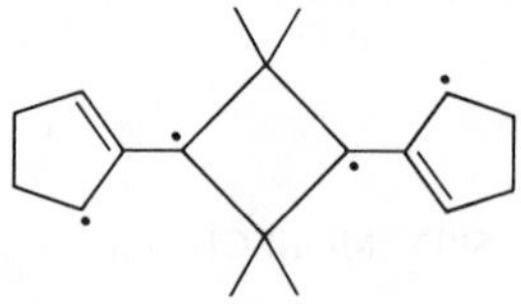   [27]

of the corresponding bis(azo)-compound is in a $S = 2$ ground state (Novak et al., 1989). Two recent examples appear, however, not to fit the above concept of dissecting high-spin molecules into the spin centres and ferromagnetic coupling units. When two ferrocenium units were connected through the p- or m-phenylenediyl unit, the interaction between the two Fe(III) centres was always antiferromagnetic although the exchange coupling through the latter unit was very small (Manriquez et al., 1989). Both the anion and cation of [3] were found to have low-spin ground states although the observed high-spin quartet state was higher in energy only by 4.5 cm^{-1}

in the anion (Matsushita *et al.*, 1990). The physical meaning of these results is not yet clear, but they seem to suggest that, as far as the spin states of those molecules are concerned, their electronic structures must in principle be considered as a whole rather than combining the dissected parts.

Seeger and Berson found a new photorearrangement of bicyclic ketones and took advantage of this reaction to generate tetraradical [2] in a quintet ground state. In good agreement with the topological consideration, the other isomer [2'] is not in a ground quintet state (Seeger and Berson, 1983; Seeger *et al.*, 1986).

(*m*-Nitrenophenyl)methylene [28], a new member of the class of non-Kekulé molecules, was generated photochemically from (*m*-azidophenyl)-diazomethane [28a] in 2-MTHF glasses at cryogenic temperatures. The signal pattern and temperature dependence of the signal intensities of the esr spectra in the range 19–69 K indicate that the ground state of [28] is a

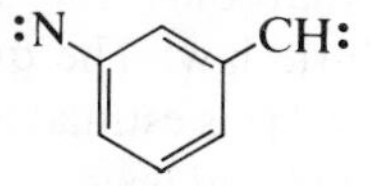

[28]

quintet with $D = 0.124$ and $E = 0.002 \, \text{cm}^{-1}$ (Tukada *et al.*, 1987). Similarly, 1,3-dinitreno-2,4,6-trifluorobenzene has been generated by photolysis of the corresponding diazide, although the spectra have not been fully interpreted (Haider *et al.*, 1989).

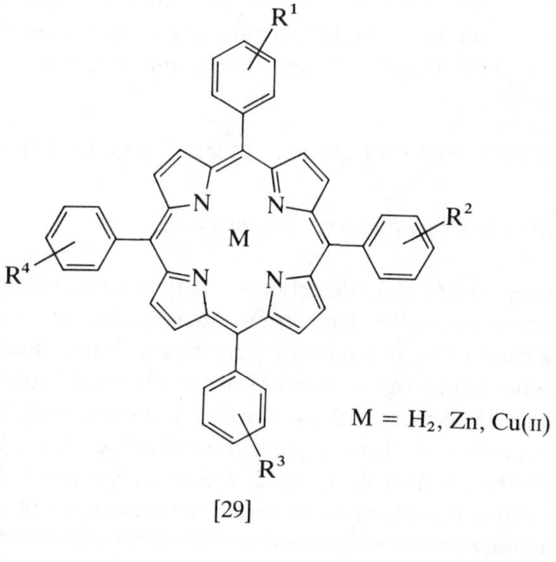

$M = H_2$, Zn, Cu(II)

[29]

Strictly speaking, conjugated molecules and polymers containing hetero-atomic groups are non-alternant systems and, therefore, the Coulson–Rushbrooke theorem (Coulson and Longuet-Higgins, 1947; Coulson and Rushbrooke, 1940) does not hold in general. Energy levels of NBMOs are no longer degenerate due to heteroatom substitution. However, when certain conditions are fulfilled, as discussed by Tyutyulkov (Tyutyulkov and Polansky, 1987), the theorem can be generalized for some classes of non-classical heteronuclear systems.

It is not straightforward to predict the ground-state multiplicity of tetraphenylporphyrins (TPP) [29] and their metal complexes carrying one to four phenylmethylenyl groups one each at the p-position of the *meso*-phenyl groups, since the porphyrins are far from AH. Both regioisomers of the dicarbene showed esr signals at 270 and 400 mT, characteristic of quintet species. The temperature dependence of the signal intensities in the range 10–50 K showed, however, that only the *syn*-isomer ($R^1 = R^2 = Ph\dot{C}$-, $R^3 = R^4 = H$) followed a Curie law. The quintet state of the *anti*-isomer ($R^1 = R^3 = Ph\dot{C}$-, $R^2 = R^4 = H$) is estimated from the temperature dependence to the thermally populated and lie at *ca.* 54 cal mol^{-1} (19 cm^{-1}) above a ground singlet state. The two carbene units can interact magnetically with each other through the conjugation of the porphyrin ring. The interaction can become ferro- or antiferro-magnetic depending on the position of the carbene units in the 18 π-electron aromatic system of the ring. Whereas the zinc complexes gave similar results, the copper complex of TPP–mono-carbene gave esr signals quite different from those of Cu–TPP and TPP–monocarbene ($R^1 = Ph\dot{C}$-, $R^2 = R^3 = R^4 = H$). The spectrum was simulated by $g = 2.03$, $D = 0.075$, $E = 0$ cm^{-1} and $S = 3/2$. The quartet state was found to arise from the ferromagnetic interaction between Cu(II) at the centre of the TPP ligand and the carbene unit in the side chain (Koga and Iwamura, 1989).

POLY(ACETYLENES) AND OTHER POLYMERS HAVING PENDENT RADICAL CENTRES

Model experiments on high-spin polymers

Poly(acetylenes) [16], poly(diacetylenes) [17], poly(phenylenes) [18] and poly(phenylenevinylenes) [19] have been much studied from the viewpoint of preparing conducting polymers or polymers with non-linear optical properties of some technological use (Alcacer, 1987; Gillespie and Day, 1986; Skotheim, 1985; Wegner, 1979). Guiding principles have been advanced to solve the question of where to place the radical centres in the side chains so as to align the electron spins, as discussed in Section 3. Here we describe some experiments performed to verify the theoretical predictions by using model oligomers.

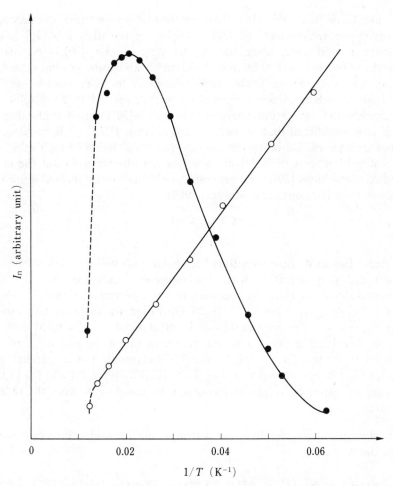

Fig. 16 Temperature dependence of the epr signal intensities due to m,m'- (●) and m,p'- (○) isomers of dicarbene [20]. Only the latter obeys the Curie law.

The isomeric stilbene dicarbenes [20] were generated in 2-MTHF matrices in an epr cavity at 16 K (Murata *et al.*, 1987; Iwamura, 1988). The spectra obtained by the photolysis of the diazo-compounds [20a], precursors to m,p'-[20] and m,m'-[20], at 16 K exhibited conspicuous signals at *ca.* 250 mT, characteristic of quintet species. The signals of m,m'-[20] showed a dramatic temperature dependence (Fig. 16). First, their intensity increased as the temperature was raised, reaching a maximum at 50 K, and then decreased somewhat and eventually irreversibly at above 65 K. In contrast, the intensity of the strong signal at *ca.* 250 mT and some weak signals due to m,p'-[20] decreased linearly with the reciprocal of the temperature as dictated by the Curie law (Fig. 16). The contrasting temperature dependences between the regioisomers of [20] strongly suggests that m,p'-[20] has a quintet ground state, while the quintet state of m,m'-[20] is populated thermally but lies ca. $200\,\text{cal mol}^{-1}$ ($70\,\text{cm}^{-1}$) above the ground state [see (12)]. The two remote triplet diphenylcarbene moieties linked together through a carbon–carbon double bond are concluded to have sufficient interaction to form a ground quintet state in m,p'-[20] as predicted by theory. It is now established that the head-to-tail polymer [16b][1] of diphenylacetylenes with the radical centres at the m,p'-positions should be high spin.

Similarly, when diphenyldiacetylenes having nitreno-groups at the m,p'- and m,m'-positions [30i] were generated, only the former showed a ground quintet state (Iwamura and Murata, 1989).

$$X-C\equiv C-C\equiv C-Y$$

[30]

According to Wegner, substituted 1,3-butadiynes undergo topochemically controlled polymerization when the molecules stack side by side, and minimal atomic motions are necessary to effect polymerization as shown in Fig. 17(a) (Wegner, 1969, 1977, 1979). Operationally, there are four modes conceivable for the polymerization in crystals of a substituted phenyldiacetylene [30c] carrying a radical centre at the m- or p-position of the phenyl ring: these are $1,1'/4',4''$ and $1,4'$ polymerization in parallel and antiparallel stacks as shown in Fig. 17(a–d). Only types (a) and (d) in this Figure are expected to lead to ferromagnetic coupling between the radical centres.

Attempts at constructing high-spin polymers carrying radical centres in the side chains

Poly(acetylenes) [16]. There are several catalysts available for polymerization of substituted acetylenes. Whereas Ziegler–Natta catalysts are quite effective for polymerization of acetylene itself and simple alkylacetylenes, they are not active towards other substituted acetylenes, e.g. phenylacetylenes. Olefin-metathesis catalysts (Masuda, 1985; Masuda and Higashimura, 1984, 1986) and Rh(I) catalysts (Furlani et al., 1986; Tabata, 1987) are often employed. In our experience, however, many persistent radicals and typical nitrogen-containing functional groups serve as good poisons for these catalysts. Therefore, radical centres have to be introduced after construction of the polymer skeletons. Fortunately, the polymers obtained with these catalysts are often soluble in one or other organic solvent. For example, methyl p-ethynylbenzoate can be polymerized to a brick-coloured amorph-

[1] See the Appendix on p. 245 of suffixes to structural formula numbers.

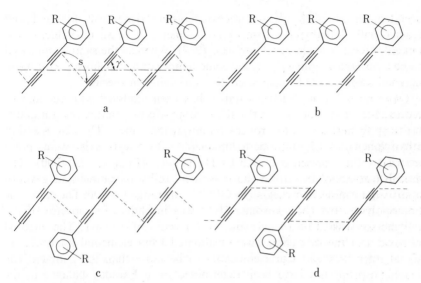

Fig. 17 Schematic diagram showing the molecular packing and topochemical polymerization of diacetylenes [30] in crystals. There are two idealized packing arrangements either (a), (b) or (c), (d) in crystals and each has two modes of polymerization. The polymerization in the solid state is said to occur smoothly when $s = 240 \sim 400$ pm and $\gamma = 45°$ (Baughman, 1974; Baughman and Yee, 1978).

ous polymer of $\overline{M}_w = 7.4 \times 10^4$ and $\overline{M}_n = 3.2 \times 10^4$, and then converted almost quantitatively to the corresponding poly(p-ethynylbenzaldehyde) (Iwamura and McKelvey, 1988). Chemical conversions of the formyl groups into verdazyl and other stable radicals have, however, been only partly successful; the polymer samples obtained contain only 10–24% of the theoretical radical concentrations. Their properties are mostly paramagnetic. A phenylacetylene carrying a p-(3,5-di-t-butyl-4-hydroxyphenyl) group is polymerized by using a catalyst system $W(CO)_6$–Ph_4Sn. When phenoxyl radicals [16e] are generated from the polymer of average molecular weight of 4×10^4, a paramagnetic sample is obtained that has a spin concentration of only 4.4×10^{19} spins g^{-1} (Iwamura et al., 1987). Similar discouraging results have been reported for [16f] (Nishide et al., 1988). When m- and p-bromoethynylbenzenes are treated with WCl_6–Ph_4Sn in toluene at 40°C, polymers with $\overline{M}_w = 5.2 \times 10^4$ and 2.9×10^4, respectively, are obtained. The latter, when treated with butyllithium in ether, allowed to react with 2-methyl-2-nitrosopropane dimer, and oxidized, yielded a powder sample [16g] that showed a spin concentration of 1.3×10^{21} spins g^{-1} (Ishida and Iwamura, 1989). Again, the concentration was not as high as designed and the polyradical sample [16g] was mostly paramagnetic, although a part of it responded to a magnet at room temperature. A polymer of bromophenyl-

acetylene cross-linked with *m*-phenylenebis(acetylene) was lithiated and
reacted with 2-methyl-2-nitrosopropane dimer to give an insoluble, cross-
linked polymer (Ishida and Iwamura, 1989). Whereas the sample appeared
to be a normal paramagnet at cryogenic temperatures, it showed saturation
magnetization at 100 K, characteristic of a high-spin species.

Quite recently, the Furlani catalyst has been modified with the aim of
reducing the Lewis acidity of the Rh(I) catalyst centre and making it active
for the polymerization of nitrogen-containing monomers. Thus 2-(*m*- and *p*-
ethynylphenyl)-4,4,5,5-tetramethylimidazoline-3-oxide-1-oxyls were polymerized in the presence of $Rh(COD)(NH_3)Cl$ (Fujii *et al.*, 1990). The
greenish homopolymer from the *m*-isomer [16j] was soluble and gave an
approximate molecular weight (by GPC) in the range 150 000. The dark blue
homopolymer from the *p*-isomer [16k] was insoluble. Esr spectra of the
polymers consisted of a single line with a width of 0.70 mT. The nitronyl
nitroxides in the side chains were estimated from elemental analyses, esr
signal intensities and Curie constants to be more than 95% intact. The
magnetic properties have been determined on a Faraday balance in the
temperature range 2–300 K and magnetic field range 0–7.00 T. The $1/\chi$ *vs*
temperature plots gave straight lines characteristic of paramagnetic species
with very weak antiferromagnetic coupling. The magnetization *vs* magnetic
field strength data on the two isomeric samples at 1.8 K deviated slightly
downwards from the Brillouin functions with $S = 1/2$, revealing again the
presence of antiferromagnetic coupling between the $S = 1/2$ spins. The
expected ferromagnetic coupling among the radical centres in the side chains
through the conjugated main chain was not operative in these polymers. In
Ullman's nitronyl nitroxides [31] (Ullman *et al.*, 1972), one unpaired electron

[31]

is delocalized on the two equivalent NO groups. However, as approximated
by allyl radicals with three π-electrons or pentadienyl systems with seven π-
electrons, there is a node at C-2 in the singly occupied MO. The spin density
at that carbon atom, which is attached to the phenyl ring, is very small in
size and negative in sign. Therefore, the disappointing results may be due to
lower delocalization of the electron spins from the side chains into the main
chain. Some steric congestion must also be contributing to the lower
exchange interaction through the main chain since the main chain and side
chains cannot assume a planar conformation; the effective conjugation
length must be very short.

In [16e–h] it appears that, since the spin delocalization is considerable, the radicals must be kinetically less stable at sites other than the original radical centre where there is no steric protection. Therefore, it was difficult to obtain samples that contain satisfactory spin concentrations. In [16j] and [16k], by contrast, the spin distribution seems not to be extensive and so the kinetic stability of the radical centres must be well protected.

Much yet remains to be done as regards the choice of the pendent radicals. The ring-opening metathesis polymerization of cyclooctatetraenes (Ginsburg *et al.*, 1989; Klavetter and Grubbs, 1988) has not yet been applied to the construction of high-spin polyacetylenes.

Poly(diacetylenes) [17]. Attempts are in progress to prepare high-spin polymers based on the above molecular design and model experiments. There appear to be two practical limitations, however. First, since this polymerization is effected by heat or irradiation (from high-energy radiation to visible light) in the solid state and is topochemically controlled (Baughman, 1974; Baughman and Yee, 1978; Wegner, 1969, 1977, 1979), not all diacetylenes are amenable to the reaction. Secondly, the polymers are very insoluble in ordinary organic solvents; radical centres have to be introduced in advance.

1-Phenyl-1,3-butadiyne derivatives carrying *N*-t-butylhydroxyamino- [30l] and stable t-butylaminoxyl [30m] groups have been prepared (Koga *et al.*, 1990). The crystal structure of triclinic [30l] from ether, mp 140°C, was found to show that the molecules are stacked head-to-tail along the a/c diagonal direction as in Fig. 17(c,d). Since the nearest distances between C-1 and 4′ and C-4 and 4′ of neighbouring molecules are 455 and 472 pm, respectively, the polymerization is considered to take place in this direction. The one-dimensional columnar stacks are bridged by intermolecular hydrogen bonds between two hydroxy amino-groups. Whereas crystalline [30l] polymerized as expected when heated at 100°C, [30m] did not show any tendency to polymerization. Mixed crystals of [30l] and [30m] (1: < 1) did polymerize in the solid state at 120°C to give black-violet microcrystals with a metallic lustre. This polymeric sample showed a broad esr signal at *ca.* 300 mT in addition to a normal signal centred at $g = 2.006$. The magnetization curve of this sample at 4.2 K agreed with the Brillouin function with $S = 1.5$–2, suggesting ferromagnetic coupling of the doublet centres. The Curie constant of the linear part of the plots showed that *ca.* 6% of the monomer units in the sample had active aminoxyl groups.

Photolysis of 1-{*p*-(diazomethyl)phenyl}-1,3-butadiyne [30n] in MTHF at 4.6 K gave an esr fine structure characteristic of a triplet phenylcarbene: $|D| = 0.4666$ and $|E| = 0.0210 \, \text{cm}^{-1}$. The signals obeyed a Curie law in the temperature range 4.6–77 K. A microcrystalline sample of the diacetylene underwent spontaneous polymerization at room temperature to give insoluble polymer [17] which was then photolysed at cryogenic temperatures to generate triplet carbene centres. Magnetic susceptibility measurements revealed weakly coupled (Weiss temperature = −2.4 K) independent spins ($S = 1$), although the triplet phenylcarbene was estimated from the Curie constant to be generated only in a low percentage yield. Although no crystal structural data are available, the results may be due to wrong regioselectivity of the polymerization, e.g. Fig. 17(b) or (c) (Koga et al., 1990).

Diacetylenes like [30o] after being laboriously prepared, are not polymerized either by heat or ultraviolet irradiation (Iwamura, 1990). Control of molecular packing in crystals is now needed. Introduction of the technology of liquid crystals or Langmuir–Blodgett membranes (Hupfer et al., 1981; Koch and Ringsdorf, 1981) may be of help.

In 1986, Korshak and Ovchinnikov reported that black powdery materials were obtained by polymerization of a diacetylene carrying symmetrically a pair of persistent aminoxyl radicals [30p]. A part of this sample was described as showing ferromagnetic properties (spontaneous magnetization of ca. 0.022 emu G g^{-1}) (Korshak et al., 1986, 1987). The results were for some time a subject of considerable interest and controversy. The observed crystal structure does not appear to satisfy the empirical criterion for topochemical polymerization (Wiley et al., 1989). Even if the expected poly(diacetylene) skeleton is formed, there are two ways in which the structure of the polymer [17p] is not consistent with the molecular design described in Section 3. First, radical centres are not conjugated with the main chain so as to have meaningful exchange coupling; they are too far apart by three saturated carbon atoms. Even if there were exchange coupling, the interaction is expected to be antiferromagnetic since the structure is topologically symmetric. According to the latest report (Zhang et al., 1989), two polymorphs of [30p] exhibit normal paramagnetic behaviour before and after heat treatment; any ferromagnetic contribution that might be observed at higher temperatures is due to a contribution corresponding to 5 ppm Fe contamination. Similarly, a claimed spontaneous magnetization of 88.5 emu G mol^{-1} and a coercive field of 455 G due to the thermally treated product of 2,4-hexadiyne-1,6-diyl bis(2,2,5,5-tetramethyl-piperidin-1-oxyl-3-carboxylate) (Cao et al., 1988a,b) was later questioned (Wiley et al., 1989).

Torrance obtained a sample that showed ferromagnetic properties with a high Curie temperature by a reaction of 1,3,5-triaminobenzene with iodine under rather drastic conditions (Torrance et al., 1987). A charge-transfer complex of iodine with an aniline-black-type polymer [32], a heteroatom-containing analogue to [14′] (Johannsen et al., 1989), may have been

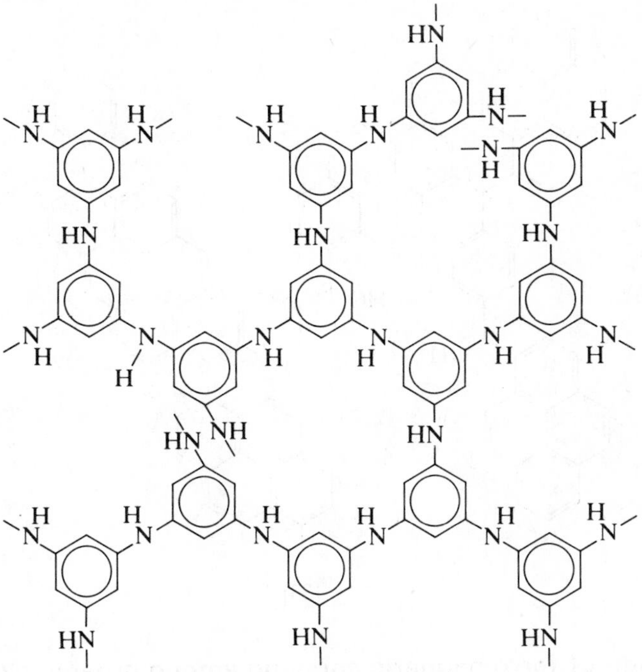

[32]

responsible for the magnetic property. Similarly, powdery material was obtained by thermal decomposition of 2,4,6-tris(diazo)phloroglucinol at 210°C. A part of this sample, which was assumed to have a two-dimensional network structure, showed a ferromagnetic property with a coercive force of *ca.* 650 G (Iwamura *et al.*, 1987). Most recently, Ota and Otani reported resinous materials called COPNA obtained by condensation of aromatic aldehydes with condensed ring aromatic hydrocarbons, e.g. pyrene. When treated under conditions of hydrogen abstraction, parts of the resulting samples were attracted by a magnet at room temperature (Ota *et al.*, 1989). The substructure as shown by [33] has been proposed.

Three pyrolytic graphite samples (Kawabata *et al.*, 1990; Murata and Ueda, 1990; Ovchinnikov and Spector, 1988) are also reported to have residual magnetization (for example, $M_S = 0.5$ and $M_r = 0.35$ emu G g^{-1}, $H_c = 600$ Oe; Kawabata *et al.*, 1990).

All these samples are problematic in one way or another. Some are not obtained reproducibly and others are difficult to characterize or prepare in quantity. It is extremely important to synthesize high-spin polymers systematically and stoichiometrically, based on the strict molecular design described in the last section.

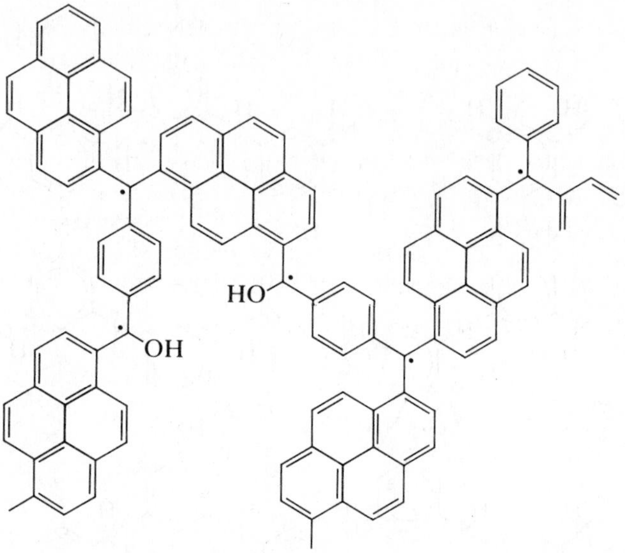

[33]

6 Design of ferromagnetic coupling among organic free radicals and high-spin molecules in molecular assemblies

The higher homologues of series [15] and other polyradicals discussed in the previous section are expected to behave at least as organic molecular superparamagnets. Long-range ordering of the electron spins in one-dimensional systems is considered to be physically unstable and can be achieved only at $T = 0$ K (Mermin and Wagner, 1966; Wannier, 1959; Stanley, 1971). However, since real molecular assemblies of the above poly(m-phenylenecarbenes) and other polyradicals are three-dimensional systems, the interchain interactions, although small, may lead to the generation of long-range magnetic order also for $T > 0$. This problem has been discussed by Tyutyulkov (Tyutyulkov and Karabunarliev, 1986). They estimated the absolute value of J between two one-dimensional chains of planar polyradicals [14] by using an expression $T_c = 2J/k_B \ln 3 = 21\,167.4 J$ K (Syozi, 1972). For J-values in the range 0.16–0.20 eV, T_c values in the range 910–1190 K have been obtained. Thus, intermolecular interactions are extremely important for spin ordering at finite temperatures.

The intermolecular interaction in molecular aggregates of some members of the series [15] has been found, however, to be antiferromagnetic (see Section 5). Exchange coupling among neighbouring free radicals or high-spin molecules is usually very small and antiferromagnetic at best in organic solids. This is the rule rather than the exception. Actually there are only a handful of examples known in the literature in which a ferromagnetic interaction has been observed in organic free radicals in well-defined crystals. Galvinoxyl [34] (Kosaki *et al.*, 1969; Mukai, 1969; Mukai *et al.*, 1967, 1982; Mukai and Sogabe, 1980); bis(2,2,6,6-tetramethylpiperidin-1-oxyl-4-yl) suberate [35] (Benoit *et al.*, 1983; Chouteau and Veyret-Jeandey, 1981) and the *p*-nitrophenyl derivative of [31] (Awaga and Maruyama, 1989) are such exceptional cases under limited conditions. Strategies leading to stabilization of ferromagnetic intermolecular coupling are very much needed.

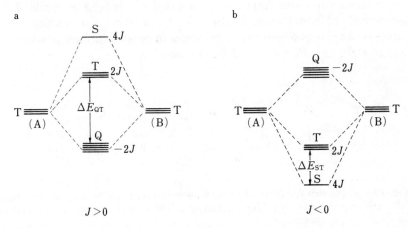

[34]   [35]

Two theories were introduced in the 1960s on how to align electron spins in parallel between open-shell molecules (McConnell, 1963, 1967).

Fig. 18 Correlation diagrams for two weakly interacting triplets in (a) ferromagnetic and (b) antiferromagnetic fashions.

McCONNELL'S HEITLER–LONDON SPIN-EXCHANGE MODEL BETWEEN SPIN DENSITIES OF OPPOSITE SIGN ON RADICALS A AND B

When two triplet species are allowed to interact weakly, singlet, triplet and quintet states are produced as shown in Fig. 18. Since the exchange integral

J is negative between open-shell organic molecules at a distance of meaningful overlap, the ground state tends to become singlet as shown in Fig. 18(b). In order to develop ferromagnetism in organic molecular assemblies, it is necessary to find conditions under with the case shown in Fig. 18(a) could result; the intermolecular interaction between high-spin molecules might then become ferromagnetic.

McConnell proposed such a model based on Heitler–London spin exchange between positive spin density on one molecule and negative spin density on another (McConnell, 1963). The theory may be paraphrased as follows when the Heisenberg Hamiltonian for the interacting spins (4) is expanded approximately as in (21). The effective exchange interaction between two free radicals can be ferromagnetic when the product of spin densities ρ_i and ρ_j at two interacting sites i and j on different radical species A and B is negative, since the valence bond exchange integral J_{ij} is negative.

$$\mathbf{H}^{AB} = -2 \sum_{ij} J_{ij}^{AB} s_i^A s_j^B = -2s^A s^B \sum_{ij} J_{ij}^{AB} \rho_i^A \rho_j^B \tag{21}$$

The theory had never been tested on a logical model system. Let us consider in detail one representative case, the superimposable stacking of the two benzene rings, one from each triplet diphenylcarbene molecule. These are considered to represent idealized modes of dimeric interaction of the aromatic ring parts of open-shell molecules in ordered molecular assemblies like crystals, liquid crystals and membranes.

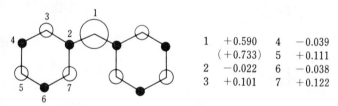

1	$+0.590$	4	-0.039
	$(+0.733)$	5	$+0.111$
2	-0.022	6	-0.038
3	$+0.101$	7	$+0.122$

Fig. 19 π-Spin-density distribution in diphenylcarbene [15; $m = 1$] determined by ENDOR experiments and MO calculations. The σ-spin density at the divalent carbon is in parentheses.

The α-spin distribution in diphenylcarbene is basically of the benzyl radical type and is shown in Fig. 19 (Brandon *et al.*, 1965). When the theory is applied to the idealized dimeric interaction modes, the signs of $\rho_i \rho_j$ at each interacting site between the two benzene rings are all negative in the *ortho*- (60°-rotated from the superimposable stacking) and *para*- (180° rotation) stacking patterns and all positive in the geminal (superimposable stacking) and *meta*- (120° rotation) modes (Fig. 20). Therefore, the *ortho*- and *para*-overlaps are predicted to give the quintet ground state [Fig. 18(a)], while the

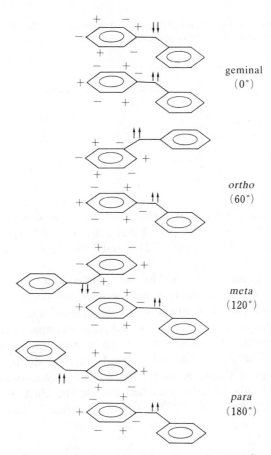

geminal
(0°)

ortho
(60°)

meta
(120°)

para
(180°)

Fig. 20 Idealized stacking modes of two diphenylcarbene molecules. Two benzene rings, one each from the two carbenes, are always in a superimposable disposition.

geminal and *meta*-overlaps would lead to the ground singlet state [Fig. 18(b)]. This theory can be regarded as a through-space version of topological symmetry if the original valence bond theory discussed in Section 3 and Fig. 7 is considered as a through-bond interaction.

Stimulated by the model experiments described on pp. 234–6, Yamaguchi *et al.* (1986a,b, 1989a) carried out APUHF STO-3G calculations for the dimeric interaction of benzyl radicals in the geminal, *ortho-*, *meta-* and *para-* conformations. The results summarized in Table 4 show that the signs of the calculated J_{ab} are positive for the *ortho-* and *para-*stackings and negative for the geminal and *meta-*stackings, in good agreement with the McConnell theory. The signs do not depend on variation of the intermolecular distance R, although the absolute values decrease with R in an exponential manner.

Table 4 Effective exchange integrals (J_{ab}/cm^{-1}) calculated by the *ab initio* APUHF method for the dimers of benzyl radicals in various parallel overlap modes at distances R.

R/pm	gem-	o-	m-	p-
340	−363	220	−225	218
360	−148	90	−91	89
380	−58	35	−35	35
400	−22	13	−13	13

The absolute values appear to increase as the basis sets are improved; $218\,cm^{-1}$ in Table 4 for the *para*-overlap at a distance of 340 pm is modified to 412 and $517\,cm^{-1}$ by using the 4-31G and APUMP 6-311G** methods, respectively.

The signs of J_{ab} can be explained by the extended McConnell model (20), (see p. 210). For example, the OO-term between the singly occupied and singly unoccupied orbitals is not zero for the geminal and *meta*-overlaps, whereas it reduces to zero in the *ortho*- and *para*-overlaps. In other words, the OO-term favours the singlet state for the geminal and *meta*-configurations. The SDP-term, on the other hand, plays a dominant role to determine the sign of the J_{ab}-value for the *ortho*- and *para*-dimers. Similar theoretical treatments have been carried out on the clustering of triplet carbenes (Yamaguchi *et al.*, 1989b) and allyl radicals (Yamaguchi and Fueno, 1989).

FERROMAGNETIC COUPLING IN DONOR–ACCEPTOR SALTS

It is usually the case that, in alternating stacks of charge-transfer salts ($\ldots D^+A^-D^+A^-D^+A^-D^+A^-\ldots$), the electron spins generated on the donor (D) and acceptor (A) components are antiferromagnetically coupled. This is as a result of admixing of the wave functions of the closed-shell components to the resultant wave function of the charge-transfer salt [Fig. 21(a)]. A second theory of McConnell predicts that, when either D or A has a triplet ground state, the electron spins developed on a pair of D and A can have ferromagnetic coupling (McConnell, 1967) [Fig. 21(b)]. In other words, configurational admixing of a virtual triplet excited state with the ground state becomes possible for a chain of alternating radical cation donors and radical anion acceptors.

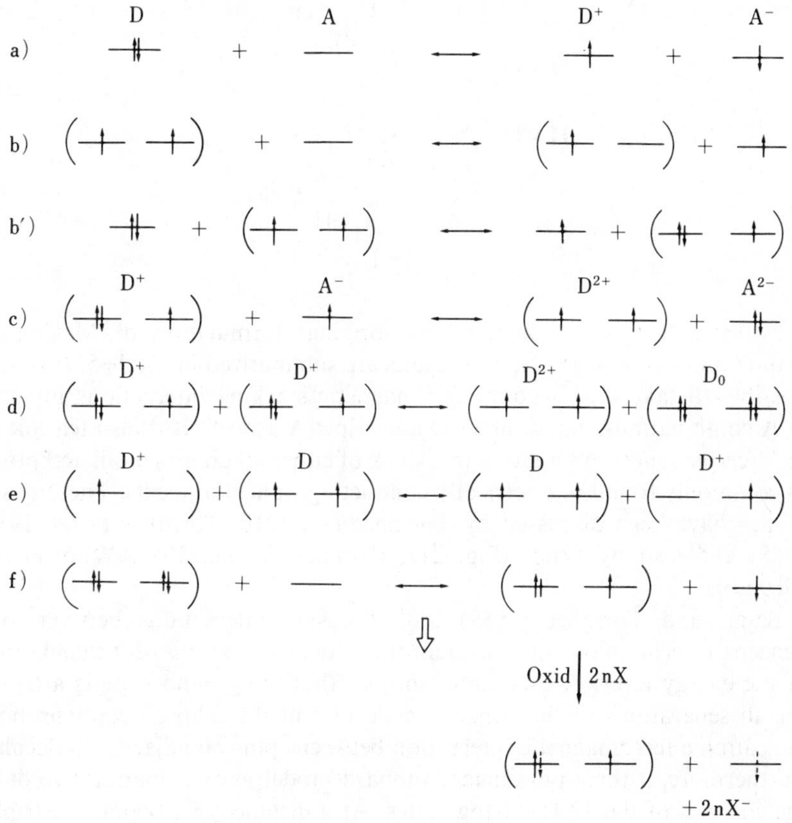

a) D + A ⟶ D⁺ + A⁻

b) (D) + ⟶ (D⁺) +

b′) + (D) ⟶ + (D⁺)

c) D⁺ + A⁻ ⟶ D²⁺ + A²⁻

d) D⁺ + D⁺ ⟶ D²⁺ + D⁰

e) D⁺ + D ⟶ D + D⁺

f) + ⟶ +

⇩

Oxid | 2 nX

( ) +

+2 nX⁻

Fig. 21 Schematic representation of strategies for spin alignment in D/A salts or complexes by application of spin conservation in different electron configurations of interacting molecular orbitals. (a) Typical D/A interaction between two closed-shell D and A, (b and b′) McConnell's proposal, (c) Breslow's extension, (d) Torrance's model, (e) Wudl's model, and (f) Chiang's model for further doping.

It was Breslow (Breslow, 1982; Breslow *et al.*, 1982) who first paid attention to this theory. Knowing that the pentachlorocyclopentadienyl cation (Breslow *et al.*, 1964; Saunders *et al.*, 1973), the hexachlorobenzene dication (Wasserman *et al.*, 1974) and the 2,3,6,7,10,11-hexamethoxy-triphenylene (HMT, [36]) dication are all ground-state triplets, in good agreement with theory, Breslow and coworkers set out on the synthesis of analogues that should have lower oxidation potentials, be chemically more stable and therefore form CT complexes more readily (Fig. 21c; Breslow *et al.*, 1982, 1984; Breslow, 1985, 1989; LePage and Breslow, 1987).

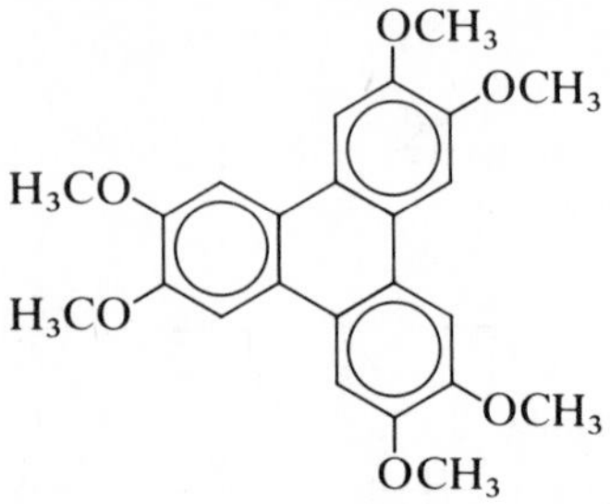

OCH₃
OCH₃
H₃CO
H₃CO
OCH₃
OCH₃
[36]

Miller and Epstein extended the original formulation of McConnell (Miller and Epstein, 1987). The results are summarized in Table 5. It is now possible to take into account ferrimagnetic stacking interactions in some D/A combinations, e.g. doublet D and tripet A as well. It is also feasible to seek ferromagnetic interaction in a stack of either all donors or all acceptors, as commonly found in electrically conducting radical-ion salts. These possibilities have been discussed by Torrance (Fig. 21d; Torrance et al., 1987, 1988) and also by Wudl (Fig. 21e; Dormann et al., 1987; Wudl et al., 1989a,b).

Bagus and Torrance (1989) took the spin interactions between two benzene radical anions into account theoretically. Ab inito MO calculations of the energy levels of this dimer showed that the ground state is a triplet for all separations of the benzene molecules in the eclipsed configuration, indicating a ferromagnetic interaction between spins on adjacent molecules. Furthermore, a three-parameter Hubbard model gave a quantitative fit to the energies of the 12 low-lying states. At a distance of 317 pm, the triplet state is estimated to be more stable than the lowest singlet by ca. 130 meV.

The idea of Chiang is a further extension of Miller's formulation. Starting from a typically antiferromagnetic alternating stack of closed-shell donors and acceptors, he generated a ferromagnetic domain by an external partial or full oxidation or reduction to induce the formation of a ground-state molecule with appreciable triplet character (Chiang et al., 1988; Chiang and Goshorn, 1989).

MODEL EXPERIMENTS AND REALIZATION OF FERROMAGNETIC INTERACTIONS

To test the first theory of McConnell as exemplified by the stacking overlap between two triplet diphenylcarbene molecules (see Fig. 20), Izuoka et al. (1985, 1987) took advantage of the [2.2]paracyclophane skeleton. It was Forrester and Ramasseul (1971a,b, 1975) who first tried to demonstrate

Table 5 Prescriptions for spin-ordered D/A charge-transfer salts.[a]

Homo-spin systems				Hetero-spin systems			
D (or A)	A (or D)	D→A	A→D	D (or A)	A (or D)	D→A	A→D
Spin $\frac{1}{2}$ systems				Spin $\frac{1}{2}$–1 systems			
s^1	s^1	AF	AF	s^1	d^2	FI	FI
				s^1	t^2	FO	FI
				s^1	t^4	FI	FO
d^1	s^1	AF	FO	d^1	d^2	FI	FO
d^3	s^1	FO	AF	d^1	t^2	FO	FO
t^1	s^1	AF	FO	d^1	t^4	FI	FI
t^5	s^1	FO	AF	d^3	d^2	FO	FI
d^1	d^1	FO	FO	d^3	t^2	FI	FI
d^3	d^1	AF	AF	d^3	t^4	FO	FO
t^1	d^1	FO	FO	t^1	d^2	FI	FO
t^5	d^1	AF	AF	t^1	t^2	FO	FO
d^3	d^3	FO	FO	t^1	t^4	FI	FI
t^1	d^3	AF	AF	t^5	d^2	FO	FI
t^5	d^3	FO	FO	t^5	t^2	FI	FI
t^1	t^1	FO	FO	t^5	t^4	FO	FO
t^5	t^1	AF	AF				
t^5	t^5	FO	FO	Spin $\frac{1}{2}$–$\frac{3}{2}$ systems			
				s^1	t^3	FI	FI
				d^1	t^3	FI	FO
Spin 1 systems				d^3	t^3	FO	FI
d^2	d^2	AF	AF	t^1	t^3	FI	FO
d^2	t^2	FO	AF	t^5	t^3	FO	FI
d^2	t^4	AF	FO				
t^2	t^2	FO	FO	Spin 1–$\frac{3}{2}$ systems			
t^2	t^4	AF	AF	d^2	t^3	FI	FI
t^4	t^4	FO	FO	t^2	t^3	FI	FO
				t^4	t^3	FO	FI
Spin $\frac{3}{2}$ systems							
t^3	t^3	AF	AF				
Spin 2 systems							
q^4	q^4	AF	AF				
Spin $\frac{5}{2}$ systems							
p^5	p^5	AF	AF				

[a] AF, FO and FI stand for the antiferro-, ferro- and ferri-magnetic interactions, respectively, expected for the D/A pair. Symbols s, d, t, q and p signify the single, double, triple, quadruple and quintuple degeneracy of the partially occupied orbitals. The superscript number corresponds to the total number of electron spins in those degenerate orbitals.

transannular interaction between two t-butyl phenyl nitroxide moieties incorporated in the [2.2]paracyclophane skeleton. While it was found that $J > a_N$, the results were interpreted in terms of direct interaction between the two nitroxyl radicals rather than spin exchange between the facing sites of the two spin-containing benzene rings.

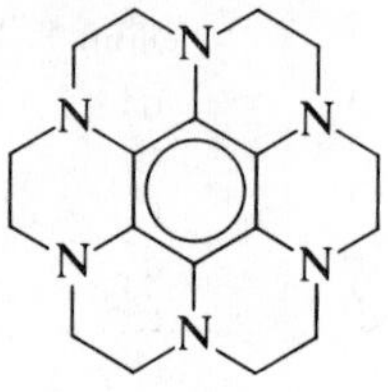

[37] 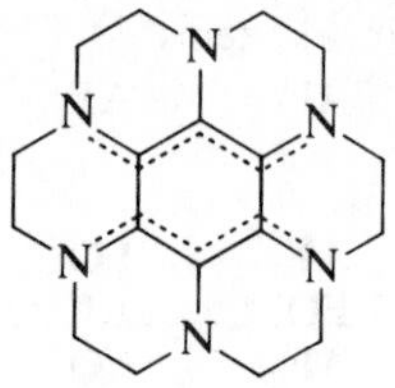   [37']

Among the three isomers of bis(phenylmethylenyl) [2.2]paracyclophanes [38], pseudo-*ortho*- and pseudo-*para*-isomers (*o*-[38] and *p*-[38], respectively) satisfy McConnell's condition to give quintet ground states. They were produced by photolysis of the corresponding bis(α-diazobenzyl)[2.2]para-cyclophanes [38a][1] in 2-MTHF at cryogenic temperatures, and their esr fine structures were studied.

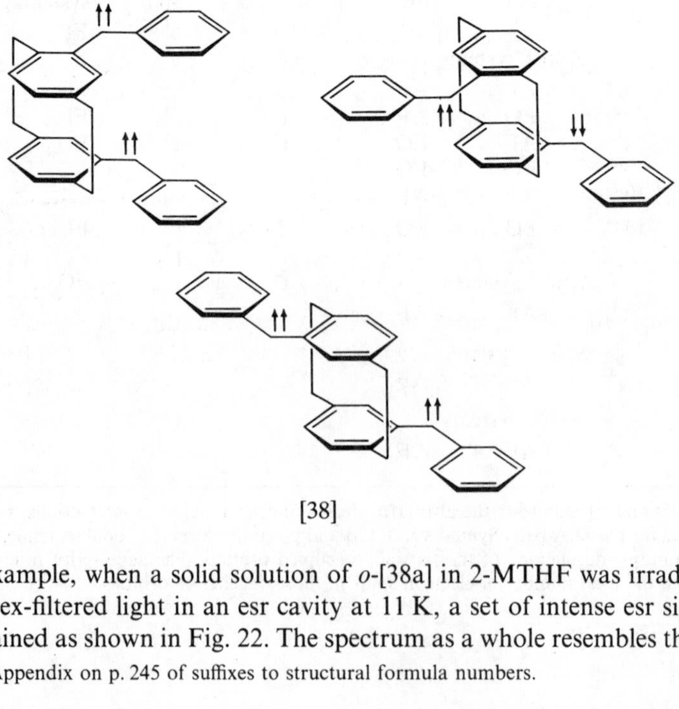

[38]

For example, when a solid solution of *o*-[38a] in 2-MTHF was irradiated with Pyrex-filtered light in an esr cavity at 11 K, a set of intense esr signals was obtained as shown in Fig. 22. The spectrum as a whole resembles that of

[1] See the Appendix on p. 245 of suffixes to structural formula numbers.

a)

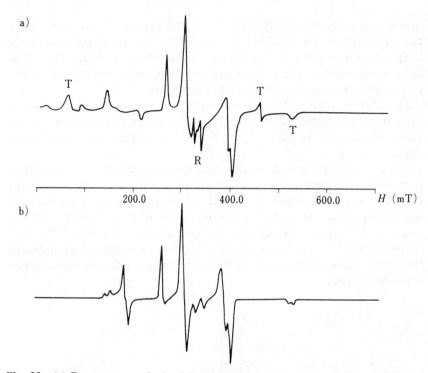

b)

Fig. 22 (a) Epr spectrum obtained by photolysis of the didiazo-compound o-[38a] of a [2.2]paracyclophane. (b) The signals due to unoriented dicarbene o-[38] in the quintet state simulated by a perturbation calculation.

m-phenylenebis(phenylmethylene) [15; $m = 2$] in the quintet ground state. The resonance positions and intensities of the signals are reproduced well by a second-order perturbation calculation as a quintet species (Q) with zero-field splitting parameters $|D| = 0.0624\,\mathrm{cm^{-1}}$ and $|E| = 0.0190\,\mathrm{cm^{-1}}$. These values are well interpreted in terms of (11). In addition to these quintet signals, triplet signals were detected at 62.8, 463.0, 526.7 and 732.0 mT and assigned to a monocarbene, T_G ($|D| = 0.3730\,\mathrm{cm^{-1}}$ and $|E| = 0.0156\,\mathrm{cm^{-1}}$). Another triplet species with small zero-field splitting parameters ($|D| = 0.0134\,\mathrm{cm^{-1}}$ and $|E| = 0.000\,\mathrm{cm^{-1}}$) was detected in the $g = 2$ region of the spectrum. This is assigned to a triplet diradical (DR) formed fortuitously from dicarbene o-[38].

The signal intensities of the quintet were found to obey the Curie law in the temperature range 11–50 K, showing that o-[38] has a quintet ground state. The other signals due to T_G and DR also obeyed the Curie law in the temperature range 11–85 K and 11–50 K, respectively. At above 50 K, the latter signal intensity increased irreversibly at the expense of Q, suggesting the thermal generation of DR from Q. At temperatures higher than 20 K, a new signal appeared at 104.0 mT. The intensity of this signal increased as the temperature was elevated and reached a maximum at 55 K. The signal decayed irreversibly at temperatures higher than 50 K. The thermal behaviour is characteristic of a thermally populated species (T_T).

By reference to Fig. 18 and by assuming a Boltzmann distribution of electron spins among the three states, the temperature dependence of the signal intensity for T_T gave $\Delta E_{QT} = 61\,cm^{-1}$ ($= 175\,cal\,mol^{-1}$) and J was calculated to be $+16\,cm^{-1}$.

In the case of the pseudo-*meta*-isomer (*m*-[38]), signals due to a quintet state were not observed. Instead, a set of signals due to the thermally populated triplet T_T appeared at 25 K. The temperature dependence data were analysed in terms of an equation similar to (13) [3 exp ($-\Delta E/RT$) in place of 5 exp ($-3\Delta E/RT$) in the numerator] to give $\Delta E_{ST} = 98\,cm^{-1}$ ($= 280\,cal\,mol^{-1}$). The behaviour of *p*-[38] was more like *o*-[38], although the former was more reactive and did not allow a satisfactory temperature-dependence study.

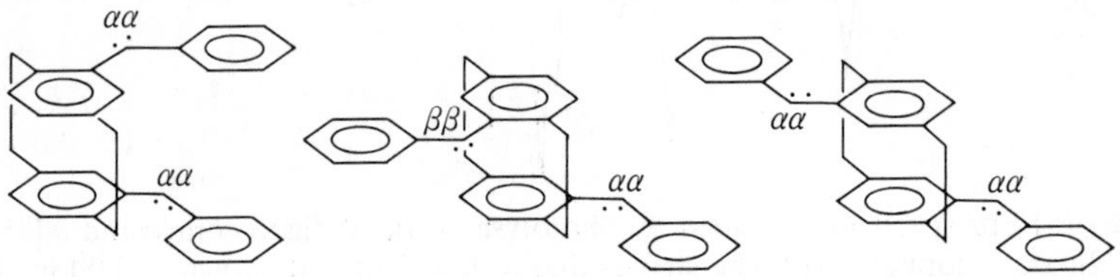

ortho-[38]   meta-[38]   para-[38]

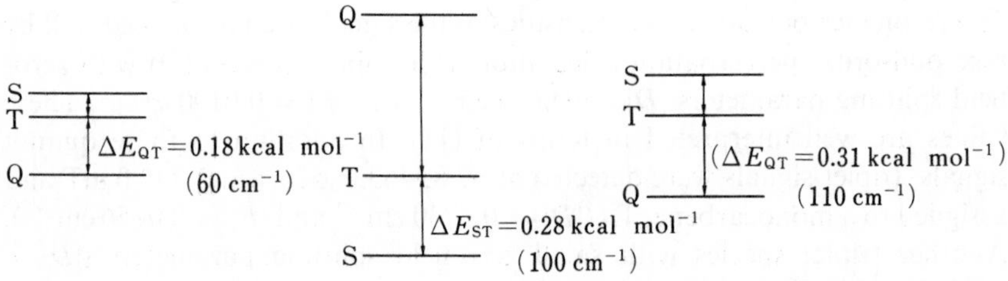

Fig. 23 The observed order of states for the isomeric [2.2]paracyclophane-dicarbenes [38], in good agreement with the McConnell theory.

All the results are summarized in Fig. 23 and serve as the first positive operational test for the McConnell theory. It is also noted that the orientational mode of stacking of spin-containing benzene rings can be useful in aligning spins parallel or antiparallel between high-spin aromatic molecules.

Similar experiments have been performed with the [3.3]paracyclophane isomers to obtain parallel results (Shinmyozu et al., 1990).

It is now well established that, if assemblies of high-spin molecules can be arranged so that the interaction between the spin densities of opposite sign could become most important between the neighbouring radicals or carbenes, the polyradicals might show longer range magnetic order and exhibit ferromagnetism as a bulk property. The conclusion was supported by inspection of epr fine structures and magnetic susceptibilities of a series of partially photolysed microcrystals of diazo-compounds. Crystalline samples of azibenzil have been shown to exhibit fine structure due to a quintet state when partially illuminated in an esr cavity at cryogenic temperatures (Murai et al., 1980a,b). It was not clear, however, if the quintet is a ground or excited state of the two interacting triplets. p,p'-Dimethoxy- and dichloro-derivatives of [15a; $m = 1$] constitute two extreme cases in that the corresponding diphenylcarbenes generated side by side are ferro- and antiferro-magnetically coupled. If the crystal structures of the precursors are assumed to be held in the partly photolysed part of the samples, we note that the nearest-neighbour interactions satisfy the McConnell ferro- and antiferro-magnetic conditions, respectively (Sugawara et al., 1986b).

Furthermore, a partial success of molecular design has been obtained by aligning the diphenylcarbene moieties by taking advantage of the van der Waals attraction between long alkyl side chains. When polycrystalline samples of bis(p-octyloxyphenyl)diazomethane [39a] were partially photo-lysed, interaction between the carbene molecules [39] generated in the immediate neighbourhood was presumably governed by the topochemical disposition of the precursor [39a] in crystals; actually, they showed ferro-magnetic coupling among four molecules on average (Sugawara et al., 1985).

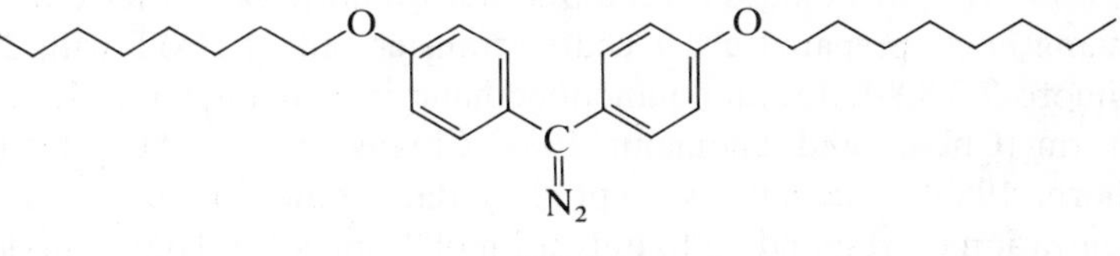

[39a]

Kinoshita and his coworkers noted the old report by Mukai and co-workers (Mukai et al., 1967, 1982; Mukai and Sogabe, 1980) that mixed crystals of [34] and hydrogalvinoxyl showed esr fine structure due to high-dependence of the esr signal intensity and magnetic susceptibility (Awaga et al., 1986a,b,c, 1988). They found that a first-order phase transition into an antiferromagnetically coupled state ($2J = -45$ meV) that takes place at 85 K in neat crystals of galvinoxyl disappears in the mixed crystals; the intermolecular exchange interaction among a few galvinoxyl radicals diluted in hydrogalvinoxyl becomes ferromagnetic with $2J = 1.5\,\text{meV}$. These results may be accounted for in terms of different orientation of the spin-distributed, neighbouring radical species in the crystals, although a contribution of intermolecular CT interaction cannot be excluded (Kinoshita, 1989a,b). The phase transition of [34] in the solid state remains unsolved in spite of an attempted low-temperature X-ray crystal-structure analysis (Chi et al., 1989).

For the second theory of McConnell, Breslow pursued stable aromatic dications of C_3 or higher symmetry to obtain triplet molecules. Since the dication of HMT (Bechgaard and Parker, 1972) was found to be stable only below cryogenic temperatures in solution, further extensions were given to its amino-counterpart, 2,3,6,7,10,11-hexaaminotriphenylenes and 2a,4a,6a,8a,10a,12a-hexaazaoctadecahydrocoronene (HOC, [37]), the latter being prepared by an unexpectedly elegant route (Thomaides et al., 1988). The corresponding dication HOC^{2+} gave a triplet spectrum ($D = 0.067 \pm 0.002\,\text{cm}^{-1}$ and $E < 0.0002\,\text{cm}^{-1}$) that obeyed a Curie law in CH_2Cl_2/CH_3CN (1.2:1), but the triplet state was $0.9\,\text{kcal mol}^{-1}$ above a ground singlet state in CH_3CN alone. Good charge-transfer salts for developing ferromagnetism were not obtained either (Breslow, 1989).

Miller and coworkers studied the cations of HOC in great detail (Dixon et al., 1989a,b; Dixon and Miller, 1989; Gabe et al., 1989; Miller et al., 1988a, 1990). Cyclic voltammetry revealed four reversible one-electron oxidation processes. The ground state structures for HOC and HOC^{4+} are "aromatic". The structures for HOC^+ and HOC^{3+} are doublets and show Jahn–Teller distortions. The structures for HOC^{2+} in the solid is a Jahn–Teller distorted, closed-shell singlet with a canonical structure [37'], as represented by cyanine/p-phenylenediammonium fragments. The one-to-one complex of HOC^+ $TCNE^-$ was indeed synthesized but found to be antiferromagnetic.

Chiang et al. prepared a 2:1 charge-transfer salt of HMT with 2,3,5,6-tetrafluoro-7,7,8,8-tetracyanoquinodimethane in a reddish purple, crystalline form (Chiang and Thomann, 1989; Chiang et al., 1989; Chiang and Goshorn, 1989). The static susceptibility data, however, indicated a spin concentration corresponding to only 0.1 mol% spins 1/2. High spin-density, solid samples were obtained only by a subsequent doping process; they obtained samples that have the formulae $(HMT)_2TCNQF_4(As_2F_{11})_y$ ($y = 1.0 \sim 4.2$), although with an anomalously small interspin coupling.

Miller and coworkers noted that the charge-transfer salts of decamethylferrocene [40] and TCNQ showed field-dependent magnetization; antiferromagnetic ($T_{\text{Neel}} = 2.55\,\text{K}$) for $H < 1.6\,\text{kG}$ and ferromagnetic for $H > 1.6\,\text{kG}$ (Candela et al., 1979). Encouraged by this discovery of a molecular metamagnet, they sought to prepare a molecular ferromagnet. The primary tactic was to utilize a smaller radical anion {as the $[Fe(C_5Me_5)_2]^+$ salt} for

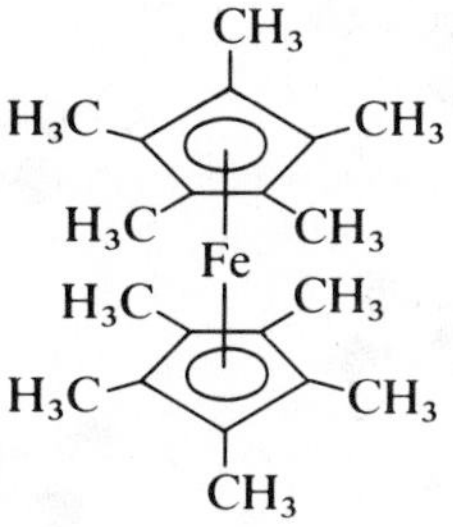

$$CH_3$$

[40]

greater spin interactions. The susceptibility of polycrystalline samples of the [40]–TCNE complex obeyed the Curie–Weiss expression with $\theta = +30$ K for $T > 60$ K, suggesting dominant ferromagnetic interactions. Below 60 K, a substantial departure from Curie–Weiss behaviour is evident and χ can be fitted by an $S = 1/2$, one-dimensional Heisenberg model (Baker et al., 1964) with ferromagnetic exchange ($J = +21$ cm^{-1}). Below 15 K, three-dimensional ordering sets in and the magnetization is no longer linearly proportional to the magnetic field. A spontaneous magnetization up to 1.1×10^4 emu G mol^{-1} was observed below the Curie temperature of 4.8 K (Fig. 13) (Miller et al., 1986, 1987). Spin alignment between chains is ascribed to the TCNE$^-$ residing in an adjacent chain equally proximal to the Fe(III) and the intrachain TCNE$^-$ moieties. This interpretation was justified by the X-ray structure (Fig. 24) which showed that the intrachain Fe—N separations range from 563 to 647 pm, whereas comparable Fe—N separations (567–571 pm) are present between the chains. Exchange mechanisms in these systems have been theoretically treated (Soos and McWilliams, 1989).

Since Fe(III) in the ferricenium ions is low spin ($S = 1/2$) and the spins on the acceptor TCNE appear to be contributing to the bulk magnetic properties as well, the D/A salt may be considered as a molecular magnetic material. It is so far the only example in which a ferromagnetic intermolecular interaction has led to a bulk ferromagnetic property (Miller et al., 1988b,c,d). Additional bulk ferromagnetic metallocene–acceptor complexes are being sought via the selective (i) replacement of the metal ions, (ii) replacement of methyl substituents with hydrogens, (iii) use of C$_6$-ring ligands instead of C$_5$-ring ligands and (iv) replacement of TCNE with other acceptors (Miller and Epstein, 1989).

The pursuit of suitable organic donors and acceptors that possess a rotation axis of order $n > 3$ or have D$_{2d}$ symmetry is in progress (Miller et al., 1989; Sugimoto et al., 1989; Wudl et al., 1989a,b; Yoshida and Sugi-

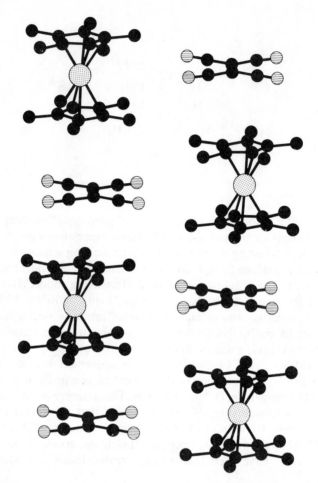

Fig. 24 X-Ray crystallographic structure for decamethylferrocene–TCNE D/A salt.

moto, 1988). Such organic donors and acceptors are shown in [41]–[46]. Whereas such studies have their own merit of finding novel multiredox systems, only a few of them showed ground triplet states and none of the CT complexes or radical salts proved to have ferromagnetic interactions.

The author believes it is not wise to take advantage of only geometrical symmetry; synthetic efforts should be directed towards triplet and quartet D or A molecules due to topological symmetry, e.g. [47] and [48] (Azuma *et al.*, 1974; Kuhn *et al.*, 1966). Of special interest is a stable perchloro-derivative of the Schlenck hydrocarbon [11]. Due to restricted rotation of the aryl groups, the biradical exists in two stereoisomeric forms: *meso* and *dl*. They

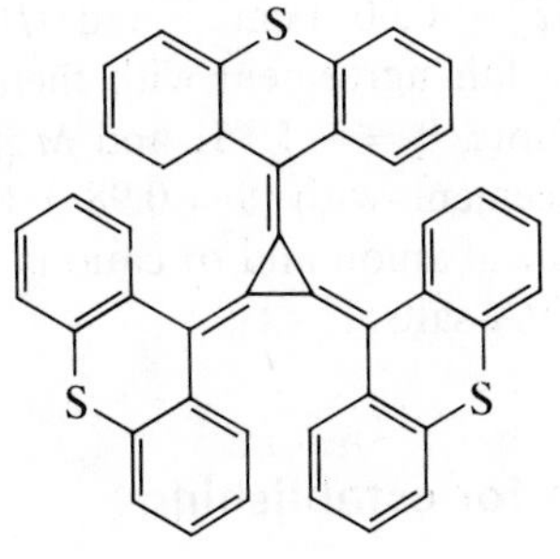

[41]

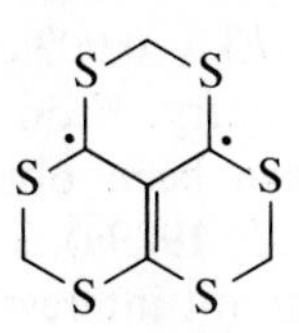

[42]

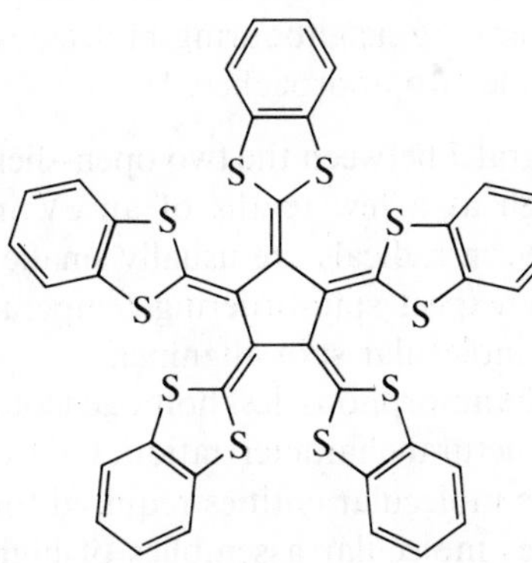

[43]

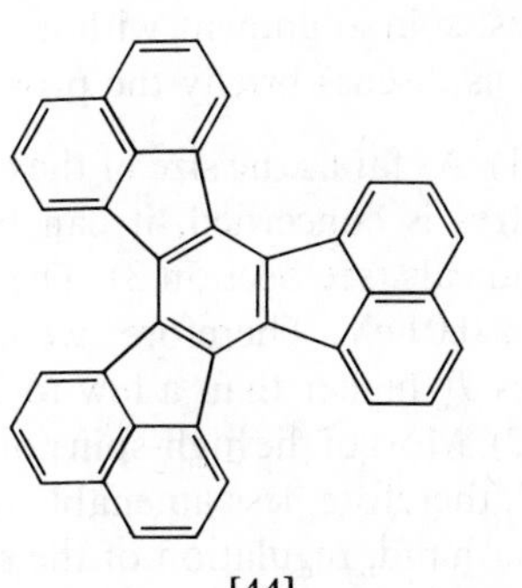

[44]

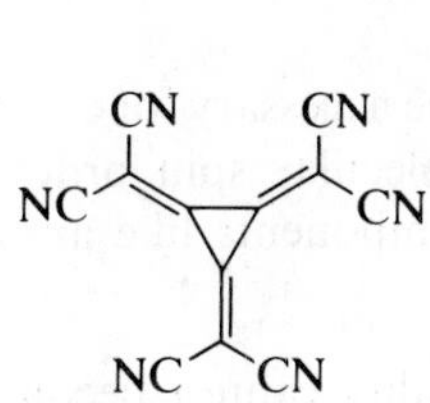

[45]

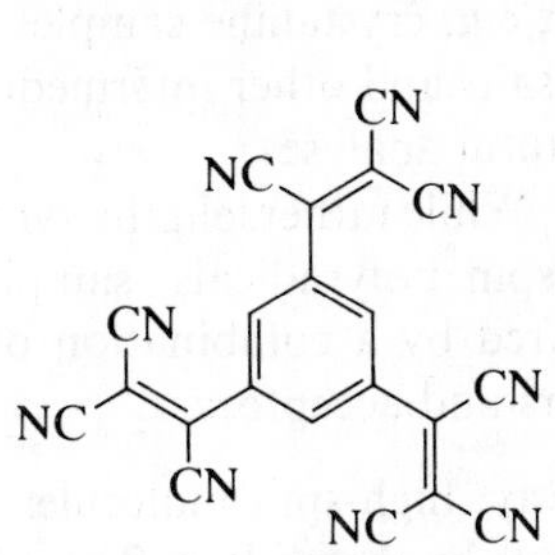

[46]

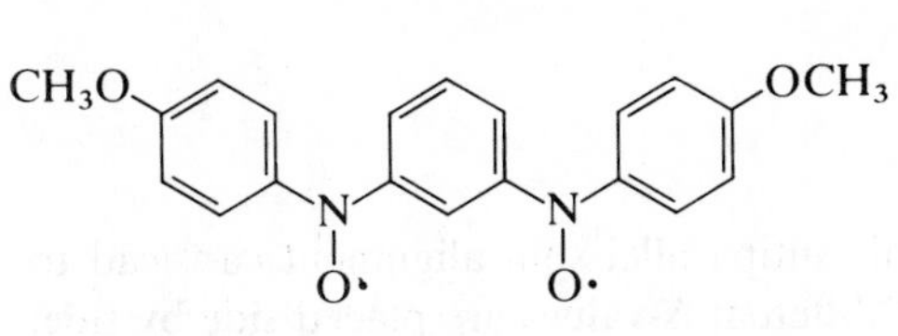

[47]

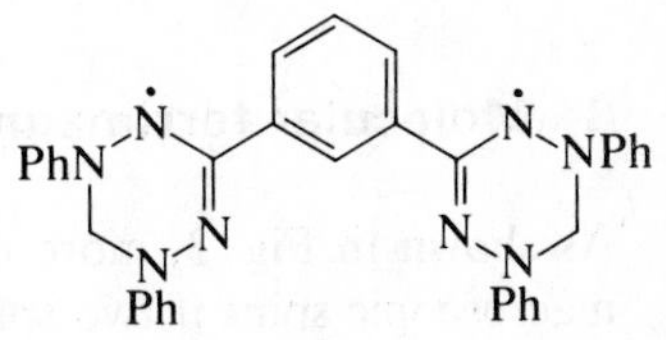

[48]

give esr fine structures $|D| = 0.0152$ and $|E| = 0.0051\,\mathrm{cm}^{-1}$ and $|D| = 0.0085$ and $|E| < 0.003\,\mathrm{cm}^{-1}$, respectively, in fair agreement with their C_s and C_2 symmetry. From the inverse χ vs T plots, $\theta = -1.1\,\mathrm{K}$ and $M_{eff} = 2.78\,M_B$ have been obtained, in good agreement with $S = 0.98 \pm 0.01$ (Veciana et al., 1989a). The expected monoradical anion and/or cations will be of paramount interest as components of D/A salts.

7 Comparison of the two approaches for establishing macroscopic spins

As discussed above, there are two approaches for constructing macroscopic spins: spin alignment within a molecule or among neighbouring molecules. Let us discuss briefly the pros and cons of the two approaches.

(1) As far as the size of the exchange integral J between the two open-shell centres is concerned, it can become as high as a few tenths of an eV in diradicals (see Section 3). The J-values between radicals are usually smaller than 0.01 eV. Therefore, we cannot readily expect spin-ordering temperatures T_c higher than a few to 20 K for intermolecular spin alignment.

(2) Most of the high-spin polymers would be amorphous, less homogeneous and, therefore, less amenable to rigorous structural characterization. On the other hand, regulation of the stacking of the molecular entities required for spin ordering between molecules necessitates molecular assemblies of high order, e.g. crystalline samples. These would be more readily obtained from D/A salts and other intermediate-size molecules, and are amenable to fuller structural analyses.

(3) While rather lengthy synthetic efforts may be necessary for constructing high-spin polyradicals, samples having intermolecular spin order can be prepared by a combination of rather simple components like appropriate donors and acceptors.

Anyway, high-spin molecules of low dimensionality cannot become ferromagnets by themselves. The additional stability at finite temperature should be derived from intermolecular interaction. Therefore the interplay of the two factors appears to be most desirable.

8 Molecular ferrimagnets

As shown in Fig. 3, more natural, antiparallel spin alignment can lead to macroscopic spins if two spins of different S-values are placed side by side. Tricarbene [49] has been shown by Itoh and coworkers to have a triplet

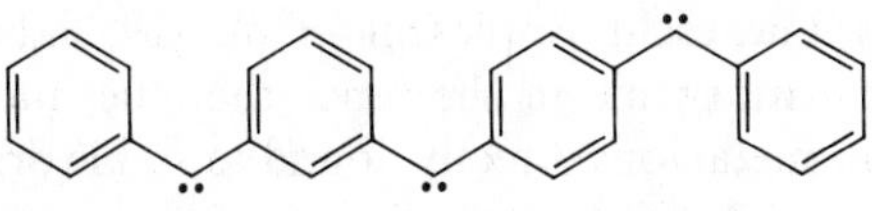

[49]

ground state; a typical example of ferrimagnetic intramolecular spin interaction. The interaction through the O and CH_2 units is said to be due to
superexchange and hyperconjugation, respectively (Itoh *et al.*, 1989; Takui
et al., 1989).

The idea of stacking monoradicals with $S = 1/2$ and di- or poly-radicals
with $S > 1$ in mixed crystals was proposed by Buchachenko (Buchachenko,
1979). If the chemical structures are almost identical, the two component
radicals will have a chance to stack in alternation and the two spin
sublattices having $S = 1/2$ and $S > 1$, respectively, will be formed with
opposite orientation.

The author cannot conclude this article without referring at this point to
extensive work by Kahn *et al.* (Georges and Kahn, 1989; Kahn 1987; Kahn
et al., 1989a,b; Lloret *et al.*, 1988; Pei *et al.*, 1986a,b, 1987; Verdaguer *et al.*,
1983) and Drillon *et al.* (Beltran *et al.*, 1982; Coronado *et al.*, 1986, 1989;
Drillon *et al.*, 1983), although it is considered to be within the realm of
inorganic rather than organic chemistry. They made arrays of transition-
metal ions bridged by closed-shell organic ligands. For example, spins 5/2 of
Mn(II) and 1/2 of Cu(II) have been arranged alternately in a ferrimagnetic
fashion by taking advantage of the antiferrimagnetic superexchange interaction (see p. 187) among the neighbouring metal ions through the multiden-
tate ligands. In one of the polycrystalline samples MnCu(obbz)H_2O [50],
long-range order aligns the spins at $T_c = 14$ K; below T_c it exhibited a
hysteresis loop characteristic of a soft ferromagnet (Gatteschi *et al.*, 1989;
Lloret *et al.*, 1988; Nakatani *et al.*, 1989).

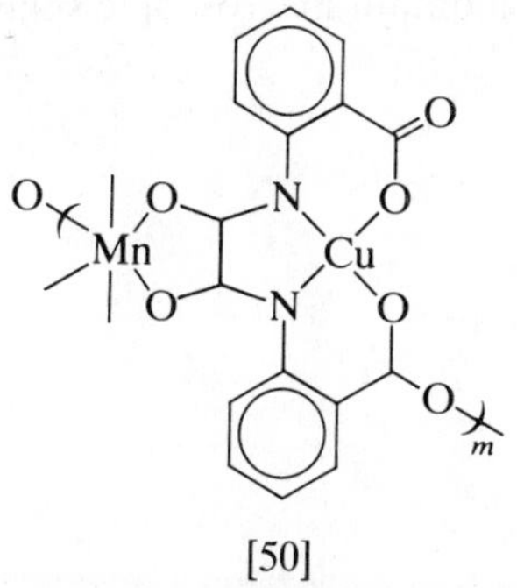

[50]

An approach by Gatteschi (Caneschi *et al.*, 1989a,b) is intermediate between organic and inorganic in the sense that the interacting magnetic centres are transition-metal ions directly bound to stable organic radicals, i.e. [31] (Ullman *et al.*, 1972). The nitroxyls coordinate with such transition-metal ions as Cu(II), Ni(II), Co(II) and Mn(II) as equatorial ligands. Under these conditions, the magnetic coupling is usually antiferromagnetic with J in the range $-150 \sim -500 \, cm^{-1}$. Therefore, when the radicals bind with the two oxygen atoms to two different metal ions, chains can be formed that are one-dimensional ferrimagnets [51]. Mn(hfac)$_2$[31; R = i-Pr] is such an example.

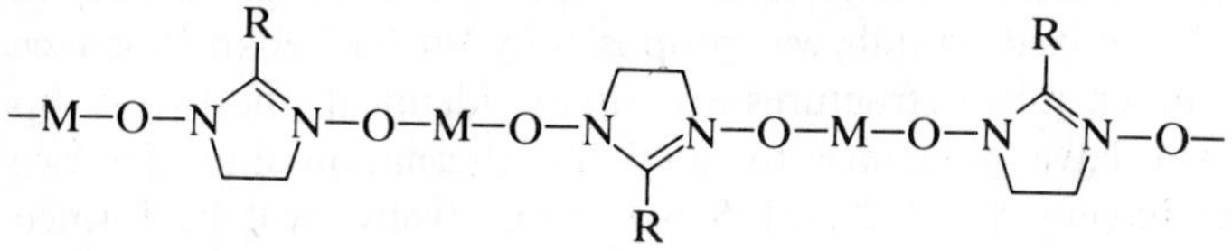

[51]

9 Conclusion

On the basis of rigorous molecular design guided by theory and model experiments, a number of high-spin organic molecules have been synthesized and spin alignment in molecular assemblies has been advanced with reasonable success. We find the two approaches, i.e. spin alignment within a molecule and among molecules, are complementary to each other in expanding our understanding of chemical bonds and organic molecules and in developing novel organic magnetic materials.

Realization of purely organic ferromagnets remains of multidisciplinary scientific interest. They are not yet usable realities, but it is already possible to state an overall conceptual framework for exploiting the magnetic properties of organic compounds. It is highly necessary to establish synthetic methods which enable us to obtain macroscopic spins almost stoichiometrically as designed.

Appendix

Suffix	X	Y	Suffix	X	Y
a	N_2	in place of $:$	j	H	
b	Ar*	Ar			
c	H	Ar			
d	H		k	H	
e	H		l	H	
f	H		m	H	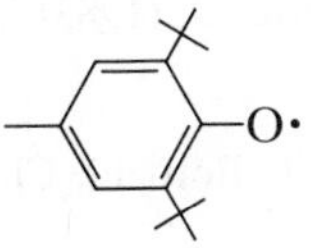
g	H		n	H	
h	H		o	CH_2OH	
i			p		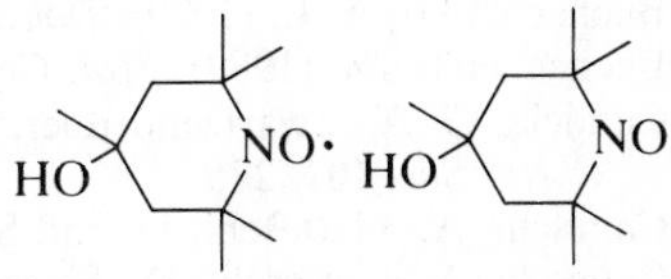

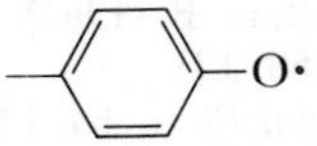

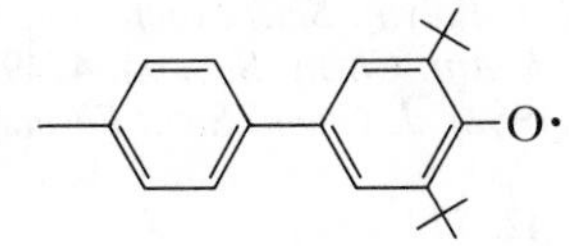

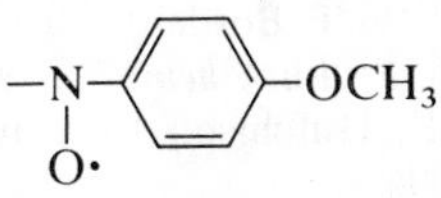

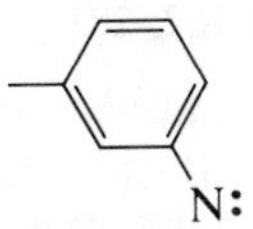

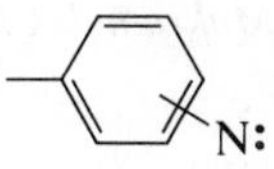

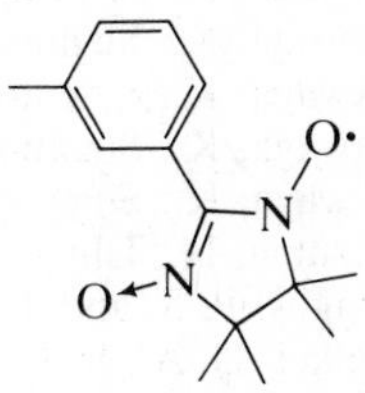

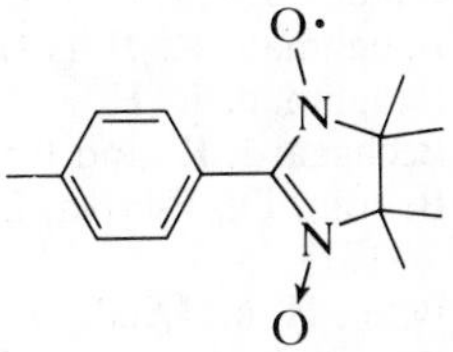

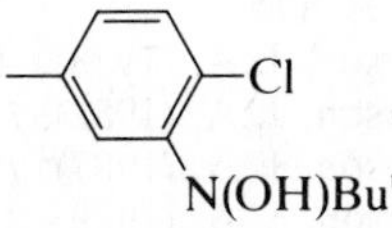

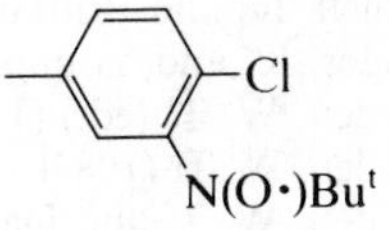

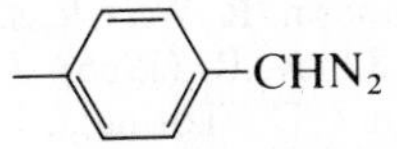

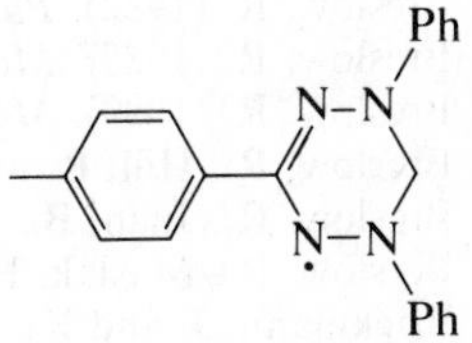

Ar* denotes a substituted phenyl.

References

Alcacer, L. (ed.) (1987). "Conducting Polymers". Reidel, Dodrecht
Alexander, S. A. and Klein, D. J. (1988). *J. Am. Chem. Soc.* **110**, 3401
Awaga, K. and Maruyama, Y. (1989). *J. Chem. Phys.* **91**, 2743
Awaga, K., Sugano, T. and Kinoshita, M. (1986a). *Solid State Commun.* **57**, 453
Awaga, K., Sugano, T. and Kinoshita, M. (1986b). *Chem. Phys. Lett.* **128**, 587
Awaga, K., Sugano, T. and Kinoshita, M. (1986c). *J. Chem. Phys.* **85**, 2211
Awaga, K., Sugano, T. and Kinoshita, M. (1988). *Synth. Met.* **27**, B631
Azuma, N., Ishizu, K. and Mukai, K. (1974). *J. Chem. Phys.* **61**, 2294
Bagus, P. S. and Torrance, J. B. (1989). *Phys. Rev.* **B39**, 7301
Baker, G. A., Jr, Rushbrooke, G. S. and Gilbert, H. E. (1964). *Phys. Rev.* **135**, A1272
Baudet, J. (1971). *J. Chim. Phys.-Chim. Biol.* **68**, 191
Baughman, R. H. (1974). *J. Polym. Sci., Polym. Phys. Ed.* **12**, 1511
Baughman, R. H. and Yee, K. C. (1978). *J. Polym. Sci., Macromol. Rev.* **13**, 219
Bechgaard, K. and Parker, V. D. (1972). *J. Am. Chem. Soc.* **94**, 4749
Beltran, D., Escriva, E. and Drillon, M. (1982). *J. Chem. Soc., Faraday Trans. 2* **78**, 1773
Benk, H. and Sixl, H. (1981). *Mol. Phys.* **42**, 779
Benoit, A., Flouquet, J., Gillon, B. and Schweizer, J. (1983). *J. Mag. Mag. Mat.* **31–34**, 1155
Berson, J. A. (1978). *Acc. Chem. Res.* **11**, 446
Berson, J. A. (1982). *In* "Diradicals" (ed. W. T. Borden), Ch. 4. Wiley, New York
Berson, J. A. (1987). *Pure Appl. Chem.* **59**, 1571
Berson, J. A. (1989). *Mol. Cryst. Liq. Cryst.* **176**, 1
Bethell, D. and Brinkman, M. R. (1973). *Adv. Phys. Org. Chem.* **10**, 53
Binder, K. and Young, A. P. (1986). *Rev. Mod. Phys.* **58**, 801
Borden, W. T. (ed.) (1982a). "Diradicals". Wiley, New York
Borden, W. T. (1982b). *In* "Diradicals" (ed. W. T. Borden), Ch.1. Wiley, New York
Borden, W. T. and Davidson, E. R. (1977). *J. Am. Chem. Soc.* **99**, 4587
Brandon, R. W., Closs, G. L., Davoust, C. E., Hutchison, C. A. Jr, Kohler, B. E. and Silbey, R. (1965). *J. Chem. Phys.* **43**, 2006
Braun, D., Tormala, P. and Wittig, W. (1981). *Makromol. Chem.* **182**, 2217
Breslow, R. (1982). *Pure Appl. Chem.* **54**, 927
Breslow, R. (1985). *Mol. Cryst. Liq. Cryst.* **125**, 261
Breslow, R. (1989). *Mol. Cryst. Liq. Cryst.* **176**, 199
Breslow, R., Hill, R. and Wasserman, E. (1964). *J. Am. Chem. Soc.* **86**, 5349
Breslow, R., Juan, B., Klutz, R. Q. and Xia, C. Z. (1982). *Tetrahedron* **38**, 863
Breslow, R., Maslak, P. and Thomaides, J. S. (1984). *J. Am. Chem. Soc.* **106**, 6453
Brickmann, J. and Kothe, G. (1973). *J. Chem. Phys.* **59**, 2807
Buchachenko, A. L. (1979). *Dokl. Akad. Nauk. SSSR.* **244**, 1146
Buchachenko, A. (1989). *Mol. Cryst. Liq. Cryst.* **176**, 307
Candela, G. A., Swartzendruber, L. J., Miller, J. S. and Rice, M. J. (1979). *J. Am. Chem. Soc.* **101**, 2755
Caneschi, A., Gatteschi, D. and Sessoli, R. (1989a). *Acc. Chem. Res.* **22**, 392
Caneschi, A., Gatteschi, D., Sessoli, R. and Rey, P. (1989b). *Mol. Cryst. Liq. Cryst.* **176**, 329
Cao, Y., Wang, P., Hu, Z., Li, S., Zhang, L. and Zhao, J. (1988a). *Solid State Commun.* **68**, 817

Cao, Y., Wang, P., Hu, Z., Li, S., Zhang, L. and Zhao, J. (1988b). *Synth. Met.* **27**, B625

Chi, K.-M., Calabrese, J. C., Miller, J. S. and Khan, S. I. (1989). *Mol. Cryst. Liq. Cryst.* **176**, 185

Chiang, L. Y. and Goshorn, D. P. (1989). *Mol. Cryst. Liq. Cryst.* **176**, 229

Chiang, L. Y. and Thomann, H. (1989). *J. Chem. Soc., Chem. Commun.* 172

Chiang, L. Y., Johnston, D. C., Goshorn, D. P. and Bloch, A. N. (1988). *Synth. Met.* **27**, B639

Chiang, L. Y., Johnston, D. C., Goshorn, D. P. and Bloch, A. N. (1989). *J. Am. Chem. Soc.* **111**, 1925

Chouteau, G. and Veyret-Jeandey, C. (1981). *J. Phys. (Paris)* **42**, 1441

Coronado, E., Drillon, M., Fuertes, A., Beltran, D., Mosset, A. and Galy, J. (1986). *J. Am. Chem. Soc.* **108**, 900

Coronado, E., Sapina, F., Beltran, D., Burriel, R. and Carlin, R. L. (1989). *Mol. Cryst. Liq. Cryst.* **176**, 507

Coulson, C. A. and Longuet-Higgins, H. C. (1947). *Proc. R. Soc., Ser A* **191**, 39; **192**, 16

Coulson, C. A. and Rushbrooke, G. S. (1940). *Proc. Cambridge Phil. Soc.* **36**, 193

Dixon, D. A. and Miller, J. S. (1989). *Mol. Cryst. Liq. Cryst.* **176**, 211

Dixon, D. A., Calabrese, J. C. and Miller, J. S. (1989a). *Angew. Chem., Int. Ed. Engl.* **28**, 90

Dixon, D. A., Calabrese, J. C., Harlow, R. L. and Miller, J. S. (1989b). *Angew. Chem., Int. Ed. Engl.* **28**, 92

Dormann, E., Nowak, M. J., Williams, K. A., Angus, R. O. Jr and Wudl, F. (1987). *J. Am. Chem. Soc.* **109**, 2594

Dougherty, D. A. (1989). *Mol. Cryst. Liq. Cryst.* **176**, 25

Dowd, P. (1966). *J. Am. Chem. Soc.* **88**, 2587

Dowd, P. (1972). *Acc. Chem. Res.* **5**, 242

Dowd, P. (1979). *J. Am. Chem. Soc.* **92**, 1066

Dowd, P., Chang, W. and Paik, Y. H. (1986). *J. Am. Chem. Soc.* **108**, 7416

Dowd, P., Ham, S.-W., Chang, W. and Partian, C. J. (1989). *Mol. Cryst. Liq. Cryst.* **176**, 13

Drillon, M., Coronado, E., Fuertes, A., Beltran, D. and Georges, R. (1983). *Chem. Phys.* **79**, 449

Du, P. and Borden, W. T. (1987). *J. Am. Chem. Soc.* **109**, 930

Dunkin, I. R and Thomson, P. C. P. (1982). *J. Chem. Soc., Chem. Commun.* 1192

Eaton, D. R. and Phillips, W. D. (1965). *Adv. Magn. Reson.* **1**, 103

Epstein, A. J. and Miller, J. S. (1989). *Mol. Cryst. Liq. Cryst.* **176**, 359

Forrester, A. R., Hay, J. M. and Thomson, R. H. (1968). "Organic Chemistry of Stable Free Radicals". Academic Press, London

Forrester, A. R. and Ramasseul, R. (1971a). *J. Chem. Soc. (B)* 1638

Forrester, A. R. and Ramasseul, R. (1971b). *J. Chem. Soc. (B)* 1645

Forrester, A. R. and Ramasseul, R. (1975). *J. Chem. Soc., Perkin Trans 1* 1753

Fujii, A., Ishida, T., Koga, N. and Iwamura, H. (1990). *Macromolecules*, in press

Fujita, I., Teki, Y., Takui, T., Kinoshita, T., Itoh, K., Miko, F., Sawaki, Y., Izuoka, A., Sugawara, T. and Iwamura, H. (1990). *J. Am. Chem. Soc.* **112**, 4074

Fukutome, H., Takahashi, A. and Ozaki, M. (1987). *Chem. Phys. Lett.* **133**, 34

Furlani, A., Napoletano, C., Russo, M. V. and Feast, W. J. (1986). *Polymer Bull.* **16**, 311

Gabe, E. J., Morton, J. R., Preston, K. F., Krusic, P. J., Dixon, D. A., Wasserman, E. and Miller, J. S. (1989). *J. Phys. Chem.* **93**, 5337

Gatteschi, D., Guillou, O., Zanchini, C., Sessoli, R., Kahn, O., Verdaguer, M. and Pei, Y. (1989). *Inorg. Chem.* **28**, 287

Georges, R. and Kahn, O. (1989). *Mol. Cryst. Liq. Cryst.* **176**, 473

Gillespie, R. J. and Day, P. (ed.) (1986). Electrical and Magnetic Properties of Low Dimensional Solids: Proceedings of a Royal Society Discussion Meeting, The Royal Society, London

Ginsburg, E. J., Gorman, C. B., Marder, S. R. and Grubbs, R. H. (1989). *J. Am. Chem. Soc.* **111**, 7621

Goodman, J. L. and Berson, J A. (1985). *J. Am. Chem. Soc.* **107**, 5409.

Haider, K. W., Platz, M. S., Despres, A., Lejeune, V., Migirdicyan, E., Balley, T. and Haselbach, E. (1988). *J. Am. Chem. Soc.* **110**, 2318

Haider, K. W., Migirdicyan, E., Platz, M. S., Soundararajan, N. and Despres, A. (1989). *Mol. Cryst. Liq. Cryst.* **176**, 85

Hay, P. J., Thibéault, J. C. and Hoffmann, R. (1975). *J. Am. Chem. Soc.* **97**, 4884

Henry, W. E. (1952). *Phys. Rev.* **88**, 559

Higuchi, J. (1963a). *J. Chem. Phys.* **38**, 1237

Higuchi, J. (1963b). *J. Chem. Phys.* **39**, 1847

Huber, R. A. and Schwoerer, M. (1980). *Chem. Phys. Lett.* **72**, 10

Hughbanks, T. and Kertesz, M. (1989). *Mol. Cryst. Liq. Cryst.* **176**, 115

Hupfer, B., Ringsdorf, H. and Schupp, H. (1981). *Makromol. Chem.* **182**, 247

Hutchison, C. A., Jr and Kohler, B. E. (1969). *J. Chem. Phys.* **51**, 3327

Ishida, T. and Iwamura, H. (1989). The 1989 Internat. Chem. Congr. Pacific Basin Scc., December 17–22, Honolulu, Abstract of Papers, Macro 93

Itoh, K. (1967). *Chem. Phys. Lett.* **1**, 235

Itoh, K. (1971). *Bussei* **12**, 635

Itoh, K. (1978). *Pure Appl. Chem.* **50**, 1251

Itoh, K., Takui, T., Teki, Y. and Kinoshita, T. (1989). *Mol. Cryst. Liq. Cryst.* **176**, 49

Iwamura, H. (1986). *Pure. Appl. Chem.* **58**, 187

Iwamura, H. (1987). *Pure. Appl. Chem.* **59**, 1595

Iwamura, H. (1988). *J. Phys. (Paris),* Suppl. **49**, C8-813

Iwamura, H. (1989). *Oyo Butsuri* (in Japanese) **58**, 1061

Iwamura, H. (1990). Unpublished work

Iwamura, H. and Izuoka, A. (1987). *Nihon Kagaku Kaishi* (in Japanese) 595

Iwamura, H. and McKelvey, R D. (1988) *Macromolecules* **21**, 3386

Iwamura, H. and Murata, S. (1989). *Mol. Cryst. Liq. Cryst.* **176**, 33

Iwamura, H., Sugawara, T., Itoh, K. and Takui, T. (1985). *Mol. Cryst. Liq. Cryst.* **125**, 251

Iwamura, H., Izuoka, A., Murata, S., Bandow, S., Kimura, K. and Sugawara, T. (1987). IUPAC CHEMRAWN VI, Abstracts of Papers, IC05, May 17–22, Tokyo, Japan

Izuoka, A., Murata, S., Sugawara, T. and Iwamura, H. (1985). *J. Am. Chem. Soc.* **107**, 1786

Izuoka, A., Murata, S., Sugawara, T. and Iwamura, H. (1987). *J. Am. Chem. Soc.* **109**, 2631

Johannsen, I., Torrance, J. B. and Nazzal, A. (1989). *Macromolecules* **22**, 566

Kahn, O. (1987). *In* "Structure and Bonding", Vol. 68, p. 89. Springer-Verlag, Berlin/ Heidelberg

Kahn, O., Pei, Y. and Journaux, Y. (1989a). *Mol. Cryst. Liq. Cryst.* **176**, 429

Kahn, O., Pei, Y., Nakatani, K. and Journaux, Y. (1989b). *Mol. Cryst. Liq. Cryst.* **176**, 481

Kamachi, M. (1987). *Nihon Butsuri Gakkaishi* **42**, 351

Kamachi, M., Akimoto, H., Mori, W. and Kishita, M. (1984). *Polym. J.* **16**, 23

Kamachi, M., Cheng, X. Su, Aota, H., Mori, W. and Kishita, M. (1987). *Chemistry Lett.* 2331

Kato, S., Morokuma, K., Feller, D., Davidson, E. R. and Borden, W. T. (1983). *J. Am. Chem. Soc.* **105**, 1791

Kawabata, K., Mizutani, M., Fukuda, M. and Mizogami, S. (1990). *Synth. Met.* **33**, 399

Kimura, K. and Bandow, S. (1984a). *Solid State Phys.* **20**, 467

Kimura, K. and Bandow, S. (1984b). *Kotai Bussei* (in Japanese) **19**, 476

Kinoshita, M. (1989a). *Mol. Cryst. Liq. Cryst.* **176**, 163

Kinoshita, M. (1989b). *Kotai Butsuri* (in Japanese) **24**, 623

Klavetter, F. L. and Grubbs, R. H. (1988). *J. Am. Chem. Soc.* **110**, 7807

Klein, D. J. (1982). *J. Chem. Phys.* **77**, 3098

Klein, D. J. (1983). *Pure. Appl. Chem.* **55**, 299

Klein, D. J., Nelin, C. J., Alexander, S. and Matsen, F. A. (1982). *J. Chem. Phys.* **77**, 3101

Klein, D. J., Alexander, S. A. and Randic, M. (1989). *Mol. Cryst. Liq. Cryst.* **176**, 109

Koch, H. and Ringsdorf, H. (1981). *Makromol. Chem.* **182**, 255

Koga, N. and Iwamura, H. (1989). *Nihon Kagaku Kaishi* (in Japanese) 1456

Koga, N., Inoue, K., Sasagawa, N. and Iwamura, H. (1990). *Mat. Res. Soc. Symp. Proc.*, **173**, 39

Korshak, Yu, V., Ovchinnikov, A. A., Shapiro, A. M., Medvedeva, T. V. and Spector, V. N. (1986). *Pisma Zh. Eksp. Teor. Fiz.* **43**, 309

Korshak, Yu, V., Medvedeva, T. V., Ovchinnikov, A. A. and Spector, V. N. (1987). *Nature* **326**, 370

Kosaki, A., Suga, H., Seki, S., Mukai, K. and Deguchi, Y. (1969). *Bull. Chem. Soc. Jpn* **42**, 1525

Kothe, G., Denkel, K.-H. and Summermann, W. (1970). *Angew. Chem., Int. Ed. Engl.* **9**, 906

Kothe, G., Ohmes, E., Brickmann, J. and Zimmermann, H. (1971). *Angew. Chem., Int. Ed. Engl.* **10**, 938

Kuhn, R., Neugebauer, F. A. and Trischmann, H. (1966). *Monatsh. Chem.* **97**, 520

Lahti, P. M. and Ichimura, A. S. (1989). *Mol. Cryst. Liq. Cryst.* **176**, 125

Lahti, P. M., Rossi, A. R. and Berson, J. A. (1985). *J. Am. Chem. Soc.* **107**, 2273

Lahti, P. M., Ichimura, A. S. and Berson, J. A. (1989). *J. Org. Chem.* **54**, 958

Leo, M. (1937). *Ber.* **70**, 1691

Lloret, F., Nakatani, K., Journaux, Y., Kahn, O., Pei, Y. and Renard, J. P. (1988). *J. Chem. Soc., Chem. Commun.* 642

LePage, T. J. and Breslow, R. (1987). *J. Am. Chem. Soc.* **109**, 6412

Lepley, A. R. and Closs, G. L. (eds) (1973). "Chemically Induced Magnetic Polarization". Wiley, New York

Longuet-Higgins, H. C. (1950). *J. Chem. Phys.* **18**, 265

Luckhurst, G. R., Pedulli, G. R. and Tiecco, M. (1971). *J. Chem. Soc. (B)* 329

Manriquez, J. M., Ward, M. D., Calabrese, J. C., Fagan, P. J., Epstein, A. J. and Miller, J. S. (1989). *Mol. Cryst. Liq. Cryst.* **176**, 527

Masuda, T. (1985). *Yuuki Gousei Kagaku Kyokaishi* (in Japanese) **43**, 744

Masuda, T. and Higashimura, T. (1984). *Acc. Chem. Res.* **17**, 51
Masuda, T. and Higashimura, T. (1986). *Adv. Polymer Sci.* **81**, 121
Mataga, N. (1968). *Theor. Chim. Acta* **10**, 372
Matsushita, M., Momose, T., Shida, T., Teki, Y., Takui, T. and Itoh, K. (1990). *J. Am. Chem. Soc.* **112**, 4700
McBride, J. M. (1983). *Acc. Chem. Res.* **16**, 304
McConnell, H. M. (1963). *J. Chem. Phys.* **39**, 1910
McConnell, H. M. (1967). *Proc. R. A. Welch Found. Chem. Res.* **11**, 144
Mermin, N. D. and Wagner, H. (1966). *Phys. Rev. Lett.* **17**, 1133
Migirdicyan, E. and Baudet, J. (1975). *J. Am. Chem. Soc,* **97**, 7400
Miller, J. S. and Epstein, A. J. (1987). *J. Am. Chem. Soc.* **109**, 3850
Miller, J. S. and Epstein, A. J. (1989). *Mol. Cryst. Liq. Cryst.* **176**, 347
Miller, J. S., Calabrese, J. C., Epstein, A. J., Bigelow, R. W., Zhang, J. H. and Reiff, W. M. (1986). *J. Chem. Soc., Chem. Commun.* 1027
Miller, J. S., Calabrese, J. C., Rommelmann, H., Chittipeddi, S. R., Zhang, J. H., Reiff, W. M. and Epstein, A. J. (1987). *J. Am. Chem. Soc.* **109**, 769
Miller, J. S., Dixon, D. A. and Calabrese, J. C. (1988a). *Science,* **240**, 1185
Miller, J. S., Epstein, A. J. and Reiff, W. M. (1988b). *Science* **240**, 40
Miller, J. S., Epstein, A. J. and Reiff, W. M. (1988c). *Acc. Chem. Res.* **21**, 114
Miller, J. S., Epstein, A. J. and Reiff, W. M. (1988d). *Chem. Rev.* **88**, 201
Miller, J. S., Krusic, P. J., Epstein, A. J., Zhang, J. H., Morand, J. P., Brzezinski, L., Lapouyade, R., Garrigou-Lagrange, C., Amiell, J. and Delhaes, P. (1989). *Mol. Cryst. Liq. Cryst.* **176**, 241
Miller, J. S., Dixon, D. A., Calabrese, J. C., Vazquez, C., Krusic, P. J., Ward, M. D., Wasserman, E. and Halow, R. L. (1990). *J. Am. Chem. Soc.* **112**, 381
Mukai, K. (1969). *Bull. Chem. Soc. Jpn.* **42**, 40
Mukai, K. and Sogabe, A. (1980). *J. Chem. Phys.* **72**, 598
Mukai, K., Nishiguchi, H. and Deguchi, Y. (1967). *J. Phys. Soc. Jpn* **23**, 125
Mukai, K., Ueda, K., Ishizu, K. and Deguchi, Y. (1982). *J. Chem. Phys.* **77**, 1606
Murahashi, S.-I., Yoshimura, N., Yamamoto, Y. and Moritani, I. (1972). *Tetrahedron* **28**, 1485
Murai, H., Torres, M. and Strausz, O. P. (1980a). *J. Am. Chem. Soc.* **102**, 5104
Murai, H., Torres, M. and Strausz, O. P. (1980b). *J. Am. Chem. Soc.* **102**, 7391
Murata, K. and Ueda, H. (1990). The 59th Annual Meeting of the Chemical Society of Japan, April 1–4, Yokohama, Japan. Abstract of Papers 4E 144–147
Murata, S., Sugawara, T. and Iwamura, H. (1987). *J. Am. Chem. Soc.* **109**, 1266
Nakatani, K., Carriat, J. Y., Journaux, Y., Kahn, O., Lloret, F., Renard, J. P., Pei, Y., Sletten, J. and Verdaguer, M. (1989). *J. Am Chem. Soc.* **111**, 5739
Nasu, K. (1986). *Phys. Rev. B; Solid State* **33**, 330
Nishide, H., Yoshioka, N., Inagaki, K. and Tuchida, E. (1988). *Macromolecules* **21**, 3119
Novak, J. A., Jain, R. and Dougherty, D. A. (1989). *J. Am. Chem. Soc.* **111**, 7618
Ota, M., Otani, S., Kobayashi, K. and Igarashi, M. (1989). *Mol. Cryst. Liq. Cryst.* **176**, 99
Ovchinnikov, A. A. (1978). *Theor. Chim. Acta* **47**, 297
Ovchinnikov, A. A. and Spector, V. N. (1988). *Synth. Met.* **27**, B615
Palacio, F., Lazaro, F. J. and Van Duyneveldt, A. J. (1989). *Mol. Cryst. Liq. Cryst.* **176**, 289
Pei, Y., Kahn, O. and Sletten, J. (1986a). *J. Am. Chem. Soc.* **108**, 3143
Pei, Y., Verdaguer, M., Kahn, O., Sletten, J. and Renard, J. P. (1986b). *J. Am. Chem. Soc.* **108**, 7428

Pei, Y., Verdaguer, M., Kahn, O., Sletten, J. and Renard, J. P. (1987). *Inorg. Chem.* **26**, 138

Platz, M. S. (1982). *In* "Diradicals" (ed. W. T. Borden), Ch. 5. Wiley, New York

Pranata, J., Marudarajan, V. S. and Dougherty, D. A. (1989). *J. Am. Chem. Soc.* **111**, 2026

Roth, W. R., Langer, R., Bartmann, M., Stevermann, B., Maier, G., Reisenauer, H. P., Sustmann, R. and Muller, W. (1987). *Angew. Chem., Int. Ed. Engl.* **26**, 256

Salem, L. (1982). *In* "Electrons in Chemical Reaction: First Principles", Ch. 7. Wiley, New York

Saunders, M., Berger, R., Jaffe, A., McBride, J. M., O'Neill, J., Breslow, R., Hoffman, J. M., Jr, Perchonock, C., Wasserman, E., Hutton, R. S. and Kuck, V. J. (1973). *J. Am. Chem. Soc.* **95**, 3017

Schmauss, G., Baumgartel, H. and Zimmermann, H. (1965). *Angew. Chem., Int. Ed. Engl.* **4**, 596

Schuster, G. B. (1986). *Adv. Phys. Org. Chem.* **22**, 311

Schwoerer, M., Huber, R. A. and Hartl, W. (1981). *Chem. Phys.* **55**, 97

Seeger, D. E. and Berson, J. A. (1983). *J. Am. Chem. Soc.* **105**, 5144

Seeger, D. E., Lahti, P. M., Rossi, A. R. and Berson, J. A. (1986). *J. Am. Chem. Soc.* **108**, 1251

Shinmyozu, T., Inazu, T., Izuoka, A., Murata, S., Sugawara, T. and Iwamura, H. (1990). Unpublished work

Sixl, H., Mathes, R., Schaupp, A., Ulrich, K. and Huber, R. (1986). *Chem. Phys.* **107**, 105

Skotheim, T. A. (ed.) (1986). "Handbook of Conductive Polymers", Vols 1 and 2. Dekker, New York

Snyder, G. J. and Dougherty, D. A. (1989). *J. Am. Chem. Soc.* **111**, 3927

Soos, Z. G. and McWilliams, P. C. M. (1989). *Mol. Cryst. Liq. Cryst.* **176**, 369

Stanley, H. E. (1971). *In* "Introduction to Phase Transitions and Critical Phenomena". Clarendon Press, Oxford

Sugawara, T., Bandow, S., Kimura, K., Iwamura, H. and Itoh, K. (1984). *J. Am. Chem. Soc.* **106**, 6449

Sugawara, T., Murata, S., Kimura, K., Iwamura, H., Sugawara, Y. and Iwasaki, H. (1985). *J. Am. Chem. Soc.* **107**, 5293

Sugawara, T., Bandow, S., Kimura, K., Iwamura, H. and Itoh, K. (1986a). *J. Am. Chem. Soc.* **108**, 368

Sugawara, T., Tukada, H., Izuoka, A., Murata, S. and Iwamura, H. (1986b). *J. Am. Chem. Soc.* **108**, 4272

Sugimoto, T., Misaki, Y., Yoshida, Z. and Yamauchi, J. (1989). *Mol. Cryst. Liq. Cryst.* **176**, 259

Syozi, J. (1972). *In* "Phase Transitions and Critical Phenomena" (eds C. Domb and M. S. Green), p. 269. Academic Press, New York

Tabata, M. (1988). *Nikkei New Materials* (in Japanese) 82

Takui, T. and Itoh, K. (1973). *Chem. Phys. Lett.* **19**, 120

Takui, T., Kita, S., Ichikawa, S., Teki, Y., Kinoshita, T. and Itoh, K. (1989). *Mol. Cryst. Liq. Cryst.* **176**, 67

Teki, Y., Takui, T., Itoh, K., Iwamura, H. and Kobayashi, K. (1983). *J. Am. Chem. Soc.* **105**, 3722

Teki, Y., Takui, T., Yagi, H., Itoh, K. and Iwamura, H. (1985). *J. Chem. Phys.* **83**, 539

Teki, Y., Takui, T., Itoh, K., Iwamura, H. and Kobayashi, K. (1986). *J. Am. Chem. Soc.* **108**, 2147

Teki, Y., Takui, T., Kinoshita, T., Ichikawa, S., Yagi, H. and Itoh, K. (1987a). *Chem. Phys. Lett.* **141**, 201

Teki, Y., Takui, T. and Itoh, K. (1987b). *Chem. Phys. Lett.* **142**, 181

Thomaides, J. S., Maslak, P. and Breslow, R. (1988). *J. Am. Chem. Soc.* **110**, 3970

Torrance, J. B., Oostra, S. and Nazzal, A. (1987). *Synth. Met.* **19**, 709

Torrance, J. B., Bagus, P. S., Johannsen, I., Nazzal, A. I., Parkin, S. S. P. and Batail, P. (1988). *J. Appl. Phys.* **63**, 2962

Trahanovsky, W. S., Chou, C.-H., Fischer, D. R. and Gerstein, B. C. (1988). *J. Am. Chem. Soc.* **110**, 6579

Tukada, H., Mutai, K. and Iwamura, H. (1987). *J. Chem. Soc., Chem. Commun.* 1159

Tyutyulkov, N. and Karabunarliev, S. (1986). *Int. J. Quantum Chem.* **29**, 1325

Tyutyulkov, N. and Polansky, O. E. (1987). *Chem. Phys. Lett.* **139**, 281

Tyutyulkov, N., Schuster, P. and Polansky, O. E. (1983). *Theor. Chim. Acta* **63**, 291

Tyutyulkov, N., Polansky, O. E., Schuster, P., Karabunarliev, S. and Ivanov, C. I. (1985). *Theor. Chim. Acta* **67**, 211

Tyutyulkov, N., Ivanov, Ts., Shopov, I. S., Polansky, O. E. and Olbrich, G. (1988). *Intern. J. Quantum Chem.* **34**, 361

Tyutyulkov, N., Karabunarliev, S. and Ivanov, Kh. (1989). *Mol. Cryst. Liq. Cryst.* **176**, 139

Ullman, E. F., Osiecki, J. H., Boocock, D. G. B. and Darcy, R. (1972). *J. Am. Chem. Soc.* **94**, 7049

Veciana, J., Rovira, C., Armet, O., Domingo, V. M., Crespo, M. I. and Palacio, F. (1989a). *Mol. Cryst. Liq. Cryst.* **176**, 77

Veciana, J., Vidal, J. and Jullian, N. (1989b). *Mol. Cryst. Liq. Cryst.* **176**, 443

Verdaguer, M., Julve, M., Michalowicz, A. and Kahn, O. (1983). *Inorg. Chem.* **22**, 2624

Wannier, G. H. (1959). "Solid State Theory". Cambridge University Press, Cambridge

Wasserman, E., Snyder, L. C. and Yager, W. A. (1964). *J. Chem. Phys.* **41**, 1763

Wasserman, E., Murray, R. W., Yager, W. A., Trozzolo, A. M. and Smolinsky, G. (1967). *J. Am. Chem. Soc.* **89**, 5076

Wasserman, E., Schueller, K. and Yager, W. A. (1968). *Chem. Phys. Lett.* **2**, 259

Wasserman, E., Hutton, R. S., Kuck, V. J. and Chandross, E. A. (1974). *J. Am. Chem. Soc.* **96**, 1965

Wegner, G. (1969). *Z. Naturforsch.* **246**, 824

Wegner, G. (1977). *Pure Appl. Chem.* **49**, 443

Wegner, G. (1979). *In* "Molecular Metals" (ed. E. W. Hartfield), p. 209. Plenum, New York

Weltner, W., Jr (1983). "Magnetic Atoms and Molecules". Van Nostrand Reinhold, New York

Wiley, D. W., Calabrese, J. C. and Miller, J. S. (1989). *Mol. Cryst. Liq. Cryst.* **176**, 277

Wilker, W., Kothe, G., Zimmermann, H. (1975). *Chem. Ber.* **108**, 2124

Williams, D. J., Pearson, J. M. and Levy, M. (1970). *J. Am. Chem. Soc.* **92**, 1436

Wright, B. B. and Platz, M. S. (1983). *J. Am. Chem. Soc.* **105**, 628

Wudl, F., Allemand, P. M., Delhaes, P., Soos, Z. and Hinkelmann, K. (1989a). *Mol. Cryst. Liq. Cryst.* **171**, 179

Wudl, F., Closs, F., Allemand, P. M., Cox, S., Hinkelmann, K., Srdanov, G. and Fite, C. (1989b). *Mol. Cryst. Liq. Cryst.* **176**, 249

Yamaguchi, K. and Fueno, T. (1989). *Chem. Phys. Lett.* **159**, 465

Yamaguchi, K., Fueno, T., Nakasuji, K. and Murata, I. (1986a). *Chemistry Lett.* 629

Yamaguchi, K., Fukui, H. and Fueno, T. (1986b). *Chemistry Lett.* 625

Yamaguchi, K., Toyoda, Y. and Fueno, T. (1987). *Synth. Metals* **19**, 81

Yamaguchi, K., Takahara, Y., Fueno, T. and Houk, K. (1988). *Theor. Chim. Acta* **73**, 337

Yamaguchi, K., Namimoto, H. and Fueno, T. (1989a). *Mol. Cryst. Liq. Cryst.* **176**, 151

Yamaguchi, K., Toyoda, Y. and Fueno, T. (1989b). *Chem. Phys. Lett.* **159**, 459

Yoshida, Z. and Sugimoto, T. (1988). *Angew. Chem., Int. Ed. Engl.* **27**, 1573

Zhang, J. H., Epstein, A. J., Miller, J. S. and O'Connor, C. J. (1989). *Mol. Cryst. Liq. Cryst.* **176**, 271

Zilm, K. W., Merrill, R. A., Greenberg, M. M. and Berson, J. A. (1987). *J. Am. Chem. Soc.* **109**, 1567

Hydrogen Bonding and Chemical Reactivity

FRANK HIBBERT AND JOHN EMSLEY

Department of Chemistry, King's College, London, UK

1 The hydrogen bond

When hydrogen bonding is invoked in discussions of structure and reactivity by chemists, biochemists and molecular biologists, it is generally assumed that it is a weak interatomic attraction involving energies of *ca.* 20 ± 10 kJ

ADVANCES IN PHYSICAL ORGANIC CHEMISTRY
VOLUME 26   ISBN 0-12-033526-3

mol^{-1}, and that it is an order of magnitude weaker than covalent bonding. While this is often the case, it has become evident in recent years that there are hydrogen bonds which are much stronger. Some are even stronger than covalent bonds. In this review we will seek first to explore the upper limits of hydrogen bonding and then to discuss the role which hydrogen bonding has in the kinetics and mechanism of chemical processes such as proton transfer and enzyme catalysis.

For most purposes it is enough to describe a hydrogen bond $A—H \cdots B$ as an electrostatic attraction between the positive end of the bond dipole of AH, and a centre of negative charge on B. The atom or group A should be sufficiently electronegative to ensure the bond is strongly polar. The site of attraction on B is commonly identified as a lone pair of electrons. These requirements mean that both hydrogen-bond donors, AH, and hydrogen-bond acceptors, B, come chiefly from the same few groups of the periodic table, i.e. 15, 16 and 17. The elements at the head of these groups, nitrogen, oxygen and fluorine, readily form hydrogen bonds. Even carbon at the head of group 14 can participate in hydrogen bonding both as donor or acceptor. The presence of electron-withdrawing groups can make $C—H$ into a hydrogen bond donor, and aromatic π-systems can act as acceptors. Such bonding is weak and is easily explained in simple electrostatic terms.

Strong and very strong hydrogen bonds are generally found with cations or anions, so much so that strength and charge seem necessarily to be linked. "Strong" and "very strong" are terms that will be defined later with respect to the potential energy well of a hydrogen bond (see p. 269 and Fig. 3). Strong hydrogen bonds are especially prevalent between acids and their conjugate bases, $AH \cdots A^-$, or between bases and their conjugated acids, the protonated base, $B^+—H \cdots B$. The former may be hemi-salts or adducts, i.e. intra- or inter-molecularly hydrogen bonded, and the same distinction can be made for $B^+—H \cdots B$ systems.

Ionization greatly affects hydrogen bonding. If the donor is positively charged, $(A—H)^+$, there will be an increased attraction for the electron density associated with the acceptor, B, as well as a strengthening of the bond moment. By the same token, a positively charged acceptor will greatly weaken the hydrogen bonding. Conversely, a negatively charged AH will weaken the hydrogen bonding but a negatively charged acceptor will strengthen it. Proton addition, removal or transfer will likewise produce strong hydrogen-bond donors, acceptors and adducts, respectively. Neutral systems with strong hydrogen bonds are much rarer and for this reason attract interest; later in this review we shall consider the best known example of a neutral strong hydrogen-bonding system, found in the cyclic enol tautomers of β-diketones.

Marked changes occur to AH and B upon formation of a hydrogen bond;

molecular dimensions, energetics, vibrational frequencies and electron distribution are all affected. In additon there are changes which occur when the hydrogen bond is perturbed by the substitution of a deuteron for the proton. These isotopic shifts could provide the key to a deeper understanding of the nature and types of hydrogen bonding, even though their behaviour seems anomalous at first sight. The chief hydrogen bond properties and the techniques used to investigate them are given in Table 1.

Table 1   The parameters of a hydrogen bond, A—H $\cdots$ B.

	Property	Investigative technique
Dimensions	Bond length, $R_{A \cdots B}$ Proton location, R_{A-H} Bond angle, $\angle$ AHB	X-Ray or neutron diffraction
Energetics	Bond energy, $E(AHB)$	Calorimetry; ir Δv_{AH}; ion cyclotron resonance spectrometry; theoretical calculations
Vibrational modes	v_{AH} v_{AH}/v_{AD}	Ir and Raman spectroscopy
Potential energy well	δ_{AHB} $\Delta[\delta(^1H) - \delta(^2H)]$ Fractionation factor, φ	Nmr spectroscopy

As with ordinary covalent bonding, it is the bond length (here $R_{A \cdots B}$) and bond energy [here $E(AHB)$ and defined in equation (1), p. 264] which command most attention. In the early days of hydrogen bonding, the former was rarely discussed, and the latter was the main focus of attention since reasonable estimates of hydrogen-bond energy could be obtained by calorimetric methods or from shifts in the vibrational modes in the infrared (ir). The change in the frequency, Δv_{AH}, of the bond A—H in the donor molecule on hydrogen-bond formation, was thought to vary linearly with the bond enthalpy. Values of $E(AHB)$ obtained in this way were little more than first approximations because of the uncertainty in measuring the ir shift. Today, owing to the availability of X-ray determinations, it is the hydrogen-bond length, $R_{A \cdots B}$, which is seen as the fundamental parameter. This quantity can be measured to within 1%.

HYDROGEN-BOND GEOMETRY

Since there are three nuclei involved in a hydrogen bond, A—H $\cdots$ B, there are three bond distances of interest: $R_{A \cdots B}$, R_{A-H} and $R_{H \cdots B}$, and one angle

$\angle$ AHB. All but the first of these parameters involve the proton itself, but this is not the easiest of entities to locate. The heavier atoms A and B can be located with certainty and for this reason $R_{A\cdots B}$ is the dimension commonly referred to as the hydrogen-bond length. Relationships between hydrogen-bond parameters $R_{A\cdots B}$, R_{A-H}, $R_{H\cdots B}$ and $\angle$ AHB have been the subject of speculation since the 1950s, when reliable estimates of the location of the proton were first obtained from neutron-diffraction studies. The most notable relationship is the observed decrease in $R_{A\cdots B}$ as R_{A-H} increases (Nakamoto *et al.*, 1955; Iwasaki *et al.*, 1967; Hamilton and Ibers, 1968; Pimentel and McClellan, 1971). Correlations between these various parameters were examined in detail by Olovsson and Jönsson (1976). A plot of $R_{O\cdots O}$ *vs* R_{O-H} (Fig. 1) shows three regions (Ichikawa, 1978a). First, there are the longer and weaker bonds in which $R_{O\cdots O} > 260$ pm. In these, the proton stays close to its parent atom and R_{O-H} remains constant at *ca.* 95 pm. Secondly, there are the shorter, stronger bonds with $R_{O\cdots O}$ in the range 245–260 pm. Here the tendency is for the proton to move towards the acceptor oxygen as the bond shortens. Finally, there are the very short, very strong hydrogen bonds with $R_{O\cdots O} < 245$ pm in which the proton is centred ($R_{O-H} \simeq R_{O\cdots H}$) giving a linear relationship of $R_{O\cdots O}$ *vs* R_{O-H} with a positive slope.

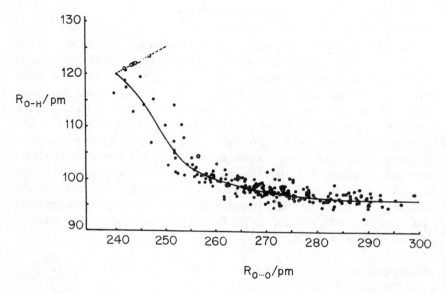

Fig. 1 Variation of $R_{O\cdots O}$ and R_{O-H} for hydrogen bonds (adapted from Ichikawa, 1978a).

Pinpointing the exact position of an atom by X-ray diffraction depends on the electron density around the atom, and so will be least easy for hydrogen. This is especially true for a hydrogen bond, because the electron density around the proton will be even more depleted. In earlier X-ray communications the position of hydrogen was not reported, but with modern techniques locating the proton is less of a problem. However, even greater accuracy is possible with neutron diffraction. Because, with this technique the exact location of the proton can be found, neutron-diffraction studies have been invaluable in settling the debate as to whether there can be a completely symmetrical and linear hydrogen bond, with the proton at the mid-point between the heavier atoms.

The deviation from linearity of a hydrogen bond could be expected to provide qualitative information about the nature of the interaction. A hydrogen bond involving a purely electrostatic interaction would be expected to be linear. If the proton is attracted towards a lone pair on the acceptor molecule, non-linear hydrogen bonds to oxygen as the acceptor would be common. However, because hydrogen bonds are relatively easy to deform compared to other bonds in crystal lattices, non-linear hydrogen bonds are generally found and the value of $\angle$ AHB as a parameter for drawing conclusions about the nature of the hydrogen bond remains questionable.

Hydrogen bonds between the elements N, O and F span the range $R_{A \cdots B} = 225$–310 pm, and towards the longer end it becomes a moot point whether a hydrogen bond is indeed present. The most reliable criterion of hydrogen-bond formation is that based on van der Waals radii. If the hydrogen-bond length $R_{A \cdots B}$ is less than the sum of the van der Waals radii of these atoms, $\Sigma_w(A + B)$, then it seems reasonable to assume that a hydrogen bond is responsible for this close proximity. Indeed, in the early days of X-ray diffraction when protons were rarely located with accuracy, this criterion was introduced as a way of identifying hydrogen bonds, and it has held good ever since.

The van der Waals radii of the atoms which most commonly engage in hydrogen bonding are N = 155, O = 152 and F = 147 pm (Bondi, 1964). Other atoms which form hydrogen bonds have the following radii: C = 170; S = 180; Cl = 175; Br = 185 and I = 198 pm. The van der Waals radius of hydrogen is uncertain, ranging between 100 and 120 pm. (This is invoked only for very weak hydrogen bonds when the dimension $R_{H \cdots B}$ is sometimes used as a fall-back criterion for hydrogen bonding where $R_{A \cdots B} < \Sigma_w(A + B)$ fails to meet the test.) If $R_{H \cdots B} < \Sigma_w(H + B)$, then a hydrogen bond is assumed to be present.

The difluoride ion, bond length $R_{F \cdots F} = 226$ pm, represents the interpenetration of van der Waals radii to the extent of reducing the distance between the two fluorines by 68 pm. Such a decrease for OHO hydrogen bonds would fix a lower limit of 236 pm, but some are shorter than this. For NHO bonds the corresponding lower limit would be 239 pm and for NHN bonds, 242 pm. In these hydrogen bonds the shortest distances reported to date are 246.5 pm for the NHO bond in [1] (Camilleri et al., 1989) and 247.4 pm for the NHN bond in [2] (Alder et al., 1983a). In [1] the structure itself guarantees a short hydrogen bond if the molecule adopts a planar arrangement to maximize delocalization.

The inside-protonated diamine [5.4.2] cage, [2], is only one of several such compounds which have been made by Alder et al. In this cage $R_{N \cdots N}$ is only 247.4 pm (White et al., 1988e), but in this case the hydrogen bond is not linear and $\angle NHN = 132°$, suggesting that the hydrogen bond is again short due to structural factors and is accommodating itself to an $N \cdots N$ distance which is determined by the geometry of the molecule.

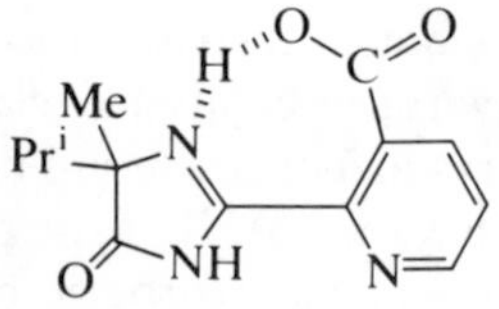

[1] 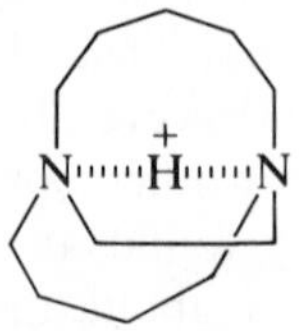   [2]

Oxygen-to-oxygen hydrogen bonds are the most common and a large number have been uncovered in crystal-structure determinations. Theoretical calculations suggest that the shape of a potential energy well depends only on the distance apart of atoms A and B, and, for $R_{O \cdots O}$, calculations showed that at 240 pm the double minimum potential energy well would become a single minimum (Peinel, 1979). This bond length has taken on a special significance since a radical change to the nature of OHO hydrogen bonding occurs for bonds with length 245–240 pm. Several reviewers have compiled lists of very short OHO bonds. In one such review (Olovsson and Jönsson, 1976) the lower limit of $R_{O \cdots O}$ was thought to be around 240 pm, and 17 compounds having bonds in the range 247.6–239.8 pm were listed.

More recently, a compilation of neutron-diffraction data gave 23 compounds with hydrogen-bond lengths in the range 250–239.1 pm (Joswig et al., 1982). A notable feature of this collection was that in only three of the 23 bonds was hydrogen centred, and these were among the shortest bonds, with $R_{O \cdots O}$ less than 240 pm. Since then, other short OHO bonds have been found, such as that of [(DMSO)$_2$H$^+$], 240.5 pm (Cartwright et al., 1988), [(PYO)$_2$H$^+$], 240.6 pm, where PYO is pyridine N-oxide (Hussain and Al-Hamoud, 1985), and [Ni(BnAOH)]I·H$_2$O [3], 241.7 pm, where BnAOH is 3,3'-(1,4-butanediylamino)bis(3-methyl-2-butanone) dioxime (Pal et al.,

1986). Another anion pair held together by a short hydrogen bond is $[H\{VO(O_2)_2(bipy)\}_2]^-$ [4] in which two peroxo complex anions are linked, with $R_{O\cdots O} = 245.6$ pm (Szentivanyi and Stomberg, 1984). An example of an uncharged system with a very short hydrogen bond is that of 2-(N,N-diethylamino-N-oxymethyl)-4,6-dichlorophenol [5], with $R_{O\cdots O} = 240$ pm (Koll $et~al.$, 1986).

[3]

[4]

[5]

Table 2 Very short OHO hydrogen bonds.

Compound	$R_{O\cdots O}$ pm	H[a]	Method[b]	Reference
2,3-Pyridinedicarboxylic acid	239.8	nc	N	c
Hydrogen cis-diacetyl	239	no	X	d
tetracarbonylrhenate[e]	239.8	nc	N	f
Tetrafluoroboric acid				
–methanol (1/2)	239.4	—	X	g
D-Quinolinic acid	239.3	nc	N	h
Imidazole hydrogen maleate	239.3	c	N	i
Copper hydrogen phthalate	239.1	nc	N	j
Lithium phthalate hydrate (1/1)	239.0	$ca.$ c	N	k
$[Ni(C_{13}H_{21}N_5O_4]^l$	239.0	c	N	m
Cobalt(2-aminoethanol) complex[n]	239.0	c	N	o
Lithium phthalate–methanol	238.8	c	N	k
cis-Dichloro(hydrogendisulphito)-				
platinate(II)	238.2	c	X	p
Lithium hydrogen phthalate	236.6	nc	N	q
Copper(thiourea)$_3$ hydrogen				
phthalate	235.1	nc	X	r
[HOHOH]$^-$	229	c	X	s

[a] Location of the hydrogen bonding proton: c = centred, nc = non-centred, $ca.$ c = almost centred; no = not observed; [b] X = X-ray methods, N = neutron methods; [c] Kvick $et~al.$ (1974); [d] Lukehart and Zeile (1976); [e] See text [6], [f] Schultz $et~al.$ (1984); [g] Mootz and Steffen (1981); [h] Takusagawa and Koetzle (1979); [i] Hsu and Schlemper (1980); [j] Bartl and Küppers (1980); [k] Küppers $et~al.$ (1981); [l] Nickel amine complex; [m] Hussain $et~al.$ (1981). [n] In this complex there are two short hydrogen bonds, the shorter one (239.0 pm) is centred, and the longer one (242.9 pm) is almost centred ($R_{O-H} = 120.4$ and 122.8 pm), see text [7]. [o] Jones $et~al.$ (1986a); [p] Kehr $et~al.$ (1980); [q] Bartle and Küppers (1978); [r] Biagini-Cingi $et~al.$ (1977); [s] Abu-Dari $et~al.$ (1979).

Most OHO bonds that are described as short are in the range 240–245 pm, but several are shorter (see Table 2). These are invariably found in charged systems. The shortest is that of the hydroxide-hydrate, $[HOHOH]^-$, which at 229 pm is almost as short as that of the bifluoride ion. The $[HOHOH]^-$ species, which is isoelectronic with $[FHF]^-$, turned up as the counter ion of a chromium complex (Abu-Dari et al., 1979). The bond length is exceptional since, apart from $[HOHOH]^-$, the shortest OHO bond is one of 231 pm between N,N'-bis(2-hydroxyethyl)-2,4-pentanediimine ligands attached to copper in the Cu(II) complex (Bertrand et al., 1976). This structure determination, however, has been questioned (Jones et al., 1986a). When hydrogen bonding involves ligands to metals, then very short bonds can result such as those in [6] and [7] with lengths of 239.8 pm and 239.0 pm, respectively (Table 2).

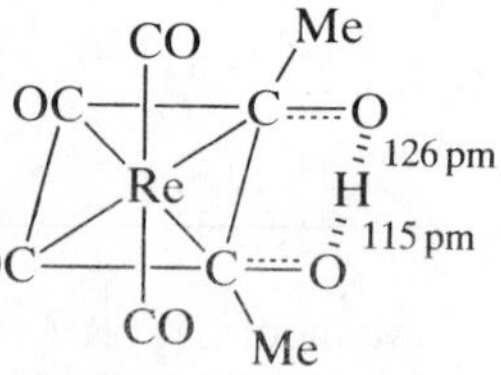

$N \frown O = NH_2CH_2CH_2OH$

[6]

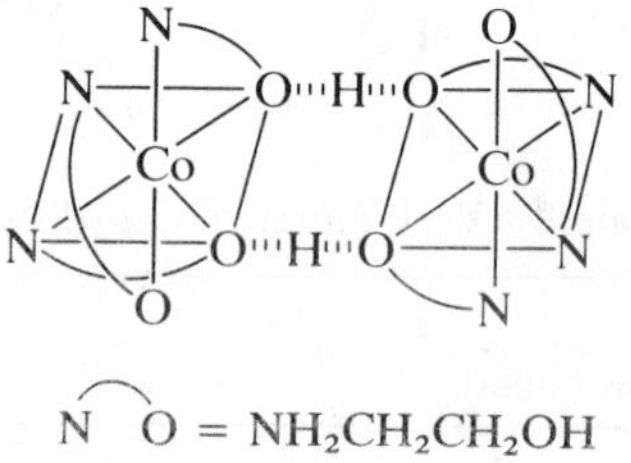

[7]

If a hydrogen bond is found by neutron diffraction to be centred, then we can reasonably conclude that it is indeed strong and the proton is above the central energy barrier. The longest OHO bond for which a centred proton has been observed appears to be potassium hydrogen malonate with $R_{O \cdots O} = 246.8$ pm, which has the proton located exactly midway between the oxygens (Schuster, 1976). It thus seems likely that in other OHO bonds shorter than this, the proton will be above the energy barrier, yet it can still be unsymmetrically placed, as some of the compounds in Table 2 with shorter $R_{O \cdots O}$ values show.

Many crystal structures have been carried out on bicarboxylates, and these can be divided into those which exhibit crystallographically symmetrical hydrogen bonds and those which are unsymmetrical (Speakman, 1972). An analysis of $R_{O \cdots O}$ values reveals that symmetrical hydrogen bonds are found within the range 251–243 pm, and unsymmetrical ones are within the range 257–244 pm. These ranges show considerable overlap, and while symmetry and lack of symmetry in this context does not *prove* the proton to be centred or non-centred, the implication is that the forces which are determining the symmetry or lack of symmetry of the rest of the system will also govern the position of the proton within the hydrogen bond.

To what extent the value of $R_{O \cdots O}$ and the location of the proton are determined by forces external to the hydrogen bond has been investigated by Misaki *et al.* They have studied the anions [(4-Me-$C_6H_4CO_2$)$_2$H$^-$] and [(4-Br-$C_6H_4CO_2$)$_2$H$^-$] in lattices with a variety of counter cations and observed a range of $R_{O \cdots O}$ values from 244.5 to 253.8 pm (Misaki *et al.*, 1986, 1989a,b). The hydrogen bond angle, $\angle$ OHO, was also found to vary and these researchers conclude that the lattice is responsible for the variation in length and symmetry of the hydrogen bond for [(4-Me-$C_6H_4CO_2$)$_2$H$^-$].

In 1,2,3-benzotriazolium dihydrogenphosphate, [$C_6H_4N_3H_2{}^+$][$H_2PO_4{}^-$], the lengths of the hydrogen bonds between oxygens of the phosphate ions are 245.1 pm (Emsley *et al.*, 1988c). This compound has chains of dihydrogen phosphates running through the lattice, linked by short hydrogen bonds. The sensitivity of the hydrogen bonding to crystal forces was revealed by the contrasting structure of the 1,3-benzimidazolium salt [$C_7H_5N_2H_2{}^+$]-[$H_2PO_4{}^-$]. The difference between the cations is small and does not disturb the basic structure, but its effect on the hydrogen bonding is profound: the dihydrogen phosphate anion chains are now linked by hydrogen bonds that are much weaker, the shortest being 260.8 pm.

Deuteration of a bond involving hydrogen is one of the most useful methods of probing a molecule or system.[1] However, on changing H for D in a hydrogen bond there may be unexpected consequences. On deuteration, hydrogen bonds nearly always lengthen, something that was first established by Ubbelohde and coworkers (Robertson and Ubbelohde, 1939; Ubbelohde, 1939; Ubbelohde and Woodward, 1942) but was still being debated 40 years later (Ichikawa, 1978a).

In 1969, calculations were published showing that deuteration should result in a longer bond if the system was a double minimum potential energy well, but in a shorter bond if it was a single minimum (Singh and Wood, 1969). However, the differences are often so small as to fall within the estimated standard deviations of the bond lengths. For many hydrogen bonds that are classed as strong, i.e. $R_{A \cdots B} < 245$ pm, we find $\Delta R(D - H)$ [defined as $R_{A \cdots B}(D) - R_{A \cdots B}(H)$] is indeed zero. Examples include HF$_2{}^-$ (McGaw and Ibers, 1963), KH(CF$_3$CO$_2$)$_2$ (Macdonald *et al.*, 1972) and KF(CH$_2$CO$_2$H)$_2$ (Emsley *et al.*, 1981). For some strong hydrogen bonds, however, $\Delta R(D-H)$ is positive, for example, $\Delta R(D-H) = 12$ pm for H$_5$O$_2{}^+$ (Brunton and Johnson, 1975) and $\Delta R(D-H) = 19$ pm for [Ni(bambo)$_2$]$^+$ [8] (Hsu *et al.*, 1980). In some cases of strong hydrogen bonds, $\Delta R(D - H)$ is even negative; for example, for the pyridine-2,3-dicarboxylic acid zwitterion [9] $\Delta R(D-H) = -5$ pm (Takusagawa and Koetzle, 1979). This anomaly in $\Delta R(D-H)$ is typical of the behaviour of

---

[1] Calculations have been extended to muonium bonding. For BrXBr and IXI, where X can be H, D or a muonium "atom" (which has a mass of 0.114 au and a lifetime of 2.2×10^{-1} s) the order of bond energies is BrDBr < BrHBr < BrMuBr (Clary and Connor, 1984).

other properties which appear to change in an irregular way when strong hydrogen bonds are deuterated, as we shall see.

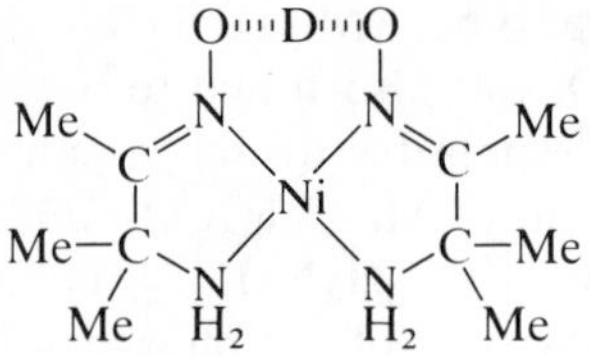

[8] 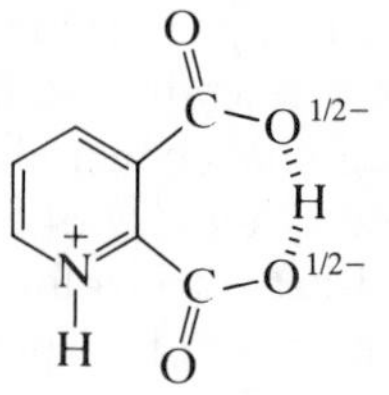   [9]

HYDROGEN-BOND ENERGY

Ideally the hydrogen-bond energy is defined with respect to isolated species (1) but obtaining this information directly has been possible for only a few gaseous systems.

$$AH(g) + B(g) = A—H \cdots B(g) \tag{1}$$

Detailed microwave spectroscopic analysis of gas-phase hydrogen-bonded adducts can yield the hydrogen-bond energy with precision, e.g. for $HCN \cdots HF$, $E(NHF) = 26.1 \pm 1.6 \, kJ \, mol^{-1}$ (Legon et al., 1980). Sadly, few systems are susceptible to this approach.

When a molecule AH forms a hydrogen bond there is a profound change in its vibrational modes, notably in the stretching vibration of the donor bond, $v(AH)$, which decreases in frequency. The bending modes δ and γ, which are vibrations in the plane and out of the plane of the system, increase in frequency. In some cases the shift Δv_{AH} can be hundreds of wavenumbers, and this has been related to the energy of the hydrogen bond. Spectra showing such changes are typical of strong hydrogen bonding and were first identified as a distinct class by Hadzi who termed them "type ii" spectra (Hadzi, 1965). Speakman referred to them as "type A" spectra (Speakman and Currie, 1970) and noted further that they are often found with hemisalts in which the hydrogen-bonding protons appear to sit at centres of inversion in the crystal or on mirror planes.

Another notable feature of systems in which hydrogen bonding is important is the broadness and increased intensity of $v(AH)$, sometimes giving continua over whole regions of the spectrum. Studies of the ir spectra of hydrogen bonds are legion and excellent reviews are available of the earlier work (Novak, 1974; Hadzi and Bratos, 1976), including the far ir (Rothschild, 1976). Extensive analysis of the ir spectra of hydrogen bonds in which there are proton-transfer equilibria, $A—H \cdots B \rightleftharpoons A^- \cdots H—B^+$, have been carried out (Zundel and Fritsch, 1986; Zundel and Eckert, 1989). These studies have been extended, recently, to biological systems (Zundel, 1988; Eckert and Zundel, 1988). For the purposes of this review, however, we will concentrate only on $v(AH)$ insofar as it provides information about the strength of the hydrogen bond.

Much of the early work on determining $E(AHB)$ was devoted to systems we now regard as weak. The methods used were chiefly calorimetric or based on ir shifts. The thermodynamic approach referred to solutions in solvents such as cyclohexane or carbon tetrachloride that were thought not to participate in hydrogen bonding. The ir methods were based on the empirical connection (first proposed by Badger and Bauer, 1937) between thermodynamic bond energies, $E(AHB)$, and shifts of the A—H stretching frequency, Δv_{AH}. Earlier texts devoted much space to hydrogen-bond energies produced in these ways (Pimentel and McClellan, 1960; Vinogradov and Linnell, 1971). *Ab initio* methods were also used to calculate the hydrogen-bond energies of the simplest hydrogen bonds, and extended to more complex adducts using model systems (e.g. see Joesten and Schaad, 1974).

For gas-phase ir studies, it has been possible to show good correlation between Δv_{AH} and $E(AHB)$ (Millen *et al.*, 1979). Even with condensed phases, this method is still fairly reliable, e.g. with alcohols (Piaggio *et al.*, 1983), and $\Delta v–\Delta H$ correlations have recently been shown to hold for hydrogen-bonded ions (McMahon and Larson, 1984b).

Most attempts to relate v_{AH} and $R_{A\cdots B}$ have concentrated on OHO hydrogen bonds. There is often a good correlation between these parameters, provided the data is limited to similar types of compound. A recent analysis of data from solid hydrates, involving 250 $O—D(H) \cdots X$ hydrogen bonds, has shown not only good agreement for $HOH \cdots O$ systems, but also when the acceptor atom X is N, S, Se, Cl, Br or I (Mikenda, 1986). This work also notes that v_{OH}/v_{OD} decreases as v_{OH} decreases, a point that will be dealt with below under hydrogen-bond properties (p. 279). However, since few water molecules in solid hydrates engage in strong hydrogen bonding the validity of ΔH *vs* v_{AH} or $R_{A\cdots B}$ *vs* Δv_{AH} over the full range of hydrogen-bond lengths could not be tested.

Strong and very strong hydrogen bonds, which mainly involve molecule–ion interactions were not accessible until the advent of ion cyclotron resonance spectroscopy (icr). The first type of hydrogen bonding to be investigated by icr was that between the fluoride ion and carboxylic acids. These studies showed the hydrogen bond to be very strong (Clair and McMahon, 1979), confirming *ab initio* calculations (Emsley *et al.*, 1977). Very strong $RCO_2H \cdots F^-$ hydrogen bonding was used to explain the curious properties of solutions of metal fluorides in carboxylic acids (Emsley, 1971; Emsley and Clark, 1974; Emsley and Hoyte, 1976). The triumph of the icr approach was its conclusive determination of the hydrogen-bond energy of the bifluoride ion at $163 \pm 4\,\text{kJ}\,\text{mol}^{-1}$ (McMahon and Larson, 1982a). Subsequent icr work led to a large range of bond-energy values for both anionic and cationic species (McMahon and Larson, 1982b, 1983, 1984a,b), several of which are given in Table 3.

Table 3 Hydrogen bond energies of ionic species determined by McMahon *et al.*

Species	Method[a]	Hydrogen bond energy/kJ mol^{-1}	Reference
$F^- \cdots H—A$ *and*			
$FH \cdots B^-$			
$F^- \cdots HF$	icr	163	b
$F^- \cdots HO_2CH$	icr	79	c
$F^- \cdots HO_2CMe$	icr	88	c
$F^- \cdots HOMe$	icr	124	c
$F^- \cdots HOEt$	icr	132	c
$F^- \cdots HOPh$	icr	82	c
$F^- \cdots HOH$	icr	97	c
$F^- \cdots H_2NPh$	icr	113	c
$F^- \cdots H_2C{=}CH_2$	ai	10	d
$FH \cdots Cl^-$	icr/ir	91	e
$FH \cdots Br^-$	icr/ir	71	e
$FH \cdots I^-$	icr/ir	63	e
$FH \cdots CN^-$	icr/ir	88	e
$B \cdots H—B^+$			
$H_5O_2{}^+$	icr	132	f
$(MeOH)_2H^+$	icr	131	f
$(Me_2O)_2H^+$	icr	128	f
$(Et_2O)_2H^+$	icr	131	f
$(Me_2CO)_2H^+$	icr	134	f
$(MeCO_2)_2H^+$	icr	124	f
Others			
$H_3F_2{}^+$	ms	135	g
$ClHCl^-$	icr/ir	97	e
$BrHBr^-$	icr/ir	84	e

[a] icr, ion cyclotron resonance; *ai, ab initio*; icr/ir, based on Δv; ms, by analogy with model system; [b] McMahon and Larson (1982a); [c] McMahon and Larson (1983); [d] McMahon and Roy (1985); [e] McMahon and Larson (1984b); [f] McMahon and Larson (1982b); [g] McMahon and Kebarle (1986).

It is taken for granted that as a hydrogen bond becomes shorter it becomes stronger. Whether these two properties behave monotonically has rarely been tested because there is as yet no way of adjusting a particular bond and measuring the change in energy. For OHO hydrogen bonds the best we can do is to compare $R_{O \cdots O}$ and $E(OHO)$ for a series of compounds. The relatively few systems where both pieces of information are known are given in Table 4 and Fig. 2. In some cases, doubt surrounds the energy values given, such as that for pentane-2,4-dione which will be discussed later in this review. Figure 2 reveals the change that occurs to OHO hydrogen bonds around 245 pm.

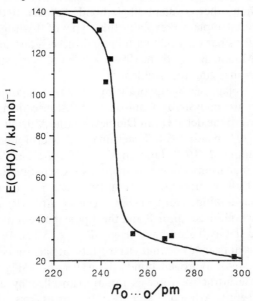

Fig. 2 Hydrogen bond energy, $E(OHO)$, vs hydrogen-bond length, $R_{O \cdots O}$

Table 4 The relationship between hydrogen-bond length, $R_{O \cdots O}$/pm, and hydrogen bond energy, $E(OHO)$/kJ mol^{-1} for O—H $\cdots$ O hydrogen bonds.

Compound	$R_{O \cdots O}$/pm	$E(OHO)$/kJ mol^{-1}	Reference
$(H_2O)_2$	297.6	21.7	[a,b]
$(MeOH)_2$	270	32	[c]
$(CH_3CO_2H)_2$	267	30.5	[c]
Pentane-2,4-dione, enol	253.5	32.76	[d,e]
$[(HCO_2)_2]^-$	244.7	135	[f,g]
3-(4'-Biphenyl)pentane-2,4-dione	244.1	117	[h]
$[(DMSO)_2H]^+$	242.0	106	[i,j]
$[(MeOH)_2H]^+$	239.4	131	[k,l]
$[HOHOH]^-$	229	135	[m,b]

[a] Odutola and Dyke (1980); [b] Joesten and Schaad (1974); [c] Vinogradov and Linnell (1971); [d] Camerman et al. (1983); [e] Buemi and Gandolfo (1989); [f] Larsson and Nahringbauer (1968); [g] Emsley et al. (1978); [h] Emsley et al. (1988b); [i] James and Morris (1980); [j] Bernander and Olofsson (1972); [k] Mootz and Steffen (1981); [l] McMahon and Larson (1982b); [m] Abu-Dari et al. (1979).

It comes as a surprise to some chemists to discover that there are hydrogen bonds stronger than covalent bonds. These so-called very strong hydrogen bonds have energies in excess of $100 \, kJ \, mol^{-1}$, and are more common than usually supposed (Emsley, 1980). The best known example is the bifluoride ion, HF_2^-, which has the proton centred between the two fluorine atoms almost as if it were divalent, with $R_{F \cdots H} = 113$ pm. This is not much longer than other covalent single bonds between hydrogen and first row elements. Attempts to explain this type of hydrogen bonding in conventional terms has only met with limited success, and terms like "semi-covalent" (McMahon and Larson, 1984c), while not being strictly definable, are useful in emphasizing the problem.

Attempts to explain the energy of the hydrogen bond go back many years to the first serious analysis by Coulson (1957). Since then, several others have contributed to the debate (van Duijneveldt and Murrell, 1967; Kollman and Allen, 1972; Schaad, 1974; Umeyama and Morokuma, 1977; Moro-kuma, 1977; Smit et al., 1979; Janoschek, 1982; Olovsson, 1982; Spackman, 1986). Some explanations invoke as many as six separate contributions to the energy. Not all theories are so complex. Allen proposed a simple but effective formula linking hydrogen-bond energy $E(AHB)$ directly to the ionization energy of the acceptor B and the dipole moment of the donor AH, and inversely to the hydrogen-bond length $R_{A \cdots B}$ (Allen, 1975).

The consensus appears to be that electrostatic attraction between AH and B accounts for weak hydrogen bonds. However, as AH and B approach more closely, the attractive force is rapidly cancelled by electron energy exchange (repulsion). Attractive second order forces must then come into play, namely polarization (i.e. induced dipole in B) and delocalization (charge transfer), and these provide the extra energy for strong and very strong hydrogen bonds. Even dispersion forces (induced dipole in B reinforcing the dipole in AH) may contribute significantly in the shortest of hydrogen bonds to counteract the electron-exchange repulsion.

HYDROGEN-BOND PROPERTIES

It has become clear in the past decade that strong hydrogen bonding has associated with it several characteristic properties. In particular, as hydrogen-bond strength changes, maxima or minima are observed in nmr chemical shifts, the isotope effect on the chemical shift $\Delta[\delta(^1H) - \delta(^2H)]$ defined on p. 271, ir ν_{AH}/ν_{AD} band ratios, and in the isotope-fractionation factor, φ.

The basic differences between hydrogen and deuterium bonding have been much debated (Buckingham and Fan-Chen, 1981). It is possible to define three kinds of hydrogen bond based on the relative positions of H and D, and these are shown schematically in Fig. 3.

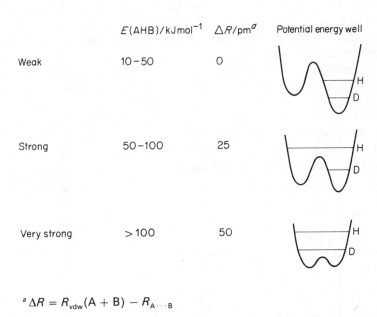

$$^a \Delta R = R_{vdw}(A + B) - R_{A \cdots B}$$

Fig. 3   The three types of hydrogen bond A—H $\cdots$ B.

For a weak hydrogen bond the potential energy barrier between the minima will be high and the proton is thus confined to the well of the parent atom to which it is covalently bonded. In the same well, the zero-point energy of the deuteron will be lower because of its increased mass, but this is often assumed to have no effect on the bond length. At the other extreme, in very strong hydrogen bonds, both H and D have zero-point energies above the barrier. Intermediate between these extremes is strong hydrogen bonding in which the zero-point energy of H may be above the central barrier with D confined to its original well. In the weak and very strong classes, the difference between D and H merely derives from the fact that the former, by virtue of its mass, sits lower in the same potential energy well.

There are cases which appear to show the difference in location of the H and D atoms quite conclusively. In the complex $[Ni(bambo)_2]^+$ [8], already referred to as one in which there is a slight increase in $R_{O \cdots O}$ on deuteration, the location of the proton is nearly centred, whereas the deuteron is asymmetrically placed. For the proton the $R_{O \cdots H}$ distances are 118.7 and

124.2 pm (Schlemper *et al.*, 1971), suggesting that it is above the internal energy barrier, while for the deuteron the $R_{O...D}$ distances are 105.8 and 139.1 pm (Hsu *et al.*, 1980), consistent with its being confined to one potential energy well. However, the issue is clouded by a change in crystal structure which occurs on deuteration. Other molecules which show marked differences in the locations of H and D in their hydrogen bonds are quinolinic acid (Takusagawa and Koetzle, 1979) and imidazolium hydrogen maleate (Hussain *et al.*, 1980).

The manner in which four characteristic properties of hydrogen bonds vary are shown diagrammatically in Fig. 4(a–d). In every case the lower limit of the hydrogen-bond length, $R_{A...B}$, is fixed by the very strong hydrogen bond of the bifluoride ion with length 226 pm, and in every case this has a value for the particular property that is more typical of a weak hydrogen bond. In each plot the property varies through a minimum (v_{AH}/v_{AD} and φ) or a maximum $\{\Delta[\delta(^1H) - \delta(^2H)]$ and $\delta\}$.

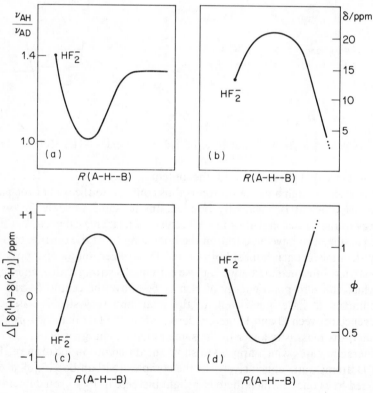

Fig. 4 The "anomalous" properties: (a) v_{AH}/v_{AD}; (b) δ_{AH}; (c) $\Delta[\delta(^1H) - \delta(^2H)]$; (d) φ.

Nmr chemical shift and the isotope effect, $\Delta[\delta(^1H) - \delta(^2H)]$

Recently, solid state nmr spectroscopy has been used to probe hydrogen-bonded materials and first reports showed that there was a direct relationship between $R_{O\cdots O}$ and σ, the anisotropic chemical shift: as bonds become shorter the chemical shift moves downfield. Twenty-four compounds were examined and some, such as potassium hydrogen malonate, had a σ-value below 20 ppm (Berglund and Vaughan 1980).

A linear correlation between $R_{H\cdots O}$, obtained from X-ray and neutron-diffraction data, and $\sigma_\parallel$ has been observed (Jeffrey and Yeon, 1986). The survey was based on 20 compounds having OHO hydrogen bonds with $R_{H\cdots O}$ varying from 123.4 pm (KH malonate) to 197.9 pm (α-form of oxalic acid dihydrate). Whether based solely on those crystals for which neutron data are available, or on X-ray and neutron data, a good linear relationship is revealed, showing that the lower $R_{H\cdots O}$ is, the more deshielded is the proton. However, the perpendicular shielding component $\sigma_\perp$ has a much stronger dependence on $R_{O\cdots O}$, and this underlies the σ-dependence through the relationship $\sigma = (2/3)\sigma_\perp + (1/3)\sigma_\parallel$. The parallel shielding component $\sigma_\parallel$ lacks correlation with $R_{O\cdots O}$ (Rohlfing *et al.*, 1983).

Unfortunately, the range of compounds for which data are available does not span the complete range of $R_{O\cdots O}$ and none is less than 243.7 pm (KH malonate), which is at the longer end of the range of very strong hydrogen bonds discussed above. Were $R_{O\cdots O}$ to decrease beyond this, then σ-values much lower than 20 ppm would be forecast for some systems. Surprisingly, the strongest of all hydrogen bonds, HF_2^-, in which $R_{F\cdots F}$ is 226 pm, has a chemical shift for the proton at *ca.* 16, as we shall see.

Following the first observation (Chan *et al.*, 1970) of a substantial difference of chemical shift in the ^{1}H- and ^{2}H-nmr spectrum of hydrogen-bonded protons and deuterons, the measurement of $\Delta[\delta(^1H) - \delta(^2H)]$ has become a useful tool to identify the hydrogen-bond type. The first theory to explain the anomalous values of $\Delta[\delta(^1H) - \delta(^2H)]$ appeared somewhat later (Gunnarsson *et al.*, 1976; Altman *et al.*, 1978). According to this theory, near-zero, positive and negative values of $\Delta[\delta(^1H) - \delta(^2H)]$ could be related to the three categories of hydrogen bond—weak, strong and very strong. Data for the chemical shift $\delta(^1H)$ and isotope effect on the shift of hydrogen-bonded species are given in Table 5. The data in Table 5 were mostly obtained from studies in organic solvents. Values of the isotope effect vary from large and positive ($+0.92$ ppm) to large and negative (-0.7 ppm). For non-hydrogen-bonded species, the values are usually within ±0.05 ppm of zero (Evans, 1982).

The deshielding effect of the two nearby nuclei in the hydrogen bond A—H $\cdots$ B is expected to cause the ^{1}H-nmr signal to occur further down-

Table 5 Isotope effects on the chemical shift of hydrogen-bonded protons.

Hydrogen bond	$\delta(^1H)$ /ppm^a	Solvent	$\Delta[\delta(^1H)-\delta(^2H)]$ /ppm	Reference

Intramolecular hydrogen bonds in diketones

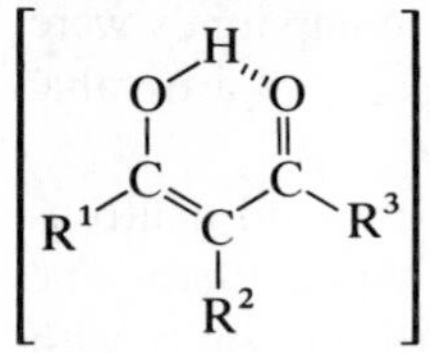

$$\left[\begin{array}{c} \underset{\displaystyle R^1-\overset{\displaystyle O}{\underset{\displaystyle \underset{R^2}{|}}{C}}{\overset{\displaystyle H}{\cdots}}\overset{\displaystyle O}{\underset{\displaystyle C}{\parallel}}-R^3} {} \end{array}\right]$$

R^1	R^2	R^3				
CH_3	H	CH_3	15.58	Neat	+0.50	b
				Neat	+0.62	c
			16.11	C_6H_6	+0.61	d
			13.64^e	C_6H_{12}	+0.58	f
			15.8	CCl_4	+0.58	g
CH_3	H	Ph	16.59	C_6H_6	+0.67	d
			14.81^h	CCl_4	+0.42	f
			16.23	$CDCl_3$	+0.64	b
Ph	H	Ph	17.61	C_6H_6	+0.72	d
			15.52^h	CCl_4	+0.45	f
H	H	H	13.99		+0.42	d
Bu^t	H	Bu^t	16.37	Neat	+0.59	b
CF_3	H	CF_3	13.1	Neat	+0.30	i
CH_3	H	CO_2CH_3	14.89	CCl_4	+0.61	b
Ph	H	CO_2CH_3	14.19	CCl_4	+0.60	b
CH_3	CH_3	CH_3	15.02^h	C_6H_{12}	+0.45	f
CH_3	Et	CH_3	16.6	Neat	+0.66	j
CH_3	Pr^i	CH_3	16.75	Neat	+0.66	j
CH_3	Pr^n	CH_3	16.75	Neat	+0.68	j
CH_3	CO_2Et	CH_3	18.01	Neat	+0.64	b
CH_3	4-MeOC$_6$H$_4$	CH_3	16.65	Neat	+0.31	k

Other O—H $\cdots$ O intramolecular hydrogen bonds

	$\delta(^1H)$ /ppm	Solvent	Δ /ppm	Reference
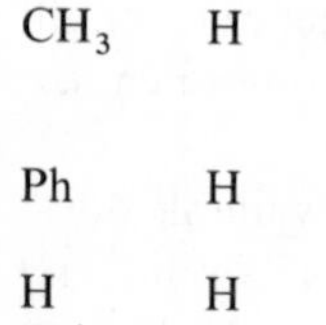 (o-hydroxybenzaldehyde structure)	11.0	CH_2Cl_2	+0.06	i
	11.03	$CDCl_3$	−0.03	b
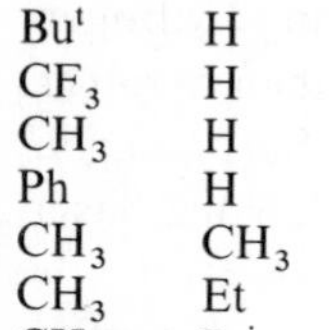 (2-hydroxyacetophenone structure)	12.2	CH_2Cl_2	+0.10	i

Table 5 Cont.

Hydrogen bond	$\delta(^1H)$ /ppm[a]	Solvent	$\Delta[\delta(^1H) - \delta(^2H)]$ /ppm	Reference
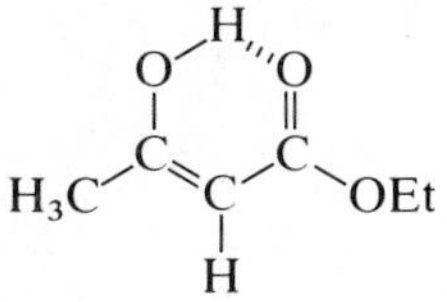	12.14 10.64[h]	Neat CCl$_4$	+0.10 −0.02	b f
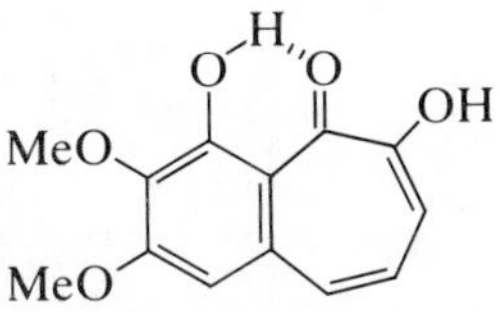	14.75	CDCl$_3$	+0.10	b
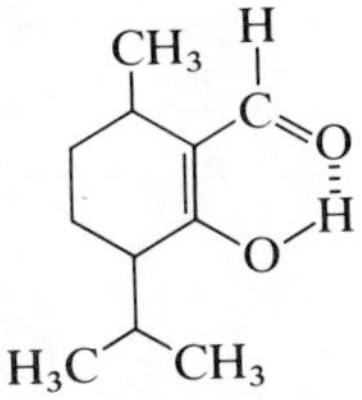	14.92	Neat	+0.44	b
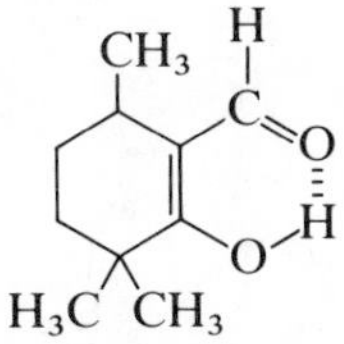	15.55	Neat	+0.60	b
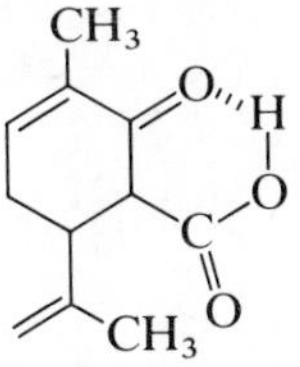	14.61	Neat	+0.60	b
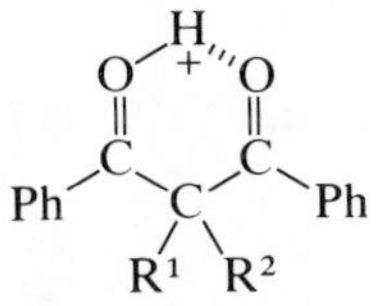				
R^1 = H, R^2 = CH$_3$	21.44	HBr/ CF$_2$Br$_2$	−0.26	l
R^1 = R^2 = CH$_3$	14.1	HBr/ CF$_2$Br$_2$	−0.7 ± 0.4	m

Table 5 Cont.

Hydrogen bond	$\delta(^1H)$ /ppm[a]	Solvent	$\Delta[\delta(^1H) - \delta(^2H)]$ /ppm	Reference
Intramolecular hydrogen bonds in carboxylates				
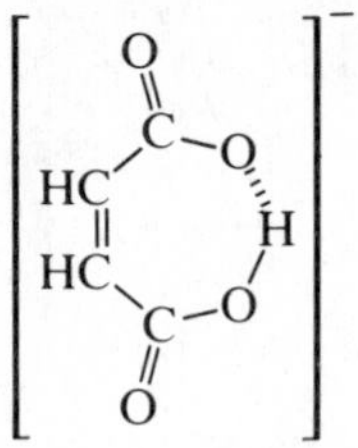	20.5	CH_2Cl_2	-0.03	i
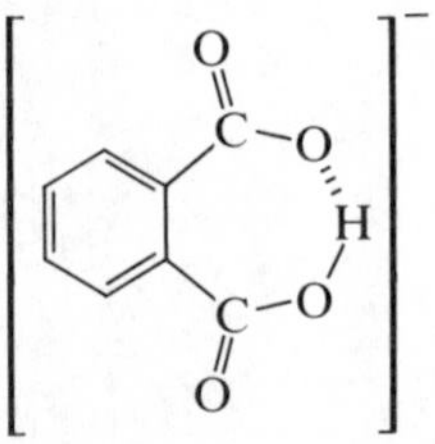	21.0	CH_2Cl_2	-0.15	i
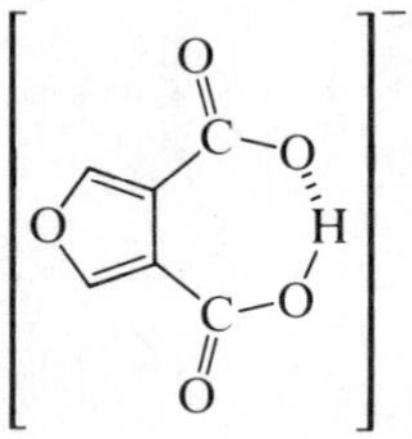	20.3	CH_2Cl_2	$+0.11$	i
Intramolecular $O—H \cdots N$ hydrogen bond				
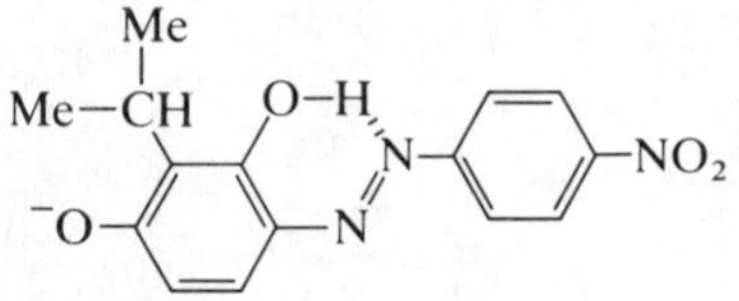	17.50	Me_2SO-H_2O	$+0.60$	n
Intramolecular hydrogen bonds in proton sponges				
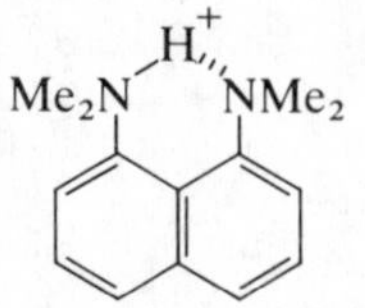	18.46	CH_2Cl_2	$+0.66$	i

| 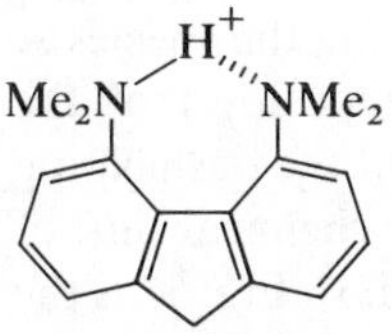 | 18.25 | Me_2SO | $+0.50$ | o |

| 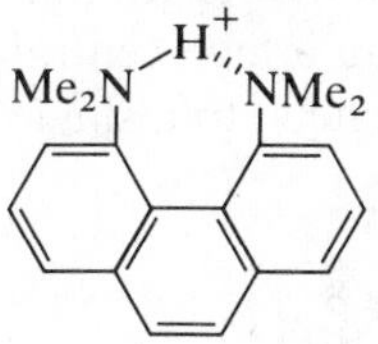 | 18.37 | Me_2SO | $+0.47$ | p |

Intramolecular hydrogen bonds in cyclic diamines

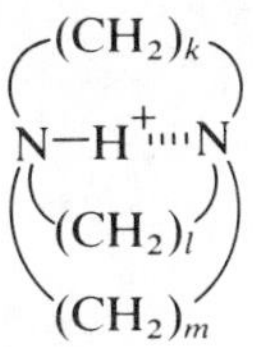

$[k.l.m]H^+$				
[6.3.3]	17.25	$CHCl_3$	$+0.76$	q
[5.4.3]	17.48	$CHCl_3$	$+0.53$	q
[4.4.4]	17.40	80% H_2SO_4	$+0.06$	q
[6.4.3]	14.62	$CHCl_3$	$+0.65$	q
[5.5.3]	17.20	H_2O	$+0.87$	q
[5.4.4]	16.19	$CHCl_3$	$+0.43$	q
[6.5.3]	13.78	$CHCl_3$	$+0.58$	q
[5.5.4]	15.30	$CHCl_3$	$+0.92$	q

| 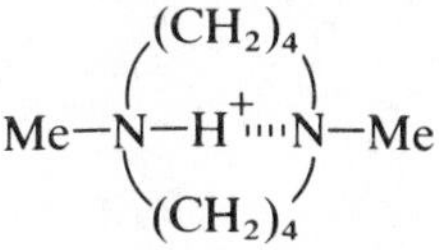 | 15.71 | $CHCl_3$ | $+0.54$ | q |

Intermolecular hydrogen bonds

| HF_2^- | 16.4 | CH_3CN | -0.30 | i |
| $(HCO_2)_2H^-$ | 14.1 | H_2O | $+0.64$ | r |

---

[a] TMS reference except where stated; [b] Shapet'ko *et al.* (1976); [c] Leipert (1977); [d] Altman *et al.* (1978); [e] δ-Value referenced relative to methyl resonance in molecule; [f] Chan *et al.* (1970); [g] Robinson *et al.* (1977); [h] δ-Value referenced relative to cyclohexane; [i] Gunnarsson *et al.* (1976); [j] Freeman (1987); [k] Emsley *et al.* (1988a); [l] Clark *et al.* (1988a); [m] Clark *et al.* (1989); [n] Hibbert and Phillips, unpublished work; [o] Staab *et al.* (1983); [p] Saupe *et al.* (1986); [q] Alder *et al.* (1983d); [r] Fenn and Spinner (1984).

field than a similar proton A—H that cannot take part in hydrogen bonding. As the hydrogen bond is made stronger by changing the species A and B, or the geometry of A—H $\cdots$ B, the equilibrium position of the proton will be closer to the mid-point between A and B and the deshielding effect will increase. It is usually assumed, therefore, that the chemical shift of the proton increases with the strength of the hydrogen bond. In Fig. 5, as the A $\cdots$ B distance is decreased, the hydrogen bond increases in strength and the potential function changes from a double minimum with a high central barrier to a double minimum with a low barrier and, ultimately, to a single minimum potential.

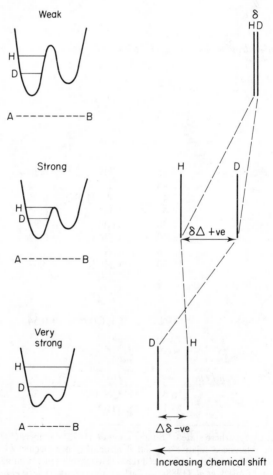

Fig. 5 The three types of hydrogen bond and their associated isotopic nmr shifts $\Delta[\delta(^{1}H) - \delta(^{2}H)]$.

For the double minimum function, the anharmonicity of each potential well close to the bottom is low and the equilibrium positions of H and D will not be very different. In this case, the difference in chemical shifts will be small. For a strong hydrogen bond, the hydrogen and deuterium are closer to the top of the barrier where the anharmonicity of the potential function is high. This will result in differences in the equilibrium positions of the two isotopes. The lighter isotope, on average, will be closer to the mid-point between A and B, and the ^{1}H shift will be more downfield. For very strong hydrogen bonds for which a single minimum potential may apply, the anharmonicity is low and the equilibrium positions of the isotopes are similar. The larger vibrational amplitude of the lighter isotope is predicted to result in an upfield shift relative to that for the heavier isotope. Thus, $\Delta[\delta(^1H) - \delta(^2H)]$ will have a value close to zero, a positive value or a negative value, depending on whether the hydrogen bond is weak, strong or very strong. The sign of the isotope effect is, therefore, of some diagnostic value in determining the potential function that best describes a particular hydrogen bond.

Values of $\Delta[\delta(^1H) - \delta(^2H)]$ for a large number of enols have been determined. The results are close to $+0.6$ ppm and point to double minimum potentials and strong hydrogen bonds. Low values are found for 1,1,1,5,5,5-hexafluoropentane-2,4-dione (0.30 ppm) and 3-(4-methoxyphenyl)pentane-2,4-dione (0.31 ppm), but these two species are of the same double minimum hydrogen-bond type and the O $\cdots$ O bond length in 3-(4-methoxyphenyl)pentane-2,4-dione ($R = 245$ pm; Emsley et al., 1988a) is similar to that for other enols.

The other examples of intramolecular O—H $\cdots$ O hydrogen bonds given in Table 5 mostly consist of neutral aldehyde- or keto-group acceptors and phenol donor groups. These include examples of weak hydrogen bonds, for example salicylaldehyde, 2-hydroxacetophenone and ethyl acetoacetate, and a few strong hydrogen bonds, for example the three alicyclic ketoaldehydes. The hydrogen bond in protonated 1,3-diphenyl-2-methylpropane-1,3-dione gives a signal at exceptionally low field ($\delta = 21.44$ ppm) and the isotope effect $\Delta[\delta(^1H) - \delta(^2H)]$ is negative (-0.26 ppm). The species is prepared under strongly acidic conditions (CBr_2F_2 containing HBr) by protonation of the diketone, and the signal for the hydrogen-bonded proton is only observed at low temperature (180 K) (Clark et al., 1988b, 1989). Other evidence is not available to confirm the type of hydrogen bond, but the negative value of $\Delta[\delta(^1H) - \delta(^2H)]$ appears to indicate a very strong hydrogen bond with a single minimum potential. In protonated 1,3-diphenyl-2,2-dimethylpropane-1,3-dione prepared in the same way from the corresponding ketone, the value of $\Delta[\delta(^1H) - \delta(^2H)]$ is somewhat uncertain but, again, appears to indicate a strongly hydrogen-bonded proton in a single minimum potential (Clark *et al.*, 1989). In this case the proton is found much further upfield, $\delta(^1H) = 14.1$ ppm. It would be of interest to use other techniques for probing the hydrogen bonds in these species.

The O—H $\cdots$ N hydrogen bond in the phenylazoresorcinol monoanion is strong, giving a large and positive value of $\Delta[\delta(^1H) - \delta(^2H)]$, characteristic of a double minimum potential (Hibbert and Phillips, 1989). Two examples of intermolecularly hydrogen-bonded species for which the isotope effect on the chemical shift has been measured are also given in Table 5.

The proton sponges are exceptionally strong bases (see Section 4, p. 321 for a discussion) and one of the reasons for this is the strong hydrogen bonds with double-minimum potentials $\{\Delta[\delta(^1H) - \delta(^2H)] \simeq 0.5\}$ present in the protonated species. The properties of bicyclic diamines are discussed in detail in Section 4. Strong hydrogen bonds are present in some of the protonated amines. In all cases the hydrogen bonds have double-minimum potentials except for inside-protonated 1,6-diazabicyclo[4.4.4]tetradecane for which the hydrogen bond is very strong and linear with a single minimum potential.

There are only a few hydrogen bonds that have a negative shift; the bifluoride ion is the best documented, but it is not alone among hydrogen bonds. A few OHO hydrogen bonds have a molecular geometry which favours close approach of the two oxygen atoms sufficient to give a negative $\Delta[\delta(^1H) - \delta(^2H)]$, e.g. the hydrogen phthalate hemi-salt [10].

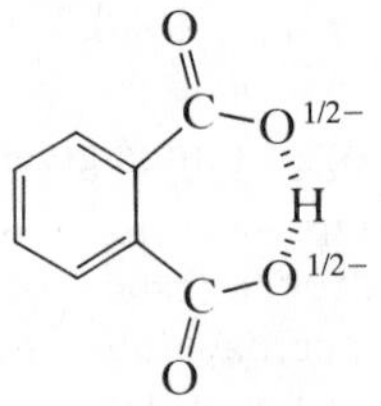

[10] 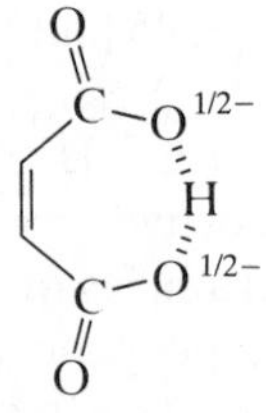   [11] 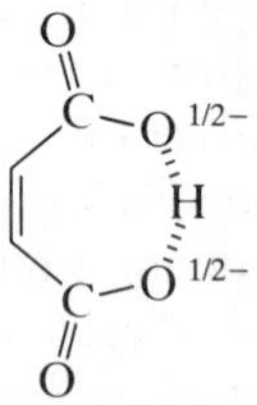   [12]

Recently, this conclusion has been challenged in studies using ^{13}C-nmr spectroscopy (Perrin and Thoburn, 1989). The ^{13}C chemical shift is sensitive to the small decrease in acidity of a —CO$_2$H group that occurs when ^{18}O replaces ^{16}O (Ellison and Robinson, 1983), which in turn affects this group's ability to act as a hydrogen-bond donor and the carboxylate group to act as an acceptor. The intramolecular hydrogen bonding of the hemi-salts, hydrogen succinate [11], hydrogen maleate [12], as well as hydrogen phthalate [10], was interpreted as consistent with a double minimum potential energy well for both H and D. It is difficult to rationalize this within the framework of weak, strong and very strong hydrogen bonds outlined above. However, the

[13]C-nmr investigations referred to aqueous solutions whereas the results in Table 5 were mostly obtained from studies in organic solvents. Competitive bond formation with the aqueous solvent may affect the potential energy well and lead to a different well from that which applies in solution in an organic solvent.

Infrared and $\Delta v(AH)$

One of the first diagnostic properties to be recorded as such was the isotope shift in the ir. "Highly anomalous" spectra were reported for some compounds that failed to show the expected ir shift on deuteration (Hadzi, 1965; Somorjai and Hornig, 1962). An early example was acetamide hemihydrochloride, $(CH_3CONH_2)_2 \cdot HCl$, with $v_{AH} : v_{AD} = 1$ (Albert and Badger, 1958). For most hydrogen bonds the ratio of the two frequencies, $v_{AH} : v_{AD}$, is about 1.4 (i.e. $\sqrt{2}$). For stronger hydrogen bonds the ratio can be less than this, and even 1.0 in some examples; in other words, there is no isotope effect. In some compounds, such as isonicotinic acid [13] and 6-hydroxy-2-pyridone [14], there is even the suggestion that v_{OH}/v_{OD} is less than unity (Spinner, 1974).

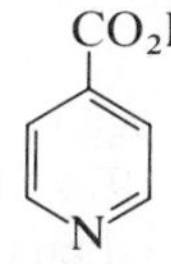

[13] 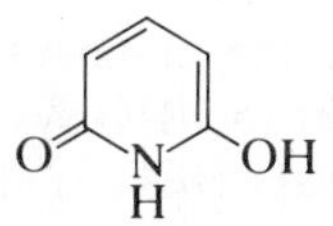   [14]

In a study of compounds for which the hydrogen bond length, $R_{O\cdots O}$, was accurately known it was observed that bonds in the range 250–260 pm had much reduced $v_{AH} : v_{AD}$ values (Novak, 1974), as Fig. 6, based on the original paper, shows. For even shorter, and stronger, hydrogen bonds the isotopic ratio eventually returned to the classical value 1.4 arising from the difference in the reduced mass of AH and AD (Guissani and Ratajczak, 1981).

This situation applies with weak hydrogen bonds at one extreme and very strong hydrogen bonds at the other with H and D confined to the same potential well. However, when the potential energy barrier has fallen sufficiently to allow the proton to escape the confines of its parent well, but leaves the deuteron trapped, then different values of the isotopic ratio can be observed (Fig. 7). The effect of isotopic exchange is now much more than merely one of doubling the reduced mass of the vibrating bond. When the proton is above the barrier, the force constant of the A—H bond, $k(A\text{—}H)$,

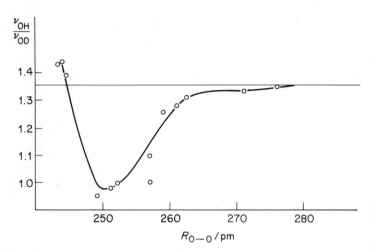

Fig. 6 Variation of the deuterium isotope-shift ratio, v_{OH}/v_{OD}, with hydrogen-bond length $R_{O \cdots O}$ (adapted from Novak, 1974).

will decrease significantly. Indeed, if $k(A—H)$ were now to be half $k(A—D)$, then the disappearance of the isotopic shift would be exactly explained, and even $v_{AH}/v_{AD} < 1$ is explicable.

Cases exhibiting the anomalous ratio, $v_{AH}/v_{AD} = 1$, are not common. Most examples refer to OHO systems (Novak, 1974; Berglund *et al.*, 1978). The ir spectra of $NaH(HCO_2)_2$ and its deuterated analogues (in which one, two or all of the hydrogens are replaced by D) have been closely analysed and v_{OH} identified as a broad intense band at *ca.* $1100 \, cm^{-1}$ with $v_{OH}/v_{OD} = 1$, (Spinner, 1983). Other examples are known (Spinner, 1974, 1980b).

Hemi-salts with symmetrical hydrogen bonds such as $KH(CF_3CO_2)_2$ and $KH(CCl_3CO_2)_2$ give normal values for v_{OH}/v_{OD} of *ca.* 1.41 (Macdonald *et al.*, 1972; Miller *et al.*, 1972; Stepisnik and Hadzi, 1972). In the case of $BaC_2O_4 \cdot H_2C_2O_4 \cdot 2H_2O$, however, $v_{OH}/v_{OD} = 1$ is observed (Hadzi and Orel, 1973). Acetamide hemihydrochloride has been investigated by X-ray and neutron diffraction (see references in Spinner, 1980b). The structure revealed a centred hydrogen bond with $R_{O \cdots O} = 241.8$ pm. The observed ratio $v_{AH}:$ $v_{AD} = 1$ means that this is a strong rather than a very strong hydrogen bond.

Isotope fractionation factors, φ

Measurement of the equilibrium distribution of deuterium relative to hydrogen atoms in a proton-exchanging site compared with the distribution in the solvent or some other standard is a particularly subtle probe that can provide important information about the nature of the environment of the

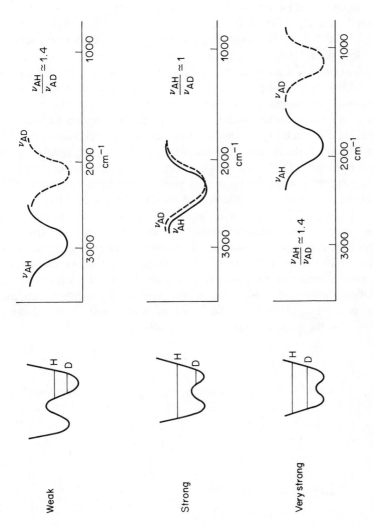

Fig. 7 Ir isotope shift associated with weak, strong and very strong hydrogen bonds.

proton. There are revealing differences between the behaviour of protons involved in a hydrogen bond and protons in non-hydrogen-bonding sites, and these differences can be used to provide details of the hydrogen bond.

The effect of isotopic substitution on the equilibrium constant for a reaction (equilibrium isotope effect) can always be expressed in terms of an isotope-exchange equilibrium (Gold, 1969; Schowen, 1972; Albery, 1975; More O'Ferrall, 1975). For example, the value of the equilibrium solvent isotope effect, $K(H_2O)/K(D_2O)$, on the equilibrium between the monoanions of a dicarboxylic and a monocarboxylic acid in aqueous solution (2), can be represented by the equilibrium constant (K_e) for the isotope-exchange equilibrium in (3). In turn, the equilibrium constant for (3) can be expressed in terms of the fractionation factors for the exchangeable sites in reactants and products. For each site (i), the fractionation factor (φ_i) is defined as the equilibrium ratio of deuterium to protium in that site, $(D/H)_i$, relative to the ratio in the solvent, $(D/H)_s$, as shown in (4). The equilibrium constant for the isotope-exchange equilibrium is then given by the ratio of fractionation factors of the proton in the monoanion of the dicarboxylic acid (φ_d) and the proton in the monocarboxylic acid (φ_m) as shown in (5). In principle, the values of the individual fractionation factors can be measured. In expressing (5) in this way, it is assumed that the equilibrium position is determined by the difference in values of the fractionation factors for the two exchanging proton sites alone. For example, it is assumed that the values of the fractionation factors for solvent molecules surrounding the reactants and products either cancel or have the value of unity; that is, the isotopic fractionation is the same as that in the bulk solvent.

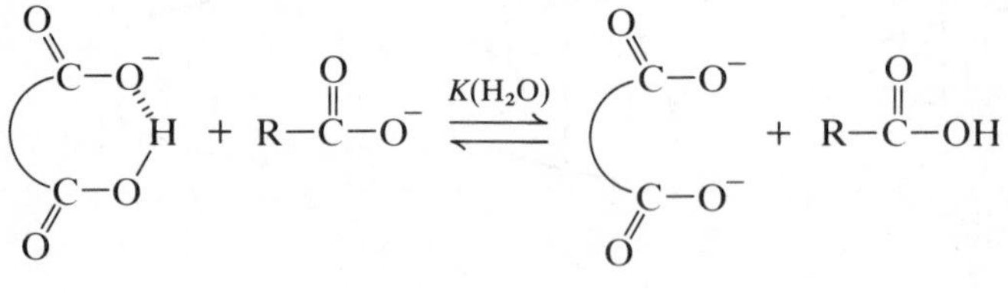

$$\text{(2)}$$

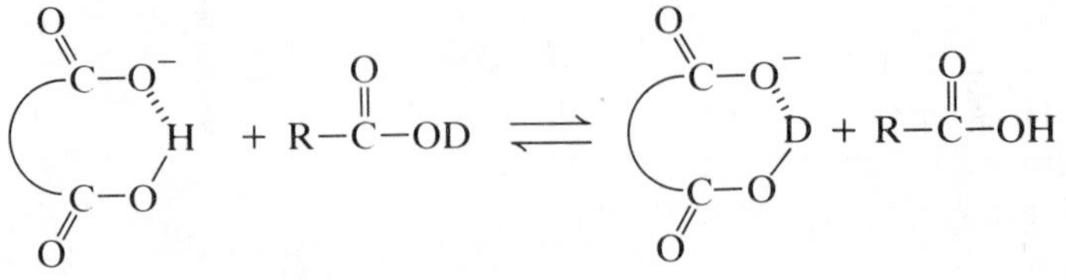

$$\text{(3)}$$

$$\phi_i = (D/H)_i/(D/H)_s \tag{4}$$

$$K_e = K(H_2O)/K(D_2O) = \phi_d/\phi_m \tag{5}$$

The value of the fractionation factor for any site will be determined by the shape of the potential well. If it is assumed that the potential well for the hydrogen-bonded proton in (2) is broader, with a lower force constant, than that for the proton in the monocarboxylic acid (Fig. 8), the value of the fractionation factor will be lower for the hydrogen-bonded proton than for the proton in the monocarboxylic acid. It follows that the equilibrium isotope effect on (2) will be less than unity. As a consequence, the isotope-exchange equilibrium will lie towards the left, and the heavier isotope (deuterium in this case) will fractionate into the monocarboxylic acid, where the bond has the larger force constant.

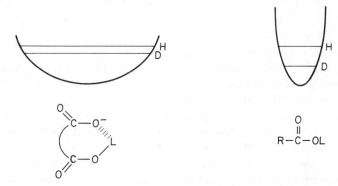

Fig. 8 Potential energy wells for the isotope-exchange equilibrium in (3), L = H or D.

The equilibrium constant for the isotope-exchange equilibrium can be expressed (6) in terms of the solvent isotope effects on the acid-dissociation constants K_a^m and K_a^d of the monocarboxylic acid and dicarboxylic acid monoanion, respectively. It follows that a lower value for the fractionation factor of the hydrogen-bonded proton means that the solvent isotope effect on the acid-dissociation constant will be lower for the dicarboxylic acid monoanion than for the monocarboxylic acid.

$$K_e = \{K_a^d(H_2O)/K_a^d(D_2O)\}/\{K_a^m(H_2O)/K_a^m(D_2O)\} \qquad (6)$$

A large number of different techniques have been used to measure isotopic fractionation factors, and any method that can be used for measuring the equilibrium position of reaction (3) is appropriate. The discussion here will be limited to the more recent methods and those that have been used for studying hydrogen-bonded species. The discussion is further limited to fractionation factors for species in solution. Recent measurements (Larson and McMahon, 1986, 1987, 1988) of φ for hydrogen-bonded species in the gas phase using ion cyclotron resonance give results that differ in a number of respects from solution values. Values of φ for species in solution are usually assumed to result from zero-point energy effects (Fig. 8), whereas rotational contributions are considered to be important in determining values of φ for gaseous species.

The most widely used method is nmr spectroscopy and various approaches are applicable. The first method was used for measuring the

$$
\begin{array}{c}
\text{(structure)} \quad \text{H} + \text{H}_2\text{O} \quad \xrightleftharpoons{\text{proton exchange}}
\end{array}
\tag{7}
$$

fractionation factor of the hydronium ion in aqueous solution (Gold, 1963; Kresge and Allred, 1963). For example, if the exchange between the hydrogen-bonded acid and solvent in (7) occurs rapidly on the nmr time-scale, a single ^1H peak will be observed with a chemical shift that is the concentration-weighted average of the chemical shifts of the acid and water protons that would be observed if exchange occurred slowly. If in a mixed solvent of H_2O and D_2O, the isotope is distributed unevenly between the two sites, the weighted average of the chemical shifts of the acid proton (δ_a) and the solvent proton (δ_s) will be different from the average that would be obtained if there were an even distribution of isotopes. The variation in chemical shift as a function of the mole fraction (c) of the acid in the medium is given by (8) in which δ is the observed shift for a solution containing the acid and δ_d is the difference in chemical shifts of the two sites. In a solvent containing an atom fraction n of deuterium $(n = \text{D}/[\text{H}+\text{D}])$, the corresponding equation is (9) and it follows that the ratio of gradients of plots of δ against c in H_2O and in the $\text{H}_2\text{O}/\text{D}_2\text{O}$ solvent has a value equal to $1 - n + n\varphi_a$, from which a value for φ_a can be obtained. Thus, the procedure simply involves the measurement of the variation of the chemical shift of the time-averaged peak with concentration of the hydrogen-bonded acid (or any other species whose fractionation factor is required) in two solvents containing different atom fractions of deuterium, for example one consisting of pure H_2O and the other a $50:50$ $\text{H}_2\text{O}/\text{D}_2\text{O}$ mixture.

$$
\delta = c\delta_a + (1 - c)\,\delta_s = c\delta_d + \delta_s
\tag{8}
$$

$$
\delta = c\delta_d/(1 - n + n\phi_a) + \delta_s
\tag{9}
$$

Using this method, the result $\varphi = 0.69$ (Gold, 1963; Kresge and Allred, 1963) was obtained for the fractionation factor of each of the three protons of H_3O^+. The fractionation factors of the further solvating water molecules around H_3O^+ were found to have values of unity. The low value of φ (usually given the symbol l for the hydronium ion) may be the result of a strong intermolecular hydrogen bond between each of the three protons and a solvating water molecule, as in [15]. Interestingly, a value $\varphi = 0.79$ was found (Kurz et al., 1984) for H_3O^+ in acetonitrile-water mixtures from spectrophotometric measurements of the dissociation of substituted anilinium ions (10). The method depended upon assuming values for the fractionation factors of the other species present in the equilibrium.

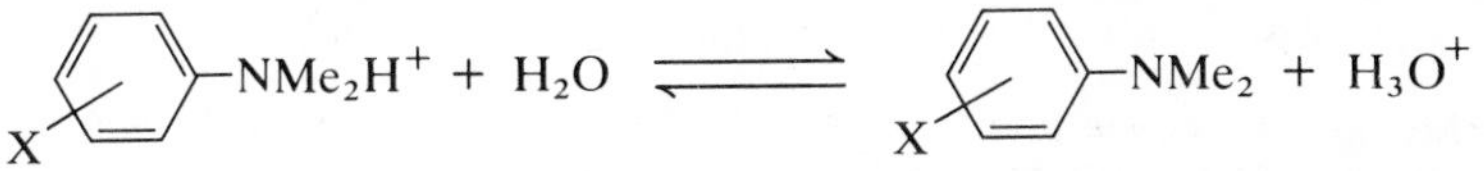

$$X\text{—}\bigcirc\text{—}NMe_2H^+ + H_2O \; \rightleftharpoons \; X\text{—}\bigcirc\text{—}NMe_2 + H_3O^+ \quad (10)$$

Using the nmr procedure, a value for the fractionation factor of hydroxide ion in water, $\varphi = 0.42$, has been determined (Gold and Lowe, 1967; Gold and Grist, 1972). This value was later explained (Gold and Grist, 1972) in terms of a number of different values for each of the solvating water molecules as in [16]. A somewhat different interpretation has also been suggested (Walters and Long, 1972).

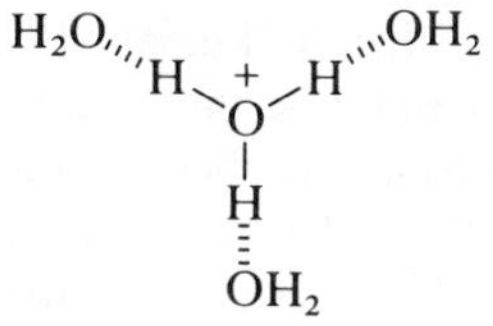

[15] 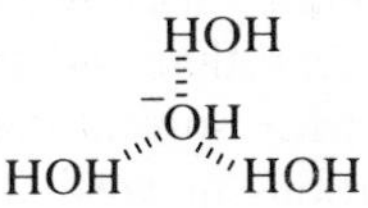   [16]

For solutions of the methoxide ion in methanol [17], $\varphi = 0.76$ was obtained (More O'Ferrall, 1969). This result was later slightly modified to $\varphi = 0.74$ (Gold and Grist, 1971) and a value for the solvated proton in methanol [18], $\varphi = 0.625$, was also measured. The result for deuterium fractionation was deduced as $\varphi = 0.7$ from observations of tritium fractionation (Al-Rawi et al., 1979) and $\varphi = 0.74$ has been obtained more recently (Baltzer and Bergman, 1982). Values for several alkoxide ions in alcohol were used to reach conclusions about the solvation of the alkoxide ions [19] in these solutions (Gold et al., 1982).

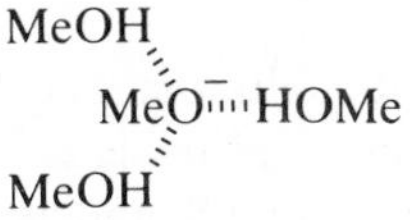

[17] 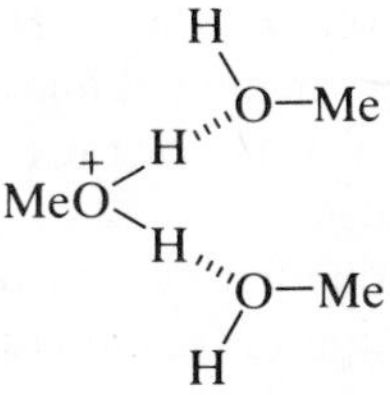   [18]

Similar measurements have given values for the fractionation factor of hydrogen-bonded complexes of the fluoride ion (Emsley *et al.*, 1986c) and the acetate ion (Clark *et al.*, 1988a) in acetic acid solution, [20] and [21]. For the chloride ion in acetic acid, the result (Emsley *et al.*, 1986c) was $\varphi = 1.26$, which means that the exchangeable sites in acetic acid molecules in the solvation sphere of the chloride ion are favoured by deuterium compared to the sites in the bulk solvent.

$RO^- \cdots (HOR)_n$	$F^- \cdots HOAc$	$AcO^- \cdots HOAc$
[19]	[20]	[21]
$R = Me, \phi = 0.74, n \simeq 3$	$\phi = 0.55$	$\phi = 0.58$
$R = Et, \phi = 0.74, n \simeq 2.7$		
$R = Pr^i, \phi = 0.75, n \simeq 2.1$		
$R = Bu^t, \phi = 0.76, n \simeq 1.4$		

The values of the fractionation factors in structures [15]–[21] are not strictly comparable since they are defined relative to the fractionation in different solvent standards. However, in aqueous solution, fractionation factors for alcohols and carboxylic acids relative to water are similar and close to unity (Schowen, 1972; Albery, 1975; More O'Ferrall, 1975), and it seems clear that the species [15]–[21] involving intermolecular hydrogen bonds with solvent have values of φ consistently below unity. These observations mean that fractionation of deuterium into the solvent rather than the hydrogen-bonded site is preferred, and this is compatible with a broader potential well for the hydrogen-bonded proton than for the protons of the solvents water, alcohol and acetic acid.

Several other nmr procedures have been used for the determination of fractionation factors. These have advantages in some systems. Instead of determining the effect of the concentration of an exchanging site on the averaged chemical shift, the effect on the averaged relaxation rate of water protons can be used in a very similar way (Silverman, 1981; Kassebaum and Silverman, 1989). For example, addition of the enzyme Co(II)-carbonic anhydrase to an aqueous solution increases the observed value of $1/T_1$ because the proton-relaxation rate is the average of that for the bulk solvent ($ca.$ $0.3\,s^{-1}$) and that for water bound to the cobalt ($ca.$ $6 \times 10^4\,s^{-1}$). The average is different in an H_2O/D_2O mixture if the bulk solvent and the Co-bound solvent have different deuterium contents, and it has been used to determine a value for the fractionation factor of Co-bound water molecules in the enzyme.

A method that can be used with relatively dilute solutions and for the determination of fractionation factors at different distinguishable sites in the same molecule has been developed by Saunders (Jarret and Saunders, 1985). The technique applies to sites which are in rapid proton exchange with the solvent and depends on the observation that introduction of deuterium into an OH, SH or NH group brings about an upfield shift of the ^{13}C peaks of the carbon atoms α, β and γ to the site of substitution (Pfeffer et al., 1978; Hansen, 1983). The magnitude of the shift for a particular carbon atom and a particular site depends on the fraction of deuterium in the exchanging site, and it can therefore be used to calculate the degree of isotope fractionation in comparison to that in the solvent. Application of the procedure to aqueous solutions of alcohols, thiols, phenols, carboxylic acids protonated amines, carbohydrates and amides at concentrations of typically ca. 0.5–1.0 mol dm^{-3} gave values of fractionation factors with uncertainties of ca. $\pm$ 0.05. Chemical shifts induced in signals of a very wide range of other nuclei were found to be equally useful for the determination of fractionation factors in different molecules. It appears that the fractionation at almost any exchangeable site in most molecules could be investigated in this way (Jarret and Saunders, 1986). In some cases, the method was used with extremely dilute solutions.

The major conclusion which was reached from the work of Jarret and Saunders was that intramolecularly hydrogen-bonded protons generally give values of φ below the value for a similar proton which is not involved in an intramolecular hydrogen bond. For example, a value of φ of 0.77 was obtained for the hydrogen-bonded site in the maleate ion [12] compared to 0.94 for acetic acid. The value for the phthalate ion [10] was higher, $\varphi = 0.95$. The value of $\varphi = 0.95$ was obtained for the salicylate ion [22] in comparison to the result for phenol, $\varphi = 1.13$.

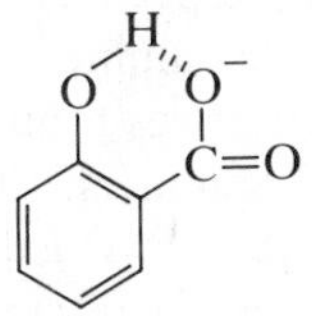

[22]

A somewhat different procedure has been used by Spinner (Fenn and Spinner, 1984) to determine the value of φ for the hydrogen-bonded complex formed between the formate ion and formic acid in concentrated solutions of the salt $NaH(HCOO)_2 \cdot 3H_2O$ in H_2O/D_2O mixtures. If there were an equal distribution of H and D between the solvent and the hydrogen-bonded complex in a 50:50 H_2O/D_2O mixture, the observed ^1H chemical shift would be an average between that of H_2O and the hydrogen-bonded proton. The deviation from this average was used to calculate the equilibrium constant for the isotope-exchange equilibrium in (11) and a value $\varphi = 0.68$ was obtained. The result is not very different from the value obtained by nmr measurements for the fractionation factor of solutions of the acetate ion in acetic acid of 0.58 (Clark *et al.*, 1988a). This value was measured relative to the distribution in the acetic acid solvent. The result differs from the value $\varphi = 0.42$ obtained by Kreevoy for the complex $CF_3COOH \cdots {}^-OOCCF_3$. Spectrophotometric measurements in acetonitrile were used in this case and the value was measured relative to the distribution in H_2O (Kreevoy *et al.*, 1977; Kreevoy and Liang, 1980).

$$HCOOH \cdots {}^-OOCH + 1/2D_2O = HCOOD \cdots {}^-OOCH + 1/2H_2O \qquad (11)$$

In some cases where proton exchange between a hydrogen-bonded proton and the solvent occurs slowly on the nmr timescale, separate peaks in the ^{1}H-nmr spectrum for the hydrogen-bonded proton and for the solvent are seen and then the fractionation factor of the hydrogen-bonded site can be measured from the signal integral in the presence of a known H_2O/D_2O ratio. This was possible for protonated 1,8-bis(dimethylamino)naphthalene [23] in aqueous solution (Chiang *et al.*, 1980), for the 2,7-dimethoxy-derivative [24] in 70% (v/v) $Me_2SO–H_2O$ (Hibbert and Robbins, 1980), and for the substituted phenylazoresorcinol [25] in 90% $Me_2SO–H_2O$ (Hibbert and Phillips, 1989). These species possess intramolecular hydrogen bonds and the trapped proton exchanges with solvent much more slowly than protons not contained in hydrogen bonds. The same procedure can be used for any weakly acidic species, for example an amide or a carbon acid, under conditions where proton exchange with the solvent is slow. The φ-values for [23] and [24], of 0.90 and 0.87, respectively, are below those measured for non-hydrogen-bonded protonated amines, for which values are generally above unity; for example, $\varphi \simeq 1.2$ was found for aliphatic ammonium ions in aqueous solution (Jarret and Saunders, 1985; More O'Ferrall, 1975). The result for [25], $\varphi = 0.97$, in $Me_2SO–H_2O$ is below the value $\varphi = 1.08$ found for phenol under these conditions (Jarret and Saunders 1985).

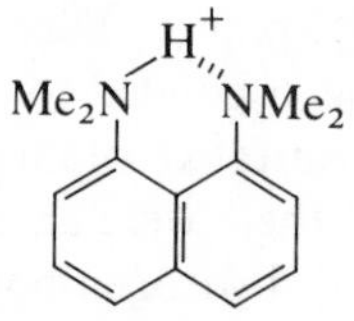

[23] 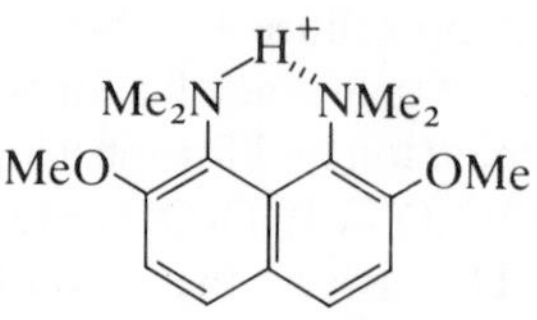   [24] 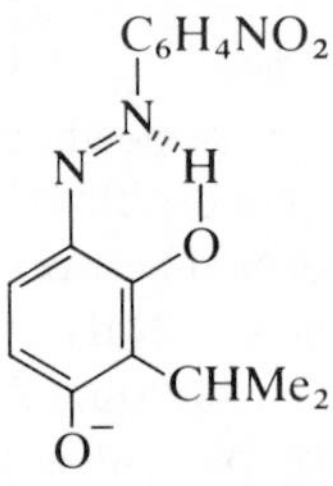   [25]

The solvent isotope effect on the dissociation constant of an acidic proton in aqueous solution has been used to deduce a value for the fractionation factor of the proton in the acid (φ_a), using (12), in which the value $l = 0.69$ for the fractionation of the protons of hydronium ion is known precisely. However, it is usually necessary to make the assumption that the water molecules solvating the undissociated and dissociated acids have identical values for their fractionation factors since these usually cannot be measured. Because of this, the reliability of values of φ determined in this way is usually low. Values of the fractionation factors for the ion HF_2^- ($\varphi = 0.60$) and for the maleate ion ($\varphi = 0.84$) have been estimated in this way (Kreevoy et al., 1977; Kreevoy and Liang, 1980), and in the latter case an independent measurement (Jarret and Saunders, 1985) gives a similar value ($\varphi = 0.77$). A lower value for the fractionation factor for a hydrogen-bonded acid in comparison with a non-hydrogen-bonded acid should mean that the values of solvent isotope effects on the acid-dissociation constants of hydrogen-bonded acids are lower. A comprehensive tabulation (Laughton and Robertson, 1969) of solvent isotope effects on acid-dissociation constants does not include sufficient data for intramolecularly hydrogen-bonded acids to be conclusive on this point.

$$K_a(H_2O)/K_a(D_2O) = \varphi_a/l^3 \tag{12}$$

$$Ph_3COD + A_1HA_2^- \rightleftharpoons Ph_3COH + A_1DA_2^- \tag{13}$$

The position of the isotope-exchange equilibrium (13), between triphenylmethanol and various intermolecularly hydrogen-bonded complexes $A_1HA_2^-$ in acetonitrile was measured directly by ultraviolet spectrophotometry (Kreevoy and Liang, 1980). The exchange is slow on the ultraviolet timescale and a measurement of the equilibrium constant gave values for the fractionation factor of the $A_1HA_2^-$ complexes, relative to the distribution of H/D in water. The results are strikingly low and some of the values are given in Table 6. According to Kreevoy, these low values can be explained by a potential well for the proton in the hydrogen bond consisting of a double minimum with the zero-point energy levels of the proton and deuteron close to the top of the barrier between the two minima (Fig. 9).

The complex formation between 2,4,6-trinitrophenol and the acetate ion in acetic acid solution (14) has been investigated by ultraviolet spectrophotometry (Clark et al., 1988a). The equilibrium constant at 298 K was found to have values of 34.8 and 41.7 dm^{-1} in acetic acid and in acetic acid–OD, respectively. From the measured value of $\varphi = 0.58$ for the fractionation factor for the solvated acetate ion under these conditions, the result $\varphi \simeq 0.7$ was deduced for the fractionation factor of the intermolecularly hydrogen-

Table 6 Fractionation factors (φ) for protons in hydrogen bonds.[a]

H-Bonded species	φ	Technique	Reference
Intermolecular hydrogen bonds			
$MeO^- \cdots (HOMe)_3$	0.74[b]	Nmr spectroscopy	c
$F^- \cdots HF$	0.60	$K_a(H_2O)/K_a(D_2O)$	d
$F^- \cdots HOAc$	0.55[e]	Nmr spectroscopy	f
$AcO^- \cdots HOAc$	0.58[e]	Nmr spectroscopy	g
$(4\text{-}NO_2C_6H_4O)_2H^-$	0.31	Uv/isotope exchange eqm	d
$(CF_3COO)_2H^-$	0.42	Uv/isotope exchange eqm	d
$[3,5\text{-}(NO_2)_2C_6H_3COO]_2H^-$	0.30	Uv/isotope exchange eqm	d
$[3,5\text{-}(NO_2)_2C_6H_3O]_2H^-$	0.36	Uv/isotope exchange eqm	d
$(C_6Cl_5O)_2H^-$	0.40	Uv/isotope exchange eqm	d
$3,5\text{-}(NO_2)_2C_6H_3OH \cdots Cl^-$	0.47	Uv/isotope exchange eqm	d
$HCOO^- \cdots HOOCH$	0.68	Nmr spectroscopy	h
$2,4,6\text{-}(NO_2)_3C_6H_2OH\text{-}$ $\cdots {}^-OAc$	0.7[e]	Uv/eqm isotope effect	g

Intramolecular hydrogen bonds

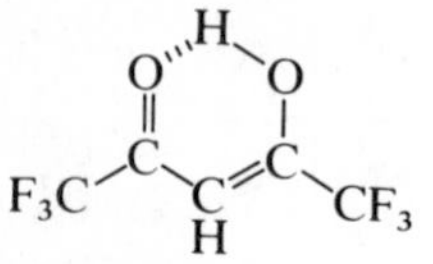	0.6	Uv/isotope exchange eqm	i

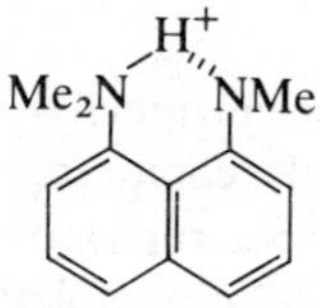	0.90	Nmr spectroscopy	j

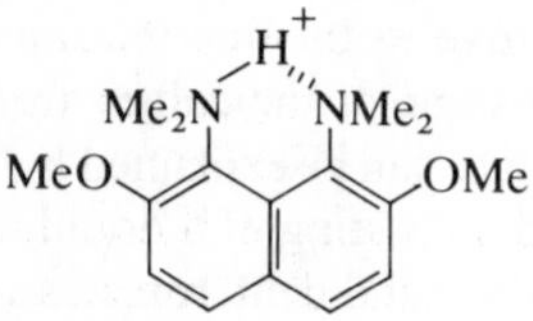	0.87[k]	Nmr spectroscopy	l

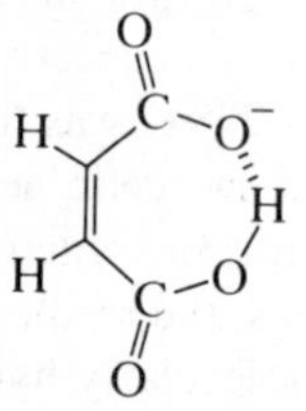	0.77, [0.84]	Nmr spectroscopy $[K_a(H_2O)/K_a(D_2O)]$	m

Table 6 Cont.

H-Bonded species	φ	Technique	Reference
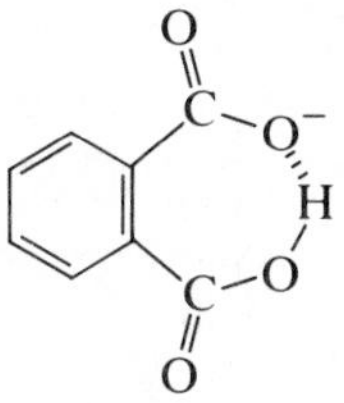	0.95	Nmr spectroscopy	[m]
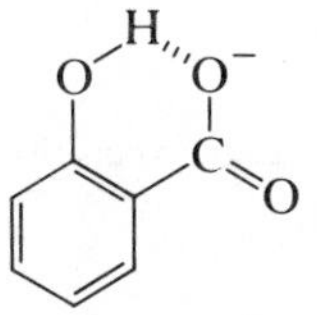	0.95	Nmr spectroscopy	[m]
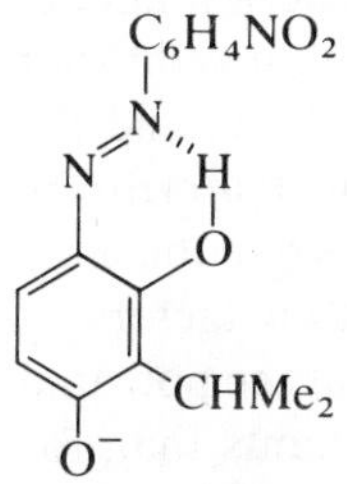	0.96	Nmr spectroscopy	[n]

[a] Reference solvent is H_2O unless otherwise stated; [b] Solvent is methanol; [c] Gold et al. (1982); [d] Kreevoy and Liang (1980); [e] Solvent is acetic acid; [f] Emsley et al. (1986c); [g] Clark et al. (1988); [h] Fenn and Spinner (1984); [i] Kreevoy and Ridl (1981); [j] Chiang et al. (1980); [k] Solvent is 90% (v/v) $Me_2SO–H_2O$; [l] Hibbert and Robbins (1980); [m] Jarret and Saunders (1985); [n] Hibbert and Phillips (1989).

bonded proton in the complex formed between the 2,4,6-trinitrophenolate ion and acetic acid. The equilibrium in (14) lies in favour of the product, but, for solutions of 4-nitrophenol or 2,4-dinitrophenol with acetate ion, complex formation is unfavourable.

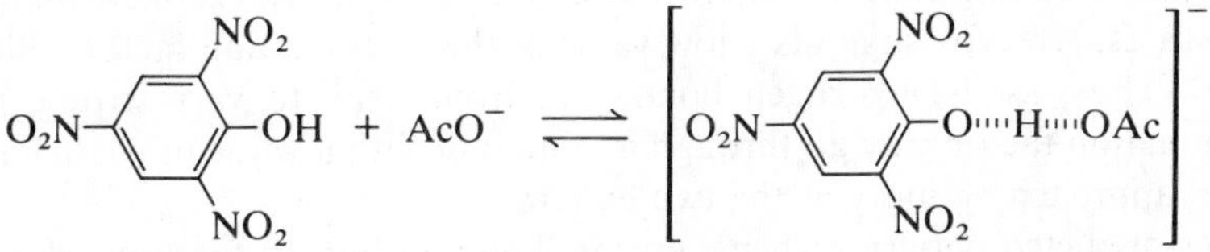

A selection of representative values of the fractionation factors of protons in hydrogen bonds is given in Table 6. The results clearly show that hydrogen bonding reduces the value of fractionation factors and the hydro-

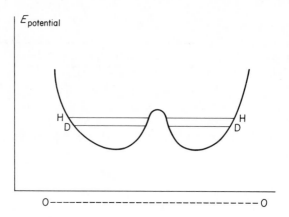

Fig. 9 Proposed potential well for the hydrogen-bonded complex between 4-nitrophenol and 4-nitrophenolate (Kreevoy and Liang, 1980).

gen bond discriminates against the heavier isotope. This conclusion applies both for inter- and intramolecular hydrogen bonds. At present, it is difficult to discern a correlation between the strength of a hydrogen bond and the magnitude of fractionation factors. However, the theoretical treatment by Kreevoy (Kreevoy and Liang, 1980) does suggest that a correlation may exist. To account for the reduced values of the fractionation factors of hydrogen-bonded protons, it seems reasonable to conclude that the potential wells for hydrogen bonds are broader with lower force constants than for proton sites not involved in hydrogen bonds.

Extremely weak hydrogen bonds will perturb the potential of the hydrogen-bond donor very little, and in this case a fractionation factor close to that expected for the hydrogen-bond donor, that is close to unity, would be expected. For very strong hydrogen bonds, a narrow single minimum potential well will be anticipated, for which a fairly normal difference between the zero-point energy levels of the H- and D-substituted bonds will be observed; again, the value of the fractionation factor will not differ greatly from unity. In the intermediate situation of hydrogen bonds of moderate strength, a broad double minimum potential well is likely and, in these cases, Kreevoy suggests a low value of the fractionation factor will be found. Thus, as the hydrogen bond goes from weak to very strong, the fractionation factor may go through a minimum with a value of ca. 0.3 and values approaching unity at the two extremes.

This predicted pattern of behaviour is shown in Fig. 10 for some chosen hydrogen-bonded species. The value of the bond length of $F^- \cdots HF$ in the solid state (see Emsley, 1980, for a discussion) and the negative value (Gunnarsson et al., 1976) of the isotope effect on the chemical shift of the hydrogen-bonded proton in solution, $\Delta[\delta(^1H) - \delta(^2H)]$, have been interpreted as meaning that this hydrogen bond is very strong with a single minimum potential well (see Section 2). Thus, a point for $F^- \cdots HF$ would appear to the left in Fig. 10. Likewise, the slightly negative value of $\Delta[\delta(^1H) - \delta(^2H)]$ for the intramolecular hydrogen bond in the maleate ion (Altman *et al.*, 1978) and the φ value of 0.77 would place a point for this species in a similar position. However, the data are too limited to support the idea of a minimum in the variation of φ with hydrogen-bond strength.

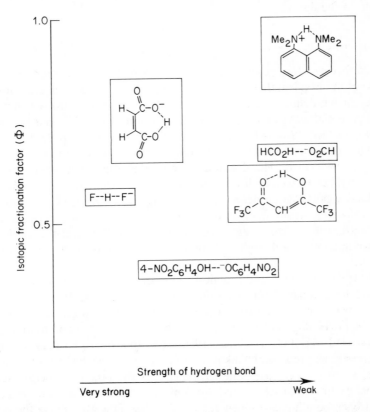

Fig. 10 Possible variation of isotopic fractionation factors with bond strength for hydrogen-bonded species, A_1AH^-.

The ir spectrum of bis(phenolate) complexes has been interpreted in terms of a double minimum potential function (Kreevoy *et al.*, 1977; Kreevoy and Liang, 1980). With values of φ of *ca.* 0.3–0.4, points for these species probably lie close to the centre in Fig. 10. The point for the enol of 1,1,1,5,5,5-hexafluoropentanedione with φ = 0.6 (Kreevoy and Ridl, 1981) is probably in a similar position. The bis(formate) complex with $\Delta[\delta(^1H) - \delta(^2H)] \simeq 0.64$ (Fenn and Spinner, 1984) and $\varphi = 0.68$ would place this species to the right in Figure 10, although the lower results obtained by Kreevoy for the bis(trifluoroacetate) and bis(3,5-dinitrobenzoate) complexes, with $\varphi = 0.42$ and 0.30, respectively (Kreevoy and Liang, 1980), would favour a point nearer the centre. Protonated 1,8-bis(dimethylamino)-naphthalene, with $\Delta[\delta(^1H) - \delta(^2H)] = + 0.66$ ppm and $\varphi = 0.90$, probably possesses a hydrogen bond of moderate strength with a double minimum potential (Altman et al., 1978) and would be represented by a point to the right in the figure.

Geometric isotope effects

There are two geometric isotope effects: $\Delta R(D - H)$, which refers to the change in bond length $R_{A\cdots B}$ of a hydrogen bond AHB on deuteration, and δR, which is defined as the change in the distance between the two minima of the potential energy well of the hydrogen bond when it is deuterated. The isotope effect, $\Delta R(D - H)$, was dealt with previously (see p. 263).

The second isotope effect, δR, requires the proton and deuteron to be accurately located. The distance between the equilibrium positions of the potential energy well of double minima, symmetrical hydrogen bonds, which Ichikawa calls $R_{H/H}$, is defined as $R_{H/H} = R_{O\cdots O} - 2R_{O-H}$. This distance can be calculated from neutron-diffraction data and it decreases as $R_{O\cdots O}$ diminishes. Early work suggested a smooth transition over the whole range down to 245 pm, although with an inflection point at ca. 260 pm (Ichikawa, 1978a).

When more neutron-diffraction data became available, $R_{H/H}$ and $R_{D/D}$ could be compared for the same crystals (Ichikawa, 1978b, 1981). The difference between these two distances, δR, is only known in a few cases such as $(NH_4)_2H_3IO_6$ (Tichy et al., 1980), $KH_3(SeO_3)_2$ (Lehmann and Larsen, 1971) and KH_2PO_4 (Nelmes, 1980); δR appears to increase as $R_{O\cdots O}$ decreases, as predicted (Ichikawa, 1981). For a weak bond δR will be zero but, as $R_{O\cdots O}$ decreases, the asymmetry of the potential energy well will affect H first, and δR should suddenly reach a maximum for the strong hydrogen-bonding case when $R_{H/H} = 0$. When D is also above the internal energy barrier $R_{D/D}$ will also be zero and the isotope effect will again disappear.

A less well-documented effect is that of the phase-transition temperature of certain crystals which are very sensitive to deuteration. Some crystals of ferroelectric and antiferroelectric materials, and in particular dihydrogen phosphates and hydrogen selenites, which are extensively hydrogen bonded, display this effect (Blinc and Zeks, 1974). For some crystals, such as caesium dihydrogen phosphate, the phase-transition temperature, T_c, was higher by 114 K for CsD_2PO_4 as compared to CsH_2PO_4. A plot of the differences, ΔT_c, defined as $T_c(D) - T_c(H)$, versus $R_{O \cdots O}$ (Fig. 11) shows a maximum at *ca.* 250 pm, with weak and very strong hydrogen bonds having ΔT_c near to zero. This property may therefore serve to distinguish between the three kinds of hydrogen bonding.

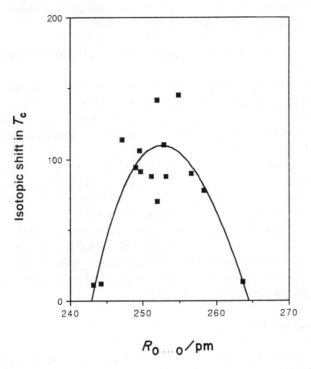

Fig. 11 Isotopic shift associated with phase change temperature and hydrogen bond length (adapted from Ichikawa, 1981).

A further curious feature of KH_2PO_4 and KD_2PO_4 has recently been reported (Endo *et al.*, 1989). In KH_2PO_4 at room temperature (which is above T_c), the hydrogen-bond length decreases initially with increasing pressure from 249.8 pm (0.0001 GPa) to 244.3 pm (1.4 GPa), then increases suddenly at the critical pressure of 2.7 GPa to 245.7 pm, and finally reaches 255.4 pm at 5 GPa. The same effect occurs with KD_2PO_4 at its critical pressure of 4.2 GPa.

Although there may be other very strong hydrogen-bond systems, it is still the case that the bifluoride ion represents the upper limit of hydrogen-bond strength. (If any system has both H and D above the internal energy barrier, this is the most likely candidate.) Despite the large amount of work expended on this species, it continues to attract a great deal of attention. The properties of the difluoride ion are reviewed here since this is the best example of a very strong hydrogen bond. Covalent bonding robs fluorine of most of its hydrogen-bonding acceptor ability. This section will end with a review of covalently bonded fluoride as an example of the weakest type of hydrogen bonding. The β-diketones of the subsequent section have been chosen as examples of strong hydrogen bonding.

2   The upper limit of hydrogen bonding (the bifluoride ion)

An excellent review of bifluorides up to 1980 is available (Gmelin, 1982). An earlier review of very strong hydrogen bonding (Emsley, 1980) assumed that HF_2^- represented the upper limit of hydrogen bonding and that the proton was in a single minimum potential well. Both assumptions have since been questioned.

The prerequisites for very strong hydrogen bonding are a large dipole bond moment in the donor A—H and a centre of high negative charge on the acceptor B. These depend primarily on the electronegativity of the atoms A and B, so that, in the case of fluorine, conditions are most favourable for the strongest hydrogen bonds. The large dipole moment of HF, 1.82 D, and the high charge/radius ratio of F^- make it likely that the difluoride ion itself will represent the best combination, an assumption that went unquestioned for many years. Recently, it has been suggested that a positively charged species might produce a stronger bond.

THE FLUORONIUM ION, $H_3F_2^+$

McMahon and Kebarle (1986) studied $(MeF)_2H^+$ as a model for $(HF)_2H^+$. They thought this to be reasonable because the hydrogen bond of a proton bound to two methanols or dimethyl ethers, e.g. $[Me_2O \cdots H \cdots OMe_2]^+$, gives cations with very similar energies to that of the hydrated oxonium ion $[H_2O \cdots H \cdots OH_2]^+$ (Grimsrud and Kebarle, 1973; Meot-Ner, 1984). An attempt was made to determine the hydrogen-bond energy of $(MeF)_2H^+$ using pulsed electron-beam, high-pressure mass spectrometry. Because of the rapid nucleophilic displacement reaction which occurs between $MeFH^+$ and MeF, the energy could not be measured directly. Instead, the product was approached from proton-bound sulphur dioxide, $(SO_2)_2H^+$, by successive displacements of SO_2 by MeF, and the value obtained was $134 \pm 8 \text{ kJ mol}^{-1}$. This is in good agreement with an *ab initio* calculation for $(HF)_2H^+$ which gave 133 kJ mol^{-1} (Del Bene *et al.*, 1985). [The hydrogen bond energy of $(SO_2)_2H^+$ was determined to be 97 kJ mol^{-1}

(McMahon and Kebarle, 1986).] In an earlier assessment of the hydrogen-bond energy of $(HF)_2H^+$, the photoionization $(HF)_3 \rightarrow (HF)_2H^+ + F^-$ was used to obtain a value of $105 \, kJ \, mol^{-1}$ (Ng et al., 1977; Tiedemann et al., 1979).

McMahon and Kebarle's results indicate that the hydrogen bonding in $(HF)_2F^+$ is about $30 \, kJ \, mol^{-1}$ weaker than the bifluoride ion. This is not the case with other systems where it would appear that positively charged species are stronger than their negatively charged counterparts; for example, the hydrogen bond of $(H_2O)_2H^+$ (Lau et al., 1982) is claimed to be stronger than that of $(HO)_2H^-$ (Arshadi nd Kebarle, 1970). However, when their hydrogen-bond lengths are measured, the former, which often turns up in acid hydrates, has a bond length of ca. 245 pm, whereas, in the one example of the latter so far discovered, the bond length is 229 pm (Abu-Dari et al., 1979).

The protonated derivatives of HF were only recently discovered in crystals obtained from the superacid system HF–SbF$_5$. Interpretation of the ir spectrum of this combination had shown that H_2F^+ was present when there was a 40% molar excess of SbF$_5$ (Bonnet and Mascherpa, 1980) and $H_3F_2^+$ existed below this concentration (Olah et al., 1985). Mootz and Bartmann (1988) were able to grow two crystal phases from this acid at temperatures between -20 and $-30°C$ and characterize them by X-ray structure analysis, showing them to be the fluoronium salts $H_2F^+Sb_2F_{11}^-$ and $H_3F_2^+Sb_2F_{11}^-$. The former lattice has hydrogen bonds between the fluoronium ions with $R_{F\cdots F} = 264$ and 278 pm, whereas the latter has discrete bifluoridonium ions with $R_{F\cdots F} = 230$ pm. The calculated theoretical value for this is 229.2 pm, with the proton centred in a C_{2h} structure (Del Bene et al., 1985). Agreement between experiment and theory could hardly be closer.

The internuclear distance in $H_3F_2^+$ represents a large overlapping of the fluorine van der Waals radii indicative of a very strong hydrogen bond, but it still falls short of the hydrogen bond length of the bifluoride ion (226 pm). The issue is not completely resolved because even in $H_3F_2^+Sb_2F_{11}^-$, the $H_3F_2^+$ ion is not an isolated entity in hydrogen bond terms. It also forms hydrogen bonds with its terminal protons to fluorine atoms on the $Sb_2F_{11}^-$ anions, and these too are quite short with $R_{F\cdots F} = 241$ pm. Until a crystal is discovered in which $H_3F_2^+$ is "free", the issue of $H_3F_2^+$ versus HF_2^- will remain unresolved. Meanwhile, there seems good evidence for HF_2^- being taken as the archetypal very strong hydrogen bond.

THE BIFLUORIDE ION, HF_2^-

Work on HF_2^- has mostly been motivated by a desire to understand the chemical bonding. Its potential as a reagent has been less well investigated but, recently, there have been examples of bifluorides being used in the synthesis of organofluorine compounds under mild conditions (Bethell *et al.*, 1977; Brown and Clark, 1985) or for the introduction of ^{18}F into a glucose molecule by a stereospecific reaction (Szarek *et al.*, 1982). The role of the hydrogen bond in these reactions is not clear.

Although the bond in HF_2^- is generally accepted as the strongest hydrogen bond, the other assumption, that it has a single minimum potential energy well, has been questioned. An interpretation of the ir spectra of $NaDF_2$ as a solute in $NaHF_2$, and $NaHF_2$ as a solute in $NaDF_2$ in terms of a double minimum potential well has been made. The key evidence was thought to be the splitting of the asymmetric stretching vibration, v_3, in the latter combination (Spinner, 1980a). However, this assignment is open to much debate as we shall see. Neutron-diffraction studies on KHF_2 placed the proton midway between the two fluorines, which suggests a single minimum (Ibers, 1964), although it was acknowledged that it would not be possible by this method to distinguish between a single and a double minimum potential energy well if the two minima were less than 16 pm apart.

Hydrogen-bond length

The chemical composition of bifluorides was known in the nineteenth century (Borodin, 1862) and their existence explained in terms of hydro-fluoric acid being dibasic (Mallet, 1881). The truth came to light in 1923 when the X-ray structure of KHF_2 was published (Borzorth, 1923); this showed $R_{F...F}$ to be 226 pm. Although it was not possible to locate the proton, it was deduced to be between the two fluorines, this being the only way in which the closeness of these two negatively charged species could be explained. Subsequent structural determinations gave different values for $R_{F...F}$ of 250 pm in $NaHF_2$ (Anderson and Hassel, 1926) and 237 pm in both NH_4HF_2 (Hassel and Luzanski, 1932) and $TlHF_2$ (Pauling, 1933). These determinations were subject to some error and fresh determinations on KHF_2 (Helmholz and Rogers, 1939) and NH_4HF_2 (Helmholz and Rogers, 1940) gave $R_{F...F} = 226$ pm.

Since these early X-ray diffraction experiments, several other crystals have been subjected to analysis by X-ray and neutron diffraction and the $R_{F...F}$ bond length has remained virtually unchanged. The location of the proton has been the point at issue. Early ir studies on KHF_2 were interpreted as evidence for a double minimum potential well (Ketelaar, 1941; Glocker and Evans, 1942), but later studies questioned this (Pitzer and Westrum, 1947; Cote and Thompson, 1951; Newman and Badger, 1951) and led to a revision of earlier opinions (Ketelaar and Vedder, 1951).

A neutron-diffraction analysis of a single crystal of KHF_2 found the proton to be centred to within 100 pm (Peterson and Levy, 1952). An early nmr investigation of the ^1H and ^{19}F spectra came to the same conclusion and narrowed the uncertainty to 60 pm (Waugh et al., 1953), and a later nmr analysis reduced it to 25 pm (Paratt and Smith, 1975). A neutron diffraction study of $NaDF_2$ showed this to have a centred deuterium atom with $R_{F\cdots F} = 226.5$ pm (McGaw and Ibers, 1963).

Table 7 The hydrogen-bond lengths of the bifluoride ion.

System	Comments	$R_{F\cdots F}$/pm	Reference
K, Rb, $CsHF_2$	Phase changes at high temperatures	226	[a]
NH_4HF_2	Variable bond length	227.5 226.9	[b]
$LiHF_2$		227	[c]
$NaHF_2$, $NaDF_2$	Neutron	226.4(H) 226.5(D)	[d]
KHF_2	Neutron	227.7	[e]
$C_7H_{10}N^fHF_2$	Symmetrical ion	227.6	[g]
$C_7H_{10}N^fHF_2$	Neutron; shows an unsymmetrical bond	226.0	[h]
$K_3[TaO_2F_5][HF_2]$	Two $HF_2{}^-$ ions	225	[i]
	$T = 290$ and 170 K	224.2, 223.3	[j]
$Na_3[NbO_2F_2][HF_2]$	Both symmetrical	228.3, 229.2	[k]
$SrF(HF_2)$	Unsymmetrical	226.9	[l]
$BaF(HF_2)$	Unsymmetrical	228.1	[l]
$Te(tu)_3(HF_2)_2{}^m$	Unsymmetrical	228.5, 229.1	[n]

[a] Kruh et al. (1956); [b] McDonald (1960); [c] Frevel and Rinn (1962); [d] McGaw and Ibers (1963); [e] Ibers (1964); [f] p-Toluidinium; [g] Denne and MacKay (1971); [h] Williams and Schneemeyer (1973); [i] Toros and Prodic (1976); [j] Stomberg (1982); [k] Stomberg (1981); [l] Massa and Herdtweck (1983); [m] tu = thiourea; [n] Beno et al. (1984).

Other structural analyses of crystals in which the bifluoride is present are listed in Table 7. One compound, p-toluidinium fluoride $[C_7H_{10}N^+][HF_2{}^-]$, is worthy of further comment. The first X-ray diffraction study reported a symmetrical anion (Denne and MacKay, 1971), but a later analysis showed that the proton was not centred between the two fluorines and $R_{F\cdots H}$ values were 102.5 and 123.5 pm (Williams and Schneemeyer, 1973). This can be explained not by a double minimum potential energy well but by asymmetry due to other forces, such as secondary hydrogen bonding between one end of the bifluoride anion and the N—H group of the cation. An alternative explanation attributes the asymmetry of the bifluoride hydrogen bond to an unsymmetrical crystal field caused by the cation (Ostlund and Bellenger, 1975).

Hydrogen-bond energy

This quantity was the subject of several indirect attempts at calculation based on physical properties as well as *ab initio* computations. A direct measurement was finally achieved by ion cyclotron experiments and gave a value of 163 kJ mol^{-1} (McMahon and Larson, 1982a) for ΔE[HF(g) + F$^-$(g) $\rightarrow$ FHF$^-$(g)]. Shortly after, *ab initio* calculations with the basis set [(14s)^8(9p)^5(2d)/(10s)^4(1p)] fell in line with a value of 169 kJ mol^{-1} (Emsley *et al.*, 1983). Prior to this, a wide range of values for the bond energy had been canvassed, often with theoretical computations providing support (see Table 8).

Table 8 Experimental, empirical and theoretical estimates of the hydrogen-bond energy of the bifluoride ion, ΔE[HF$_{(g)}$ + F$_{(g)}^-$ $\rightarrow$ FHF$_{(g)}^-$]/kJ mol^{-1}.

Method	ΔE/kJ mol^{-1}	Reference
Experimental		
ΔH[MF$_{(s)}$ + HF$_{(g)}$ $\rightarrow$ MHF$_{2(s)}$; M = K, Rb, Cs]	243	a
ΔH[Me$_4$NF$_{(s)}$ + HF$_{(g)}$ $\rightarrow$ Me$_4$NHF$_{2(s)}$; with $\Delta U = 0$]	155	b
ΔH[Me$_4$NF$_{(s)}$ + HF$_{(g)}$ $\rightarrow$ Me$_4$NHF$_{2(s)}$; with $\Delta U = 30$]	185	c
Ion cyclotron resonance	163	d
Empirical calculations		
From dipole moments	241	e
From multipole moments and Madelung energy	234	f
From temperature studies of gas-phase equilibria	125	g
From calculated lattice energies of MHF$_2$	176	c
From permanent and induced dipole moments	219	h
Ab initio *calculations*		
basis sets: 4-31G	262	i
6-31G**	239	i
(9s)^4(5p)^2/(4s)^2(1p)	222	j
(9s)^5(5p)^3/(4s)^3(1p)	214	i
(9s)^4(5p)^2(1d)/(4s)^2(1p)	213	k
(9s)^5(5p)^3(1d)/(4s)^3(1p)	214	i
(10s)^6(7p)^4(2d)/(6s)^4(1p)	179	l
(11s)^5(7p)^3/(5s)^2(1p)	168	m
(14s)^8(9p)^5(2d)/(10s)^4(1p)	169	n

a Waddington (1958); *b* Harrell and McDaniel (1964); *c* Jenkins and Pratt (1977); *d* McMahon and Larson (1982a); *e* Fyfe (1953); *f* Neckel *et al.* (1971); *g* Yamdagni and Kebarle (1971); *h* Fujiwara and Martin (1974a); *i* Clark *et al.* (1981); *j* Størgard *et al.* (1975); *k* Almlöf (1972); *l* Keil and Ahlrichs (1976); *m* Noble and Kortzeborn (1970); *n* Emsley *et al.* (1983).

Since the proton is centred in the bifluoride ion, it is possible to redefine the hydrogen-bond energy in a way comparable to the mean bond energies of covalent molecules in which all the bonds are equivalent. In this case, the total bond energy of the system is $E[\text{H}\text{—F}(g)] + \Delta E[\text{HF}(g) + \text{F}^-(g) \rightarrow \text{FHF}^-(g)]$, i.e. $566 + 163$, or 729 kJ mol^{-1}. If this is so, the mean energy per $\text{F} \cdots \text{H}$ bond is 365 kJ mol^{-1}, higher than many ordinary covalent bonds. Remarks to the effect that there is a significant covalent contribution to strong hydrogen bonds (Desmeules and Allen, 1980) seem well in accord with the bifluoride system.

Other properties

Infrared spectra. Early reports on the spectra of the difluoride salts divide into those which support (Pitzer and Westrum, 1947) or refute (Blinc, 1958) the idea of the anion having a single minimum potential energy well. This debate has rumbled on with Spinner remaining as the sole champion of the double minimum/low barrier profile, on the basis of the ir spectrum (Spinner, 1977, 1980a). A more contentious issue, however, is the assignment of the asymmetric stretching vibration, v_3.

If FHF^- is centred and linear with symmetry $D_{\infty h}$, then the symmetrical stretching frequency, v_1, will be ir inactive but Raman active; v_2, the bending mode, and v_3 will be ir active but Raman inactive. Difficulty in assigning bands in the spectra of bifluoride salts is complicated by lattice effects which can lift the degeneracy of v_2 or reduce the symmetry of the ion to $C_{\infty v}$. There has been universal agreement that v_1 comes at *ca.* 600 cm^{-1} and v_2 at *ca.* 1250 cm^{-1}, but there has been wide disagreement over the location of v_3, choices spanning the range 1284–1848 cm^{-1}. A useful tabulation of earlier assignments has been given (Kawaguchi and Hirota, 1986). These workers were the first to report a gas-phase ir spectrum of FHF^- when studying the spectrum of CF^+. Subsequently, they were able to generate the bifluoride ion by a hollow cathode discharge in a mixture of H_2 and a fluorine-containing molecule such as CF_4. Their data are given in Table 9, together with values obtained from solid-state spectra and *ab initio* calculations.

The theoretical results given in Table 9 were obtained using the highest level of theory (Janssen *et al.*, 1986). The authors of this study found that v_3 varied according to the theoretical input. Their *ab initio* methods varied from a low of "double-zeta plus polarization", which gave a calculated value of 1698 cm^{-1}, up to "triple-zeta plus triple polarization plus diffuse augmented by hydrogen d function" which gave a value of 1427 cm^{-1}. The computed values of v_1 remained steady, v_2 varied, but much less so than v_3.

The isotopic ratio $v(\text{FHF}^-)/v(\text{FDF}^-)$ has a value of unity for v_1 since the proton does not move during the course of the symmetric stretching vibration. For v_2 and v_3, this ratio should be 1.396 if the motion of the hydrogen is perfectly harmonic, with both H and D moving in a single minimum potential energy well. The ratio will also have the classical value if both H and D are above the internal barrier of a double minimum well. Several measurements of $v(FHF^-)/v(FDF^-)$ have been carried out. Details have been tabulated (Kawaguchi and Hirota, 1986) and the results are consistent with a single minimum well. The value of v_3 for the gas phase ions is a surprisingly low 1.32, whereas values obtained from the solid state give 1.42 (Chunnilall et al., 1984). Matrix isolation in Ar gave 1.41 (Ault, 1982), and theoretical calculations gave 1.45 (Janssen et al., 1986). For v_2 the isotopic ratios 1.39 (Chunnilall et al., 1984) and 1.45 have been found (Janssen et al., 1986).

Table 9 Comparison of the vibrational frequencies (cm^{-1}) of the bifluoride ion.

System	v_1	v_2	v_3	Reference
Gas phase	617	1241	1848	[a]
Solid phase				
Salts:				
$NaHF_2$	[b]	1199	1284	[c]
$NaHF_2$	629	1209	1565	[d]
$NaDF_2$	630	873	1143	[d]
KHF_2	[b]	1214	1314, 1372	[c]
$CsHF_2$	[b]	1217	1364	[c]
$C_7H_{10}NHF_2$[e]	[b]	1080, 1230	1749	[f]
Me_4NHF_2	650	1265–1255	1376	[g]
Ar matrix	[b]	1217	1364	[h]
KCl matrix[i]	[b]	1254	1563	[j]
Ab initio	617	1363	1427	[k]

[a] Kawaguchi and Hirota (1986); [b] Not reported; [c] Ault (1979); [d] Spinner (1980a); [e] p-Toluidinium bifluoride; [f] Harmon et al (1974); [g] Harmon and Lovelace (1982); [h] Ault (1978); [i] Other halide matrixes also used; [j] Chunnilall et al. (1984); [k] Janssen et al. (1986).

Harmon and Lovelace (1982) studied the ir spectrum of p-toluidinium bifluoride and tetramethylammonium bifluoride and concluded that the hydrogen bonds are different from that of KHF_2. The first is different because it is known to be unsymmetrical, but the second was claimed to be even stronger than the hydrogen bond in KHF_2. For the tetramethylammonium salts of HF_2^- and DF_2^- the isotope ratios were 1.40 (v_2) and 1.41 (v_3). Clearly, the final word on the vibrational analysis of FHF^- has still to be written. Calculations of ionic force fields based on ir data have been carried out (Matsui et al., 1986), but since these are based on v_3 it is probably unwise to read too much into them.

Nmr spectroscopy. Early ^{1}H- and ^{19}F-nmr studies were interpreted in terms of the potential energy well being a single minimum (Waugh *et al.*, 1953), and this has been supported in later nmr analyses.

What makes FHF$^-$ an attractive species to investigate by nmr spectroscopy is that it consists of three nuceli each with spin 1/2 bonded directly. Also, the proton of the strong hydrogen bond should have an unusual chemical shift. Early work failed to detect the expected ^{19}F doublet and ^{1}H triplets (e.g. Soriano *et al.*, 1969), and it was not until the importance of the solvent was appreciated that coupling was observed (Fujiwara and Martin, 1971, 1974a,b). Suitable media were found to be the dipolar aprotic solvents acetonitrile, nitromethane and dimethylformamide.

In these solvents it is possible to obtain spectra of Et_4NHF_2 at temperatures between 233 and 307 K, and peaks were found at $\delta(^1H$, ref. Me$_4$Si) = 16.37 ppm [cf. HF, 7.64 ppm] and $\delta(^{19}F$, ref. CF$_4$) = -83.35 ppm [cf. HF, -118.7 ppm] with $^1J_{HF} = 120.5$ Hz [cf. HF, 476 Hz]. In chlorinated solvents and in pure liquid $(n\text{-}C_4H_9)_4NHF_2$ the multiplets coalesce (Soriano *et al.*, 1969). What is striking about the chemical shift of the proton is that it is not as far downfield as many other hydrogen-bonded protons. This, however, is not inconsistent with the idea that a very strong hydrogen bond would be more shielded as the heavier nuclei approach each other. The critical result in assigning a single minimum potential energy well to HF$_2^-$ is the value of $\Delta[\delta(^1H) - \delta(^2H)]$, -0.30 ppm (Gunnarsson *et al.*, 1976), proving conclusively that both H and D are above the energy barrier.

Solid-state nmr spectroscopy on a single crystal of KHF$_2$ was interpreted in terms of a centred proton with $R_{H\cdots F} = 113.8$ pm (Pratt and Smith, 1975). The same conclusion was reached for the other alkali-metal bifluorides (Ludman *et al.*, 1977) and the displacement of the proton from the midpoint was estimated to be less than 60 pm. This technique cannot distinguish between a single minimum and a double minimum with fast transfer. Fluorine-19 nmr spectroscopy has been used to study solutions of HF$_2^-$ in formic acid, and the data have been interpreted in terms of equilibria involving F$^-$, HF, HF$_2^-$ and H$_2$F$_2$ (Coulombeau, 1977).

Fractionation factor φ. A value of 0.60 for φ has been calculated from the ratio of acid-dissociation constants in H$_2$O and D$_2$O by Kreevoy (Kreevoy *et al.*, 1977; Kresge and Chiang, 1973). This value is not as low as that for the many other systems which are known to have weaker hydrogen bonds.

Polyfluorides, $H_nF_{n+1}^-$ (n = 2–4)

Polyfluorides are less well known than HF$_2^-$, but they provide good examples of multiple hydrogen bonding to a single acceptor site since they consist of two or more HF molecules clustered round a central fluoride ion.

The ir spectra of $H_2F_3^-$, $H_3F_4^-$ and $H_4F_5^-$, as their potassium salts, have been reported at low temperatures (Harmon et al., 1977). These workers regard the unsymmetrical HF_2^- ion, found in the p-toluidinium salt, as the first member of the series $F(HF)_n^-$.

Temperature studies on the system Me_4NF–HF revealed four distinct phases: $Me_4NH_2F_3$, mp 110°C; $Me_4NH_3F_4$, mp 20°C (decomposes); $Me_4NH_5F_6$, mp −76°C (decomposes); and $Me_4NH_7F_8$, mp −110°C (decomposes) (Mootz and Boenigk, 1987). Most of these phases undergo solid–solid phase transitions. A crystal structure of the trifluoride shows two HFs attached to a central F^- with $R_{F...F} = 230.2$ and 231.6 pm and with an angle of 123° [Fig. 12(a)]. The hydrogen atoms remain covalently close to their parent fluorines.

In the tetrafluoride ion, $H_3F_4^-$, the central fluoride is above the plane defined by the fluorine atoms of the HF molecules [Fig. 12(b)], and the $R_{F...F}$ distances are 235.1, 235.2 and 235.7 pm. In this work by Mootz and Boenigk, there was no pentafluoride phase, but the structure of $H_4F_5^-$ has been known for some time as a distorted tetrahedral arrangement of four HF molecules around F^- with $R_{F...F} = 245$ pm (Ibers, 1964). All the above polyfluorides were the subject of ab initio computations which gave very short hydrogen bonds in configurations compatible with those observed (Clark et al., 1981b).

The hexafluoride ion, $H_5F_6^-$, was the first of the unexpected phases in the Me_4NF–HF system, and the crystal structure showed it to consist of two trifluoride units linked through a very short and centred hydrogen bond with $R_{F...F} = 226.6$ pm. An alternative way of viewing this anion is as a central FHF^- to which two HF molecules are bonded to each fluorine by asymmetrical hydrogen bonds as shown in Fig. 12(c) (Mootz and Boenigk, 1987). The octafluoride ion, $H_7F_8^-$, remains tantalizingly out of reach.

The trigonal tetrafluoride, $H_3F_4^-$, has been identified in other crystals such as NOF·3HF with $R_{F...F} = 238.9$ pm (Mootz and Poll, 1984b), KF·3HF with $R_{F...F} = 240.1$ pm (Mootz and Boenigk, 1986), and NH_3·4HF with $R_{F...F} = 240.6$ pm (Mootz and Poll, 1984a). A linear form of this anion [see Fig. 12(d)] is also known in KF·2.5HF (Mootz and Boenigk, 1986). In $H_3F_4^-$ there is again a central symmetrical FHF^- unit with $R_{F...F} = 228.1$ pm and an HF molecule attached to each end with asymmetrical FHF bonding and $R_{F...F} = 240.2$ and 244.1 pm.

The ability of the fluoride ion to act as a hydrogen-bond acceptor, even towards the very weakest of hydrogen-bond donors, such as the C—H bond, has been investigated. McMahon and Roy (1985) computed the relative stabilities of $C_2H_4F^-$ and $C_2H_2F^-$ by ab initio methods and showed that

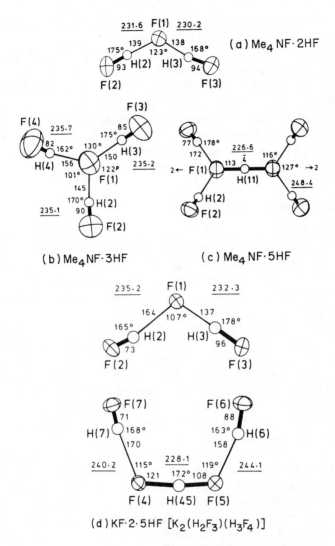

Fig. 12 Polyfluoride structures (a), (b), (c) (adapted from Mootz and Boenigk, 1987) and (d) (adapted from Mootz and Boenigk, 1986). Bond lengths are in pm.

the most stable forms of these are not the fluoroethyl or α-fluorovinyl carbanions, respectively, but a hydrogen-bonded adduct. In the case of $CH_2{=}CH_2 \cdots F^-$, the bond energy is calculated to be *ca.* $10\,kJ\,mol^{-1}$, in good agreement with the experimental fluoride-binding energy (McMahon and Larson, 1983). Similarly, the hydrogen-bond energy of the ethyne adduct, $HC{\equiv}C{-}H \cdots F^-$, was computed to be $80\,kJ\,mol^{-1}$, again in good agreement with experimental methods which gave a range between 60 and 80 kJ mol^{-1} (McMahon and Larson, 1983).

Although the fluoride ion is normally a very strong hydrogen-bond acceptor, the adduct formed between tetra-n-butylammonium fluoride and 2,6-di-t-butyl-4-methylphenol contains a weak F$^-$ ··· H—O bond with Δv_s(OH) = 216 cm^{-1} and with E(OHO) = 29 kJ mol^{-1} calculated from the temperature dependence of the ^{1}H chemical shift (Owen, 1989). Normally, however, F$^-$ forms strong hydrogen bonds. This is in sharp contrast to fluorine's hydrogen-bond acceptor properties when it is covalently bonded; then it is among the weakest acceptors.

Covalently bonded fluorides

Although the gas-phase hydrogen-bonded dimer (MeF)$_2$H$^+$ is held by a strong hydrogen bond (McMahon and Kebarle, 1986) this is a rare exception to the previous statement regarding covalently bonded fluoride. More typical are the perfluorocarbons, which are among the weakest hydrogen-bonding substances known, as their physical properties and uses clearly demonstrate.

However, not all C—F bonds are reluctant to participate in hydrogen bonding. In fluorocitrate esters, where a C—F bond is adjacent to two carboxylate groups, crystal structures have revealed C—F ··· H—N distances as short as 286.7 pm to amine groups in a nearby cation (Murray-Rust et al., 1983). Here, the attraction of ions of opposite charge is clearly of help in forming such hydrogen bonds. Fluorinated carbohydrates have the potential to form C—F ··· H—O bonds, and in these compounds hydrogen bonding plays an important role (Taylor, 1988).

One compound, 2-deoxy-2-fluoro-β-D-mannopyranosyl fluoride, whose X-ray crystal structure has been determined, shows C—F ··· H—O hydrogen bonds linking molecules in the lattice with $R_{F···O}$ = 282.7 pm (Withers et al., 1986). This internuclear distance appears short for such a bond involving a covalent fluorine, but the hydrogen bond angle is only 108°, which suggests that the shortness of the bond may owe more to lattice forces than to strong hydrogen bonding. The crystal structure was also claimed to reveal C—H ··· F—C hydrogen bonding, although with $R_{F···C}$ = 330.1 pm; this is open to debate since Σ_w(C + F) is 320 pm (185 + 135 pm). The bond energy was calculated to be ca. 13 kJ mol^{-1} (Withers et al., 1988).

Fluorinated carbohydrates have been used to great effect in studying the hydrogen bonding between sugars and the glucose-binding states in glycogen phosphorylase (Withers et al., 1986, 1989). The crystal structure of an enzyme–inhibitor complex of porcine pancreatic elastase with a peptidyl-α,α-difluoro-β-ketoamide shows the inhibitor to fit snugly within the enzyme. There are five intermolecular hydrogen bonds between the enzyme and the inhibitor, including a C—F $\cdots$ H—N bond with $R_{F\cdots N} = 276$ pm (Takahashi *et al.*, 1989). Since $\Sigma_w(F + N)$ is 302 pm, this hydrogen bond shows considerable shortening and may well justify the appellation "strong" given to it by its discoverers.

Another example of a C—F bond participating in hydrogen bonding has been found in calcium 2-fluorobenzoate dihydrate (Karipides and Miller, 1984). In the crystal structure of this compound the $R_{F\cdots O}$ distance between the fluorine and a water molecule bound to the calcium as a ligand was 299.4 pm, and the hydrogen bond angle was 170°.

A hydrogen bond between the very weak acceptor (C—F) and very weak donor (C—H) has recently been found in the diruthenium complex [26] (Howard, *et al.*, 1989). In this complex, there is hydrogen bonding between the CH_2 group bridging the two ruthenium atoms and a C—F bond of a perfluoropropene ligand. The X-ray crystal structure gave a bond length $R_{H\cdots F} = 223$ pm, which is less than the sum of the van der Waals radii of H and the acceptor F (242 pm).

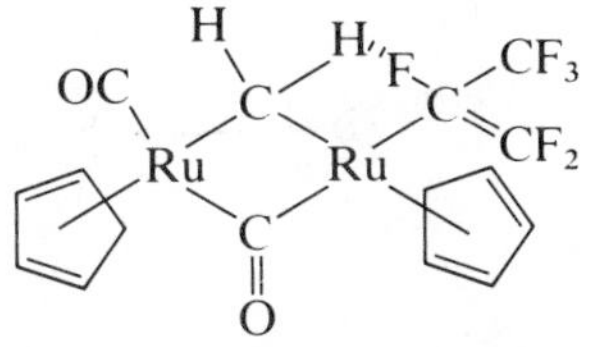

[26]

When fluorine is covalently bonded to silicon, it finds itself in a very different situation because of the affinity of the high polarity associated with the Si—F bond. In $FeSiF_6 \cdot 6H_2O$ there is hydrogen bonding between the SiF_6^- and $[Fe(H_2O)_6^{2+}]$ units with $R_{F\cdots O} = 268, 272$ pm (Hamilton, 1962). Similar inter-ion hydrogen bonding is found in $CoSiF_6 \cdot 6H_2O$ with $R_{F\cdots O}$ = 271.1 and 278.5 pm (Lynton and Siew, 1973).

When the hydrogen-bond donor is a hydrated oxonium ion, $H_5O_2^+$ or $H_7O_3^+$, hydrogen bonding to Si—F can be strong, as shown by the crystal hydrates of hexafluorosilicic acid, $H_2SiF_6 \cdot xH_2O$, where $x = 4$, 6 or 9.5 (Mootz and Oellers, 1988). The $R_{F\cdots O}$-value varies from compound to compound, but the minimum in each case is much shorter than we might expect for a covalently bonded fluorine. Figure 13 shows the hydrated oxonium ions and their hydrogen bonds in these compounds, with the shortest $R_{F\cdots O}$ given in each case.

Perhaps the best demonstration of the unwillingness of covalently bonded fluorine to participate in hydrogen bonding is hypofluorous acid, HOF. This

$H_2SiF_6 \cdot 4H_2O$ $H_2SiF_6 \cdot 6H_2O$

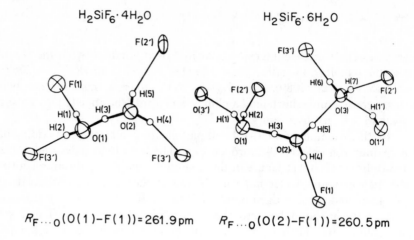

$R_{F \cdots O}(O(1)-F(1)) = 261.9 \, pm$ $R_{F \cdots O}(O(2)-F(1)) = 260.5 \, pm$

$H_2SiF_6 \cdot 9.5H_2O$

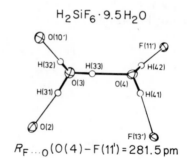

$R_{F \cdots O}(O(4)-F(11')) = 281.5 \, pm$

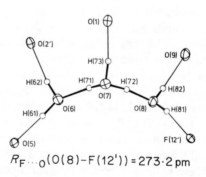

$R_{F \cdots O}(O(8)-F(12')) = 273 \cdot 2 \, pm$

Fig. 13 Hydrogen bonding between hydrated oxonium ions and SiF_6^{2-} (adapted from Mootz and Oellers, 1988).

can be isolated only at very low temperatures as an intermediate in the formation of HF and O_2 by the reaction of fluorine and water. The HOF produced by this method was investigated by ir and Raman spectroscopy and it was deduced that the condensed phase consisted of HOF molecules linked by O—H $\cdots$ F hydrogen bonds (Appelman *et al.*, 1981; Appelman and Kim, 1982; Appelman and Thompson, 1984). A later interpretation of the spectroscopic data suggested only O—H $\cdots$ O bonding, and this was supported by theoretical calculations (Christe, 1987). A single crystal of HOF, grown at $-160°$C, has now been shown to consist of layers made up of parallel strands of HOF molecules linked through O—H $\cdots$ O hydrogen bonds; $R_{O \cdots O} = 289.5$ pm (Mootz *et al.*, 1988). The O—F bonds project above and below the layers and take no direct part in the hydrogen bonding.

3 Strong hydrogen bonding in β-diketones

The keto/enol equilibrium (15) has been a spur to much research. In the absence of catalysts the equilibrium is established slowly and is very sensitive to a variety of influences, both internal, such as the nature of α- and β-substituents, and external, such as temperature and solvent. The discovery that the equilibrium was established sufficiently slowly to permit both keto and enol tautomers to be observed by ^{1}H-nmr spectroscopy allowed these several influences to be easily investigated (see Kol'tsov and Kheifets, 1971, for a review of the early work, and Emsley, 1984, for later work).

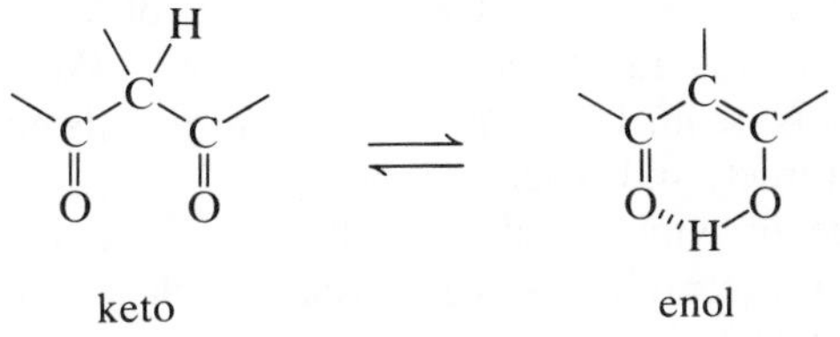

keto   enol   (15)

Investigations by ^{1}H-nmr spectroscopy also led to debate about the hydrogen bond itself, since the enol proton was found to resonate far downfield of other deshielded protons, typically in the range below 15 ppm. Pentane-2,4-dione, the best known and most investigated β-diketone, has a δ-value of 15.40 ppm. This observation has led to much speculation that the hydrogen bond might indeed be centred; it was backed up by evidence from electron-diffraction studies and supported by theoretical calculations. The enol tautomers of pentane-2,4-dione and other β-diketones are now generally considered to have strong, non-centred hydrogen bonds.

KETO/ENOL EQUILIBRIUM

The earliest estimates of the enol content of various β-diketones was made by careful distillation (Meyer and Schoeller, 1920; Meyer and Hopff, 1921). Although, as a method of studying the equilibrium, this approach has been replaced by non-intrusive methods, work on the physical separation of the tautomers has continued (e.g. Regitz and Schäfer, 1981; Vogt and Gompper, 1981).

Traditionally, the position of equilibrium of (15) has been expressed as the percentage of enol, and for neat pentane-2,4-dione this is 79%. The equilibrium constant is especially sensitive to solvents; for pentane-2,4-dione the enol content varies from 13% in water to 98% in cyclohexane, corresponding to a *ca.* 300-fold change in the value of the equilibrium constant.

Likewise, different α- and β-substituents can dramatically affect the balance at equilibrium; some compounds such as 3-biphenylpentane-2,4-dione (Emsley *et al.*, 1988b) and 1,3-diphenylpropane-1,3-dione (Allen and Dwek, 1966) exist almost entirely as the enol tautomer in CCl_4 solution, while others, such as 1,3-diphenyl-2-methylpropane)-1,3-dione (Emsley *et al.*, 1987), are present entirely as the keto tautomer in both the solid state and solution. The effect of α-substituents on the equilibrium for pentane-2,4-dione shows how the tendency for enol formation is related to the electronic effect of the substituent affecting the acidity of the CH proton. Electron-releasing groups keep this low and so favour the keto tautomer; e.g. the methyl group gives 28% enol (Allen and Dwek, 1966) while the isopropyl group gives less than 0.2% enol (Yoffe *et al.*, 1966). Electron-attracting groups increase the acidity and thus favour the enol tautomer; e.g. cyano- and carbomethoxy-substituents both give 100% enol (Wierzchowski *et al.*, 1963; Forsén and Nilsson, 1960) and halide substituents give 46% (Br) and 92% (Cl) enol (Burdett and Rogers, 1964).

The percentage of enol correlates with the dielectric constant of the medium, with few exceptions. Some solvents, such as triethylamine, were reported to give 100% enol based on the nmr spectrum and were thought to favour the enol by formation of a hydrogen bond to this tautomer (Leipert, 1977; Raban and Yamamoto, 1977). Infrared analysis, however, shows there is 8% of the keto form present, which is about the expected percentage based on the dielectric constant of triethylamine (Emsley *et al.*, 1986a). Hydrogen-bonding between the enol form and diethylamine can occur and a 2/2 adduct crystallizes from a mixture of pentane-2,4-dione and diethylamine (Emsley *et al.*, 1986b). The structure of the adduct reveals three-centre hydrogen bonds between the two oxygens of the diketone and the nitrogen of the amine (Fig. 14).

A comprehensive study of the keto/enol equilibrium for pentane-2,4-

dione in 21 solvents, for 3-methylpentane-2,4-dione in 14 solvents and for 3-ethylpentane-2,4-dione in six solvents has been carried out (Emsley and Freeman, 1987b). The percentage of enol was shown to correlate best (correlation coefficient = 0.968) with the solvent-polarity parameter, A + B, devised by Swain (Swain *et al.*, 1983). Other solvent-polarity parameters such as π^*, E_T and ε were found to relate less well. The conclusion was that hydrogen-bonding solvents have no special influence on the percentage of enol and that the cyclic intramolecular hydrogen bond of the enol remains intact in all solvents.

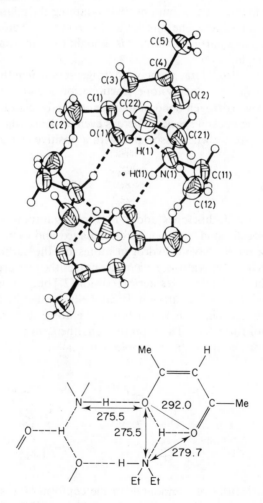

Fig. 14 Structure of the pentane-2,4-dione–diethylamine adduct (taken with permission from Emsley *et al.*, 1986b).

Since there is a large difference in the polarity of the keto and enol tautomers caused by the transfer of the proton, it has often been said that polar solvents, like water, favour the polar keto tautomer, and non-polar solvents, like cyclohexane, favour the enol form. However, theoretical calculations suggest that the keto tautomer of pentane-2,4-dione has a lower dipole moment than the enol (Buemi and Gandolfo, 1989). Consequently, these authors deduce that dipole–dipole interactions alone cannot account for the high percentage of enol found with less polar solvents. These computations were based on a solvent-cavity model and used quadrupole- and higher-moment contributions in determining the relative energies of the keto and enol forms. The answer as to where the keto/enol equilibrium comes to rest in a particular solvent is thought to lie in some unspecified way with these higher terms.

A large body of data, namely, hydrogen-bond lengths, bond energies, ir and nmr spectra, and isotope fractionation factors, is available for β-diketones. The hydrogen bond in some enols can be characterized as strong, a rarity among neutral species. There is even evidence that this bonding can become very strong when associated with a positive charge.

STRUCTURES

Although many β-diketones adopt a *cis*-configuration with an intramolecular hydrogen bond, this can be discouraged and even prevented. It can be discouraged by having hydrogen atoms on the carbonyl carbons, i.e. malondialdehyde derivatives, which encourages the two protons to adopt an arrangement which minimizes steric repulsion [27] but requires the β-diketone to adopt an s-*trans*-configuration. It can be prevented by incorporating the diketone into a cyclic system which would enforce the s-*trans*-arrangement of the dienol form [28]. These precautions then ensure that the hydrogen bonding of the enol will be intermolecular.

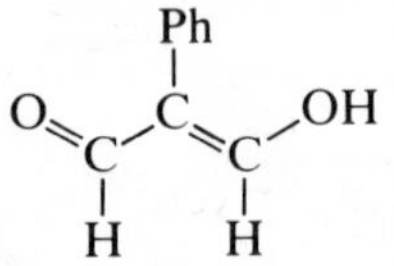

[27] 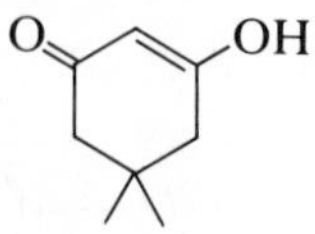   [28]

Crystal structures of compounds of the conformation shown in [27] and [28] have been reported: 2-phenylmalondialdehyde (Semmingsen, 1977) and 5,5-dimethyl-1,3-cyclohexanedione, i.e. dimedone (Semmingsen, 1974; Singh and Calvo, 1975). The former has an arrangement of infinite planar zig-zag chains with $R_{O\cdots O} = 257.7$ pm, while the latter consists of infinite helices in which $R_{O\cdots O} = 259.3$ pm. These values can be taken as 'typical' of hydrogen bonds unaffected by intramolecular steric forces within the *cis*-enols.

For the intramolecular hydrogen bond of pentane-2,4-dione, the cyclic enol ring has a bond length $R_{O\cdots O}$ of 253.5 pm, which is not significantly different from the $R_{O\cdots O}$-values for the intermolecular hydrogen bonds. The crystal structure of this β-diketone was first obtained when the molecule turned up as a solvate in a drug complex (Camerman *et al.*, 1983); the enol proton was 36 pm above the plane of the molecule. Recent gas-phase electron-diffraction measurements on pentane-2,4-dione also indicate an unsymmetrical hydrogen bond having $R_{O\cdots O} = 251.2$ pm with the proton *ca.* 45 pm out of the plane of the enol ring and $\angle$ OHO $= 137°$ (Iijima *et al.*, 1987).

Curiously, the first β-diketone structure to be reported showed a much shorter $R_{O\cdots O}$ of 246.4 pm in 1,3-bis(3'-bromophenyl)propane-1,3-dione (Williams *et al.*, 1962). Several other β-diketones proved to have similar bond lengths and these have been summarized in a previous review (Emsley, 1984). All structures showed enol tautomers to be present in the solid state. Since then, attempts have been made to tighten the grip of the carbonyl groups on the proton of the enol ring of pentane-2,4-dione by steric and electronic changes at the 3-position. Phenyl substituents produce crystalline materials and the first of these, 3-(4'-methoxyphenyl)pentane-2,4-dione showed a significant closing of the oxygen–oxygen distance to 244.9 pm (Emsley *et al.*, 1988a). Other derivatives whose crystal structures have been determined give $R_{O\cdots O}$-values as follows: 3-biphenyl, 244.1 pm (Emsley *et al.*, 1988b); 3-(4'-nitrophenyl), 244.5 pm, 3-(4'-phenoxyphenyl), 244.3 pm; 3-(4'-isopropylphenyl), 241.9 pm (Emsley *et al.*, 1989). Increasing the steric congestion with 2,4-dimethoxyphenyl and 3,4,5-trimethylphenyl substituents still resulted in bond lengths in the same range, 244.5 and 247.6 pm, respectively (Emsley *et al.*, 1990a,b). All these can be classed as strong hydrogen bonds on the basis of the $R_{O\cdots O}$-value.

Attempts to influence the hydrogen bond by changes at the β-carbons, for example, as in 1,3-diphenylpropane-1,3-dione, resulted in crystals whose structure showed them to be entirely of the keto tautomer with a *cis*-diketo conformation and a carbonyl–carbonyl dihedral angle of *ca.* 90° (Emsley *et al.*, 1987).

When the enol ring is adjacent to a cyclic moiety, then it is possible to achieve very short hydrogen bonds, as in the structure of usnic acid, a natural product found in lichens. A low-temperature X-ray diffraction analysis of this compound showed two enol moieties, one in which a carbon–carbon bond of the enol was part of a cyclohexenone ring, and this had

$R_{O \cdots O} = 240$ pm. The other enol ring, in which a carbon–carbon bond of the enol was part of a benzene ring, had $R_{O \cdots O} = 253$ pm (Norrestam et al., 1974). Also, in the molecule 2,3,4-trihydroxyacetophenone, the enol ring adjacent to the phenyl ring has $R_{O \cdots O} = 251.2$ pm (Schlemper, 1986), although in this case the OH group is also acting as a hydrogen-bonding acceptor to a neighbouring OH.

An analysis of 25 structures determined by X-ray or neutron diffraction and containing the enol grouping HO—C=C—C=O has been made with a view to finding the extent to which the hydrogen bonding in enols is "resonance assisted" (Gilli et al., 1989). In this work, attention is focused on the C—O and C—C bond lengths, the contention being that the system is resonance assisted if both C—O bonds, and both C—C bonds, are the same length. The differences in the C—O bonds was termed q_1, and this quantity was found to vary from 2 pm, showing no differentiation between the carbonyl group and the C—OH bond, up to 122 pm, where the C=O and C—OH bonds are obviously different.

The quantity q_2 was defined as the difference in bond length of the two C—C bonds in the ring. This value varied from 6 pm up to 106 pm. Since q_1 and q_2 linearly were related, as might be expected, the values of their sum, Q, was useful in discussing resonance-assisted hydrogen bonding. The value of Q was found to correlate well with $R_{O \cdots O}$-values and with other properties. However, when the Q-value of β-diketones is plotted against $R_{O \cdots O}$ there is little correlation (Emsley et al., 1990a).

The location of the enol proton is not clearly found by X-ray methods, which is why interest has focused on the other bonds of the enol ring. When the ring is symmetrical, the implication is that the proton will be centred between the two oxygens, although not in line with them. If the bonding in these compounds is to be classed as strong, then we would expect the proton to be above the energy barrier of the double minimum and to be centred. Its position will, of course, be very sensitive to crystal forces since it projects well above the plane of the molecule. This appears to be the case in those structures where some indication of the proton's location can be judged.

HYDROGEN-BOND ENERGIES

The hydrogen-bond energy, $E(OHO)$, of the cyclic enol is generally defined relative to the open form of the enol (16).

(16)

Ab initio calculations on malondialdehyde, $HCOCH_2COH$, can be made with relatively large basis sets. Nevertheless, hydrogen-bond energies vary considerably: 18.4 (Noak, 1979), 33.1 (Carlsen and Duus, 1980), 56.9 (Millefiori *et al.*, 1983), 75.0 (Emsley *et al.*, 1986b) and $37.36 \, kJ \, mol^{-1}$ (Buemi and Gandolfo, 1989). This last result was based on a semi-empirical method (AM1) and gave $E(OHO) = 32.76 \, kJ \, mol^{-1}$ for pentane-2,4-dione itself. Such a value would suggest a weak hydrogen bond in line with the longer bond length found for this β-diketone. Other β-diketones with $R_{O \cdots O}$ <245 pm would be expected to be much stronger, and their ir spectra support this.

Buemi (1990) has also carried out semi-empirical AM1 calculations on 3-(4'-biphenyl)pentane-2,4-dione whose molecular parameters have been fully determined by X-ray diffraction (Emsley *et al.*, 1988b). The most stable configuration predicted by theory was not that found in the crystal structure. The calculated hydrogen-bond length ($R_{O \cdots O} = 275.9$ pm) differed from the observed value (241.1 pm). The conclusion was reached that the short hydrogen bond observed in the crystal is a peculiarity of the solid state.

VIBRATIONAL MODES

The ir spectra of β-diketones have recently been reassessed since the shift to lower frequencies of $v(OH)$ reported in earlier assignments is inconsistent with the strength of the hydrogen bonding. The spectra of pentane-2,4-dione and its derivatives show an intense broad band lying below the region 1100–1700 cm^{-1}. Strong peaks due to carbonyl-stretching and methyl-bending vibrations protrude from this band and obscure its definition, explaining why earlier workers opted to assign a broad but weak band at *ca.* 2500 cm^{-1} to $v(OH)$ in a variety of β-diketones (Wierzchowski and Shugar, 1965; Ogoshi and Yoshida, 1971; Grens *et al.*, 1975; Tayyari *et al.*, 1979a,b). On the basis of this assignment, it was deduced that the hydrogen bonding in enol tautomers had a high internal barrier indicating that it was weak (Tayyari *et al.*, 1979b). Clearly, this interpretation is at variance with the structural and nmr data and, consequently, a reassignment of $v(OH)$ became necessary.

A simple formula has been proposed which directly relates $\Delta v(OH)$ with $R_{O \cdots O}$ (Bellamy and Owen, 1969) and this has been tested several times and found to hold good (e.g. Helder *et al.*, 1984). For $R_{O \cdots O} = 244$ pm the formula predicts that the ir band for $v(OH)$ would be at 1690 cm^{-1}, which is more in keeping with the broad continuum that underlies the spectra of the β-diketones. Inspection of the ir spectra reproduced in the literature also shows an extreme broadness around the 1600 cm^{-1} region, even though the authors chose weaker bands at *ca.* 2500 cm^{-1} for ν(OH) (Wierzchowski and Shugar, 1965; Ogoshi and Yoshida, 1971). In making this choice they may have been influenced by the ir analysis of pentane-2,4-dione itself, which with its longer hydrogen bond ($R_{O\cdots O}$ = 253.5 pm) would be expected to have ν(OH) at the assigned location of 2640 cm^{-1} (Ogoshi and Nakamoto, 1966).

The carbonyl stretching frequency of both the keto and enol tautomers can be recognized in the vibrational spectrum of pentane-2,4-dione. The enol has ν(C=O) at 1618 cm^{-1}, generally the dominant peak in the spectrum and more intense than the in- and out-of-phase ν(C=O) stretching modes of the keto form, which are found at 1727 and 1707 cm^{-1}, respectively. These are identified by their Raman counterparts at 1719 cm^{-1} (polarized) and 1697 cm^{-1} (depolarized) (Ernstbrunner, 1970). The ratio of absorbances of the enol and the out-of-phase keto bands in the ir was used as an early method of analysis of the keto/enol equilibrium in different solvents (Le Fèvre and Welsh, 1949).

The vibrational assignments for a selection of β-diketones are given in Table 10, which includes those β-diketones for which complete structural and nmr data are available.

Table 10 Hydrogen-bond lengths, chemical shifts and vibrational mode of selected enol tautomers of β-diketones.

Compound[a]	$R_{O\cdots O}$/pm	δ/ppm[b]	$\Delta\delta$/ppm[c]	ν(OHO)/ cm^{-1}	References
MeCOCH$_2$COMe	253.5	15.40	+0.61[d]	2640	[efgh]
PhCOCH$_2$COMe	248.5	16.27	+0.42[d]	2650	[gijk]
PhCOCH$_2$COPh	246.3	17.13	+0.45[d]	2620	[fgkl]
3-(2',4'-(MeO)$_2$C$_6$H$_4$)-PD[m]	246.0[n]	16.65	+0.63	1580	[o]
PD-S$_{2,3}$-PD[p]	246.1	17.06	+0.36	1500	[qr]
3-(4'-MeOC$_6$H$_4$)-PD	244.9	16.65	+0.31	1510	[s]
3-(4'-PhC$_6$H$_4$)-PD	244.1	16.72	+0.60	1485	[t]
3,3'-bis(PD)[u]	242	16.80	+0.66	1500	[v,r]

[a] Formulae given as β-diketone; [b] Measured in CCl$_4$ or CDCl$_3$; [c] $\Delta[\delta(^1H)-\delta(^2H)]$; [d] Other values for this quantity have been reported (see Table 5); [e] Camerman *et al.* (1983); [f] Lintvedt and Holtzclaw (1966); [g] Altman *et al.* (1978); [h] Emsley (1984); [i] Jones (1976b); [j] Sardella *et al.* (1969); [k] Tayyari *et al.* (1979a); [l] Jones (1976a); [m] PD = pentane-2,4-dione; [n] Average of 244.5 and 247.5 pm; [o] Emsley *et al.* (1990a); [p] Two pentane-2,4-dione moieties linked through two or three sulphur atoms at the 3,3'-positions; [q] Power and Jones (1971); [r] Emsley *et al.* (1989); [s] Emsley *et al.* (1988a); [t] Emsley *et al.* (1988b); [u] Tetraacetylethane; [v] Dewar *et al.* (1964).

NMR SPECTROSCOPY

The β-diketones have been subjected to many nmr investigations, chiefly ^{1}H and ^{13}C, although [17]O-nmr studies have also been made on one enol (Lapachev *et al.*, 1985). The chemical shift of the hydrogen-bonding proton of the cyclic enols are found to be far downfield, indicating deshielded protons. A list of these has been compiled (Emsley, 1984). The extremely deshielded signals were at one time ascribed to ring currents involving a vacant, high-energy p-orbital on hydrogen. The furthest downfield shift recorded to date is that of the protonated β-diketone, 1,3-diphenyl-2-methylpropane-1,3-dione, which in the strong acid system $HBr-CF_2Br_2$ has a δ-value of 21.44 ppm (Clark *et al.*, 1988b). This will be discussed below.

Investigation of stable simple enols has recently been undertaken with a series of β,β-dimesityl-α-ethenols [29] and [30] in which Mes = 2,4,6-trimethylphenyl. In this work, δ(OH) and $^3J_{HCOH}$ were measured for a series of compounds with various α-substituents and in a range of solvents (Biali and Rappoport, 1984; Rappoport *et al.*, 1988). The nmr parameters were shown to be linearly related to the Kamlet and Taft β-variable, a solvent property that is related to its ability to act as a hydrogen-bond acceptor (Kamlet and Taft, 1976; Taft *et al.*, 1985).

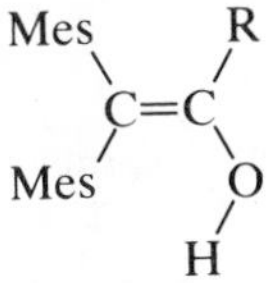

syn-planar
[29]

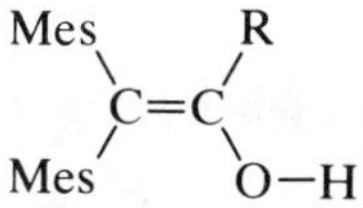

anti-clinal
[30]

When the substituent R is H, the chemical shift for these enols is solvent dependent, moving downfield from 4.74 in CCl_4 to 9.17 in DMF, and this change is taken to indicate that the *syn*-planar conformation is preferred in non-polar solvents and the *anti*-clinal conformation is more stable in polar solvents. No enol–enol, hydrogen-bonding association is found, and no intermolecular exchange of OH occurs, this being consistent with the sharp δ(OH) signals that are insensitive to concentration. In non-polar solvents there is thought to be OH $\cdots$ π hydrogen bonding in the *syn*-arrangement between the enol OH and the aromatic mesityl ring (Nadler and Rappoport, 1989).

The δ(OH)-value of the intermolecular hydrogen-bonded enol [27] is 14.2 ppm in $CDCl_3$ and 11.5 ppm in DMSO. For [28], δ(OH) has a value of 6.42 ppm in $CDCl_3$ and 10.9 ppm in DMSO (Imashiro *et al.*, 1987). In contrast, the enol proton of intramolecularly hydrogen-bonded cyclic enols are generally found downfield with δ-values > 15 ppm, and are less sensitive to solvent polarity. For example, 1,3-diphenlypropane-1,3-dione has a δ(OH)-value of 17.1 ppm in $CDCl_3$ and 17.2 ppm in DMSO.

Solid-state ^{13}C-nmr studies have been made with [27] and [28], which must be intermolecularly hydrogen bonded (Imashiro et al., 1987). By comparing the chemical shifts of the solid enols with those of their methyl ethers in solution, it has been possible to deduce the changes due to hydrogen bonding. The downfield shift of the enol carbon is always greater than the downfield shift of the carbonyl carbon, and the ratio of these shifts was shown to be 2.3 when the intermolecular hydrogen bonding produced a crystal structure that was an infinite helical array, compared to 1.5 when the intermolecular hydrogen bonding produced an infinite zig-zag arrangement. This enabled other non-cyclic enols to be categorized.

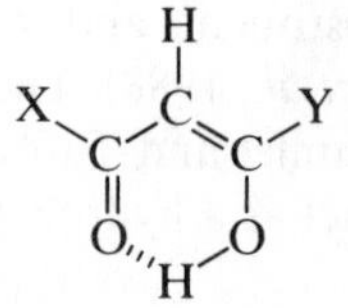

[31]

Table 11 β-Substituent factors, Δ, for calculating [1]H- and ^{13}C-nmr spectra of β-diketones.

Substituent	Δ	Substituent	Δ
δ(OHO) = Δ(X) + Δ(Y): see (31)			
Me	1.10	2-C_4H_3O	7.73
CF_3	6.54	2-$C_4H_3S^a$	8.05
Ph	8.46	4-NO_2-C_6H_4	8.20
Bu^t	8.25	2,4,6-$Me_3C_6H_2$	7.78

	Δ_a	Δ_n	Δ_d
δ(C): see (17) and (18)			
Me	0.0	0.0	0.0
CF_3	−4.1	−14.9	+3.5
Ph	−4.2	−8.5	+2.0
Bu^t	−5.1	+8.8	+0.4
2-$C_4H_3S^a$	−2.9	−12.0	−5.1
2,4,6-$Me_3C_6H_2$	+1.4	−2.1	+2.1

^{a} Thienyl.

The ^{1}H- and ^{13}C-nmr spectra of a range of enol tautomers was originally used by Shapet'ko to show that the hydrogen-bonded proton was centred (Shapet'ko 1973; Shapet'ko *et al.*, 1975). Although this thesis is no longer tenable, the idea on which it was based is still valid—that the chemical shifts of the hydrogen-bonding proton and the carbon atoms of the β-diketone skeleton are predictable from the substituent effects of groups X and Y [31], termed Δ. These substituent effects were measured relative to the methyl groups of pentane-2,4-dione and are assumed to be additive (Table 11).

For the ^{13}C nmr spectra, three equations are necessary to define the chemical shifts since three different carbon atoms are involved. Consequently, three substituent factors are required: Δ_a values to calculate the chemical shift of the α-carbon, Δ_n and Δ_d values for calculating the β-carbons, the former referring to the nearer group, the latter to the more distant β-substituent. The respective equations are (17) and (18).

$$\delta C_\alpha = 100.6 + \Delta(X)_\alpha + \Delta(Y)_\alpha \qquad (17)$$

$$\delta C_\beta = 191.5 + \Delta(X)_n + \Delta(Y)_d \qquad (18)$$

This approach has recently been extended to keto tautomers (Bassetti *et al.*, 1988). Fourier-transform nmr spectroscopy is needed to record the signals of the small amounts of the keto isomers that are present. The carbonyl carbons of the keto form are to lower field of the corresponding atoms in the enol by *ca.* 10 ppm. The data were analysed into substituent factors relative to pentane-2,4-dione. Replacing the methyl groups of this compound with 2-thienyl, phenyl and t-butyl groups caused upfield shifts to the α-carbon of -3.14, -4.4 and -6.5 ppm, respectively, when the substituents were introduced at this site.

In an asymmetric β-diketone there should be a preference by the enol proton for one of the carbonyls over the other and attempts have been made to determine which it is by ^{13}C-nmr spectroscopy (Shapet'ko *et al.*, 1975; Lazaar and Bauer, 1983). With the additional help of ^{17}O-nmr spectroscopy it has been possible to demonstrate convincingly that the enol group prefers the carbonyl with a β-group in the following order: $CF_3 > Ph > Bu^t > Me$ (Geraldes *et al.*, 1990).

Positive values of $\Delta[\delta(^1H) - \delta(^2H)]$ are obtained for the eight β-diketones given in Table 10. These were discussed earlier (Section 1, p. 271) and they confirm that the hydrogen bonding is strong. A negative value for this quantity has also been observed (Clark *et al.*, 1989) for the protonated form of 1,3-diphenyl-2-methylpropane-1,3-dione. The keto form of 1,3-diphenyl-2-methylpropane-1,3-dione is the preferred species in the solid phase. Protonation of one of the carbonyl oxygens clearly provides the impetus for the formation of an hydrogen bond ring [32] as the downfield shift of 21.44 ppm indicates. Protonation was achieved by the use of the acid HBr in CF_2Br_2. A value for $\Delta[\delta(^1H) - \delta(^2H)]$ of -0.26 ± 0.15 ppm was found, thus showing that the hydrogen bonding is very strong.

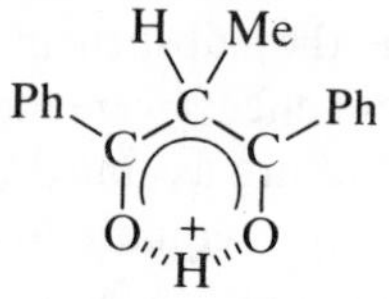

[32]

The φ-values of various β-diketones are given in Table 6 and discussed in Section 1.

4 Chemical reactivity and hydrogen bonding

The values of the equilibrium constants and rate coefficients for a variety of reactions are affected quite markedly by the presence of hydrogen bonding. The magnitude of the effect can be assessed by comparison of the values with those for similar reactions involving reactants and products in which hydrogen bonds are not present. Because this field of research is extremely large we will restrict coverage to a few important and recent examples. Some of the topics in this section have been covered in a previous review (Hibbert, 1986). In these cases, a sufficient summary is given to cover the earlier work, and the most recent developments are discussed in detail.

The position of the keto/enol equilibrium for β-diketones in comparison with monoketones is partly controlled by the possibility of hydrogen bonding in the enol form of β-diketones and details of the hydrogen bond in enols have been discussed in Section 3. The effect of isotopic substitution on equilibrium constants for reactions involving an intra- or an inter-molecularly hydrogen-bonded species provides important information about the nature of the hydrogen bond and this topic has been dealt with in Section 1 (p. 280). The effect of intramolecular hydrogen bonding on the rate coefficients and equilibrium constants for proton removal is of particular interest and will be considered first. In general, the presence of an intramolecular hydrogen bond retards the rate of proton removal. However, there are examples of chemical reactions in which rate increases are due to the development of a hydrogen bond in the transition state and products of reaction. Some examples will be given later and the consequences of hydrogen bonding on the efficiency of intramolecular catalysis discussed.

This topic is related to the role that hydrogen bonds play in enzymic reactions and recent work in this area is covered in Section 5. It will be seen that hydrogen bonds are involved in the stabilization of the enzyme, the enzyme–substrate complex, and the transition state for the reaction.

PROTON-TRANSFER BEHAVIOUR

One of the most important and well-studied effects of hydrogen bonding on chemical properties is the modification to proton-transfer behaviour. Discussion of this topic formed a major part of previous reviews (Hibbert, 1984, 1986) and, after brief summaries of the most important conclusions, coverage will be limited to the most recent developments which were not discussed previously.

Acidity and basicity of N—H · · · N species

Proton sponges. It is well established that intramolecular hydrogen bonding results in reduced acidity of the hydrogen-bonded species and enhanced basicity of its conjugate base. Some progress in this area has been achieved since our previous review. In particular, investigations of the hydrogen-bond length and likely potential function by crystallographic and spectroscopic methods for a much wider range of examples have led to attempts to correlate the properties of the hydrogen bond with the effect on acidity and basicity.

The most striking examples of the effect of hydrogen bonding on acidity and basicity are in diamines, where two identical basic sites closely positioned can accept a proton between them. The best-known examples are the proton sponges, based on amino-groups at the 1- and 8-positions of naphthalene [33] (Alder *et al.*, 1978, 1981; Hibbert and Hunte, 1983; Barnett and Hibbert, 1984). The most strongly basic proton sponge so far discovered is 1,8-bis(diethylamino)-2,7-dimethoxynaphthalene (Alder *et al.*, 1978; Hibbert and Simpson, 1987a), shown in its protonated form as [34]. With a pK_a-value of *ca.* 16.3, [34] is more than 11 units less acidic than protonated 1,8-diaminonaphthalene ($pK_a = 4.61$), and consequently the free amine is more than 11 units more basic than 1,8-diaminonaphthalene. The proton-transfer behaviour of this series of amines is well established (Alder and Sessions, 1983; Hibbert, 1986). They owe their high basicity to a severe lone pair–lone pair interaction and to steric strain in the free amines which is relieved on protonation. The protonated amines are able to adopt a relatively unhindered conformation in which the $^+$N—H · · · N system forms a moderately strong intramolecular hydrogen bond. Other factors, such as steric inhibition of resonance in the amines and the screening of the amine lone pairs from stabilization by solvation, may also contribute.

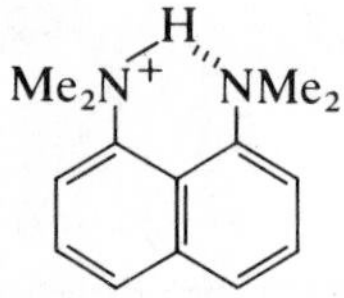

[33] pK_a = 12.1 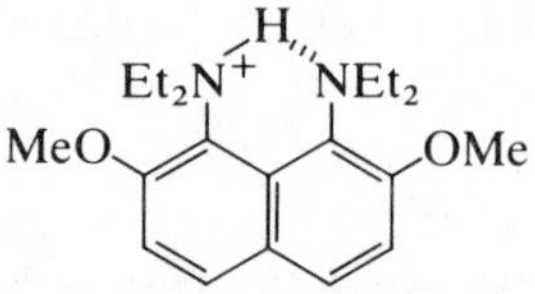   [34] pK_a = 16.3 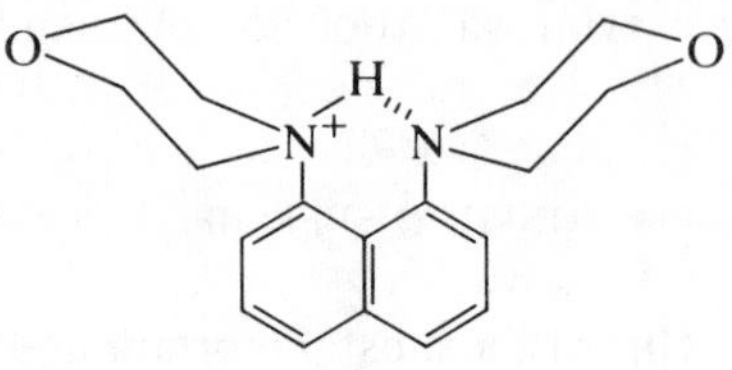   [35] pK_a = 7.49

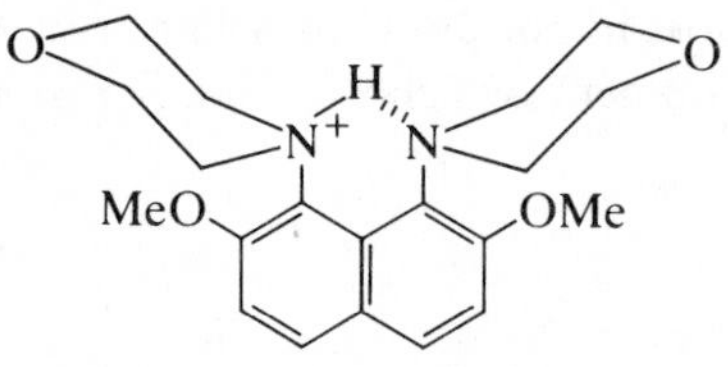

[36] pK_a = 13.0 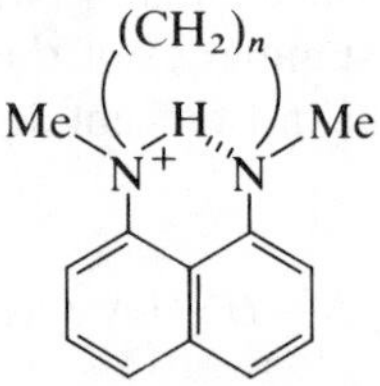   [37] $n = 2$ pK_a = 4.62   [41] pK_a = 12.8
                            [38] $n = 3$ pK_a = 10.27
                            [39] $n = 4$ pK_a = 13.6
                            [40] $n = 5$ pK_a = 13.0

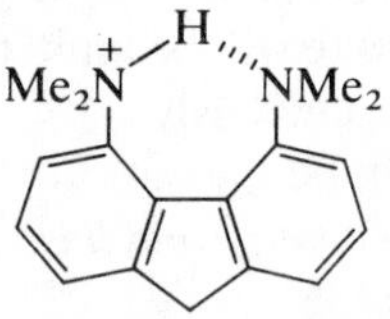

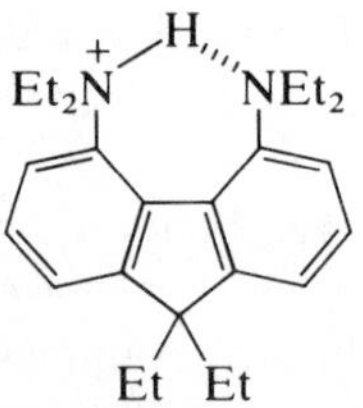

[42] pK_a = 13.6 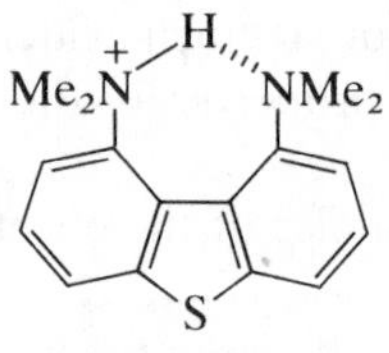   [43] pK_a = 11.9 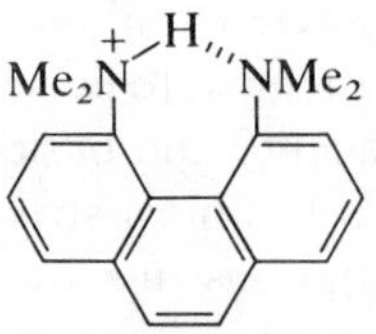   [44] pK_a = 11.5

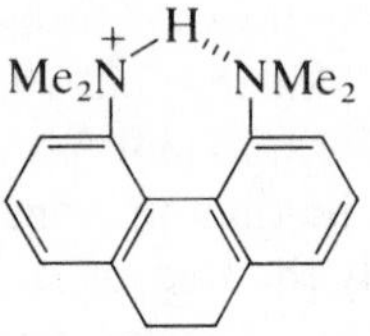

[45] pK_a = 10.9 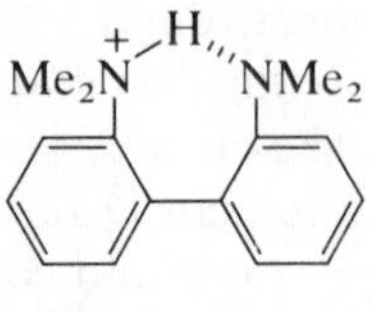   [46] pK_a = 7.9 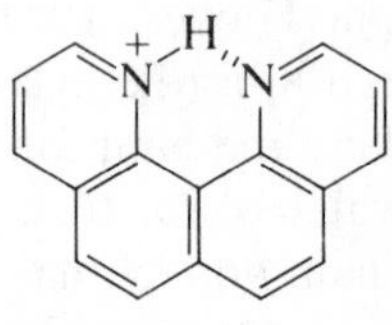   [47] pK_a = 12.8

The proton in [34] is removed by the hydroxide ion and buffer species on the timescale of minutes. The mechanism of the reaction has been established (see p. 330) as involving opening of the intramolecular hydrogen bond to give a strained open intermediate from which the proton is removed. The rates are low because formation of the non-hydrogen-bonded intermediate is unfavourable and because the attack by the hydroxide ion on the intermediate is sterically hindered. Both factors make sizeable contributions (Barnett and Hibbert, 1984).

Recent examples of proton sponges include the dimorpholino-derivatives [35] and [36]. The basicity of [36] at a pK_a-value of 13.0 is enhanced by several units, and rates of proton transfer to the free amine and from the protonated species to the hydroxide ion are in the range of milliseconds (Hibbert and Simpson, 1987a). The series of cyclic diamines [37]–[40] nicely illustrate the effect of a change in conformation on the acid–base properties (Hibbert and Simpson, 1987b). In the five-membered cycle ($n = 2$), the nitrogen lone pairs face outwards from the plane of the naphthalene ring and there is little difference in conformation between the free and protonated amine. The protonated amine is unable to form an intramolecular hydrogen bond and the acidity and rates of proton transfer are typical of an aromatic amine. Actually, the kinetics of the proton-transfer reactions of [37] were too rapid to follow and at 278 K it was only possible to estimate lower limits of $ca.$ 5×10^7 dm^3 mol^{-1} s^{-1} for rate coefficients for reaction with buffer species. With the larger cycles $n = 4$ [39] and $n = 5$ [40], the rates of proton transfer with the hydroxide ion were reduced and rate coefficients typical of those for proton sponges of moderate basicity were found. In these cases the protonated amines are less strained than the free amines since, in the larger cycle, the strain can be relieved by the formation of an intramolecular hydrogen bond.

Table 12 Structural details of the hydrogen bond in protonated proton sponges.

Proton sponge	pK_a	$R_{N \cdots N}$/pm	NHN angle	δ/ppm^a	$\Delta[\delta(^1H) - \delta(^2H)]$/ppm
[33]	12.1[b]	260.1[c]	134°[c]	18.31(19.51)[d]	+0.66[e]
[33]		265.2[f]	148°[f]		
[33]		255.4[g]	153°[g]		
[41]	12.8[h]	262.6[h]	178°[h]	18.25[h]	+0.50[h]
[43]	11.9[i]	258.7[i]	175°[i]	19.06[i]	
[44]	11.5[j]	254.4[k]	168°[k]	18.37[j]	+0.47[j]
[45]	10.9[j]			16.50[j]	
[46]	7.9[l]	265.0[l]	175°[l]	11.2[l]	
[47]	12.8[m]			19.38[m]	

[a] Me$_2$SO solution; [b] Hibbert (1974); [c] Protonated amine salt with magnesium tris(hexafluoroacetylacetonate) (Truter and Vickery, 1972); [d] CF$_3$COOH solution (Alder et al., 1968); [e] Altman et al. (1978); [f] Protonated amine salt with copper(II) tris(hexafluoroacetylacetonate) (Truter and Vickery, 1972); [g] Hydrobromide salt (Pyzalka et al., 1983); [h] Staab et al. (1983); [i] Staab et al. (1988a); [j] Saupe et al. (1986); [k] Staab and Saupe (1988); [l] Staab et al. (1988b); [m] Zirnstein and Staab (1987).

In recent years, Staab and his research group (Staab and Saupe, 1988) have prepared moderately strong aromatic bases based on the principles established for proton sponges and with closely related structures. The structures of the protonated species are shown in [41]–[47]. The amines are all less basic than 1,8-bis(diethylamino)-2,7-dimethoxynaphthalene, although many exceed the basicity of 1,8-bis(dimethylamino)naphthalene [33], the first proton sponge for which acid–base properties were established (Alder *et al.*, 1968; Hibbert, 1973, 1974). As well as basicities and rates of proton transfer, details of the hydrogen bonds present in the protonated amines have been established for a number of proton sponges. Details of the hydrogen bonds in the protonated amines are given in Table 12.

In all cases for which data are available, the $N \cdots N$ distances are much reduced in the protonated amines compared to the free amines. For example, in [33], $R_{N \cdots N}$ is reduced from 279 to *ca.* 260 pm (Truter and Vickery, 1972) and for [43] and [44] the reductions in $R_{N \cdots N}$ on protonation are 27.4 (Staab *et al.*, 1988a) and 23.9 pm (Staab and Saupe, 1988), respectively. On protonation of 2,2'-bis(dimethylamino)biphenyl to give [46], the dimethylamino-groups in the molecule change from an *anti*-conformation in which the N atoms are distant to a *syn*-conformation with a $N \cdots N$ distance of 275 pm (Staab *et al.*, 1988b).

The sign of the isotope effect on the chemical shift of the hydrogen-bonded proton in the protonated amines [33], [41] and [44] is compatible with double minimum potential functions for hydrogen bonds of moderate strength (Altman *et al.*, 1978). A double minimum potential was also suggested from the magnitude of the isotopic fractionation factor of [33] and of protonated 1,8-bis(dimethylamino)-2,7-dimethoxynaphthalene (see Section 1, p. 280). In addition, the ESCA spectrum of [33] as the tetrafluoroborate salt confirms that the nitrogen atoms are non-equivalent (Haselbach *et al.*, 1972) and this is compatible with a double minimum potential function for the hydrogen bond with the proton located in a minimum nearer to one of the nitrogen atoms.

The strength of the hydrogen bonds in the protonated amines can be increased by the use of larger and additional groups to provide further steric strain. For example, the difference in strain between 1,8-bis(dimethylamino)-naphthalene [33] and 1,8-bis(diethylamino)naphthalene leads to a pK_a-value for the ethyl derivative enhanced by 0.5 units (Hibbert, 1974), and a similar difference is also observed in the fluorene series for [41] and [42]. Further strain can be introduced into the naphthalene series by the use of 2- and 7-methoxyl groups as substituents. This is illustrated by the pK_a-values of 1,8-bis(dimethylamino)naphthalene [pK_a = 12.1 (Hibbert, 1974)] and 1,8-bis(dimethylamino)-2,7-dimethoxynaphthalene [pK_a = 16.1 (Hibbert and Simpson, 1987a)] and by the pK_a-values of [35] [pK_a = 7.49 (Hibbert and Hunte,

1981; Awwal et al., 1981)] and [36] [pK_a = 13.0 (Hibbert and Simpson, 1987a)]. Steric compression by the methoxyl groups increases the strain in the free amine and, consequently, leads to relative stabilization of the protonated hydrogen-bonded species. It will be interesting to see if a similar effect operates in the fluorene [41] and phenanthrene [44] systems.

For the series [41], [43] and [44] with almost linear $^+$N—H $\cdots$ N hydrogen bonds, the value of $R_{N \cdots N}$ in the protonated amines decreases but the base strength does not increase. This may mean that, for [44], in particular, the hydrogen bond is compressed beyond that which is most favourable energetically (Staab and Saupe, 1988).

In the free amine form of [47], solvation of the nitrogen lone pairs is much less hindered compared to the other proton sponges. The comparable basicity of [47] and the other sponges has led Staab to postulate that inhibition of solvation of the free amines plays only a minor role in determining the basicity of proton sponges (Zirnstein and Staab, 1987). The pK_a-value of [47] was estimated, however, from ^{1}H-nmr measurements of the position of the proton-exchange equilibrium between [47] and [33] and the corresponding free bases in [^{2}H$_6$]Me$_2$SO as solvent. In this solvent, of course, solvation of the type present in aqueous solution does not occur and neither [47] nor [33] nor the free amines can be stabilized by the solvent under these conditions. It would be useful to confirm that the difference in pK_a-value between [33] and [47] found in [^{2}H$_6$]Me$_2$SO is also observed in aqueous solution. There is an inherent difficulty in estimating the aqueous pK_a-values of strong bases, since it is necessary to use some solvent other than H$_2$O to bring about dissociation of the protonated amines. The dissociation of the protonated forms of the strongest proton sponges, for example [34], was observed in Me$_2$SO–H$_2$O mixtures in the presence of hydroxide ions. The dissociation was compared under the same conditions to that for 1,8-bis(dimethylamino)naphthalene [33] that has been studied in wholly aqueous solution (pK_a = 12.1), and the assumption is made that the change in solvent has the same effect on the dissociation of both amines (Hibbert and Hunte, 1983; Hibbert and Simpson, 1987a).

The proton-transfer behaviour of [47] differs in one important respect from that of the other proton sponges. It is found that the rates of equilibration of the protonated amine with the free amine in mixtures of the two in [^{2}H$_6$]Me$_2$SO at 30°C is fast on the nmr timescale and averaged proton signals are observed (Zirnstein and Staab, 1987). For mixtures of [33] and the free amine, separate proton signals are observed. Quantitative information about the extent to which the rates of proton transfer for [47] differ from those of the other proton sponges must await detailed kinetic studies.

Protons in cages.   Studies of the intramolecular hydrogen bonds and the proton-transfer behaviour of the protonated bicyclic aliphatic diamines shown in [48] have been published. The hydrogen bonds were investigated by

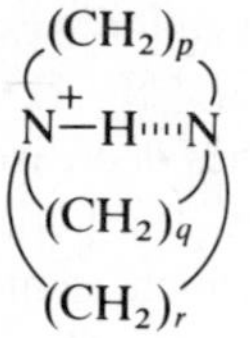

[48] 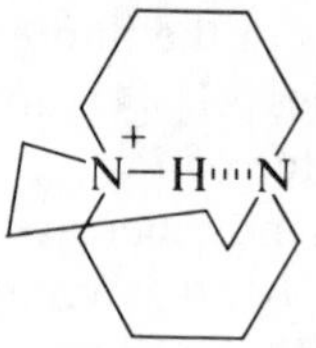   [49] $pK_a = ca.$ 25

nmr and ir techniques and, more recently, crystal structures of some of the species have been obtained. In all 20 medium-ring, bicyclic diamines were prepared (Alder *et al.*, 1983b) of which 11, containing the larger rings, could be converted to inside-protonated ions. In some cases, protonation occurred by electron-transfer mechanisms rather than by conventional proton transfers. Spectroscopic data and bond lengths for the intramolecular hydrogen bonds in the 11 inside-protonated diamines are given in Table 13.

Although X-ray data show that in several of the protonated diamines the proton is located almost centrally between the nitrogen atoms within experimental error, the values of $\Delta[\delta(^1H) - \delta(^2H)]$ prove that in only one case, 1,6-diazabicyclo[4.4.4]tetradecane [49], is the potential function of the single minimum type. It appears that the conditions for formation of a hydrogen bond with a single minimum potential function are a low $N \cdots N$ distance and a linear $^+N—H \cdots N$ bond. Thus, Table 13 contains examples of short but non-linear hydrogen bonds (the [5.4.2]-, [5.5.2]- and [5.4.3]-protonated diamines) and examples of long but almost linear hydrogen bonds (the [5.5.4]-protonated diamine), all of which have double minimum potential functions.

There is no clear correlation between the chemical shift or the ir absorption frequency and the length and linearity of the hydrogen bonds in the protonated diamines. The most reliable test for the nature of the potential function appears to be the sign of the isotope effect on the chemical shift of the hydrogen-bonded proton.

A further indication that the hydrogen bond in protonated 1,6-diaza-bicyclo[4.4.4]tetradecane [49] is strong is provided by the exceptionally low acidity; the pK_a-value is *ca.* 25 (Alder, 1989). In some of the other diamines, for which inside-protonated forms could not be obtained by conventional proton-transfer reactions, it was not known whether this was due to a low rate of protonation or to low basicity of the amines towards inside protonation. The estimated pK_a-value for [49] at least makes it clear

Table 13 Structural details of the hydrogen bonds in protonated bicyclic diamines.

Ring sizes	$R_{N \cdots N}$/pm	NHN angle	H-bond[a]	δ/ppm[b,c]	$\Delta[\delta(^1H) - \delta(^2H)]$/ppm[b]	$\nu(NH)$/cm^{-1} [b,d]	Reference[e]
[5.4.2]	247.4	132.6°	Symm.	13.5		2450	f
[6.4.2]				10.7		2450	g
[5.5.2]	255.5	134.5°	Asymm.	12.35		2475	
[6.3.3]				17.25	+0.76	2200	h
[5.4.3]	255.5	160.0°	Symm.	17.48	+0.53	1400–1900	i
[4.4.4]	252.6	Linear	Symm.	17.40	+0.06	1400–1900	j
[6.4.3]	266.3	154.8°	Asymm.	14.62	+0.65	2150	
[5.5.3]				17.20	+0.87	2100	
[5.4.4]				16.19	+0.43	1500–2300	k
[6.5.3]	261.0	154.3°	Asymm.	13.78	+0.58	2200	l
[5.5.4]	269.0	171.7°	Asymm.	15.30	+0.92	1500–2300	

[a] In X-ray crystal structure, symm. means proton placed centrally within experimental error and asymm. means proton is placed closer to one nitrogen; [b] spectral data taken from Alder et al. (1983c); [c] solvent is CHCl$_3$ or CH$_2$Cl$_2$; [d] solvent is CDCl$_3$; [e] references refer to X-ray crystal structural details; [f] White et al. (1988e); [g] White et al. (1988b); [h] White et al. (1988a); [i] Alder et al (1983d) and Schaefer and Marsh (1984); [j] Alder et al. (1988b); [k] White et al. (1988c); [l] White et al. (1988d).

that in this case the reason is a kinetic one. Where protonation of the diamines occurred by conventional acid–base chemistry, it was always found that the reaction was slow (Alder *et al.*, 1983b). For example, inside protonation of the [5.4.2], [6.4.2], [5.5.2], [6.4.3], [5.5.3] and [6.5.3] species occurs slowly on mixing with CF_3COOH in $CDCl_3$. The reactions occurred measurably slowly but were complete within 30 minutes of mixing. Inside protonation of the [5.5.4] and [6.3.3] diamines is exceptionally slow under these conditions, taking days to reach completion. However, an electron-transfer mechanism is needed to form the inside-protonated forms of the [5.4.3], [4.4.4] and [5.4.4] diamines.

The reasons for the reluctance of the diamines to undergo protonation is due to the inaccessibility of the basic sites. The high thermodynamic basicity is probably due to a combination of the formation of a strong intramolecular hydrogen bond and to unfavourable lone pair interactions in the diamines that cannot be relieved by solvation.

The basicity of the monocyclic diamines [50]–[53] has recently been investigated for comparison with that of the bicyclic species (Alder *et al.*, 1988a) and with the basicity of acyclic diamines. For 1,4-diaminobutane in aqueous solution, the results $pK_{a1} = 9.2$ and $pK_{a2} = 10.7$ were obtained (Perrin, 1972), corresponding to the dissociations in (19) and (20) respectively. These results are not very different from the pK_a-values found for monoamines; for example, 1-aminopentane gives $pK_a = 10.6$ (Perrin, 1965). In the gas phase, dissociation of monoprotonated 1,4-diaminobutane is much less favourable than dissociation of protonated monoamines, meaning that the diamine is a substantially stronger base than expected for a primary amine (Aue *et al.*, 1973, 1976; Yamdagni and Kebarle, 1973). In the gas phase, the monoprotonated diamine is stabilized by an intramolecular hydrogen bond. In aqueous solution the free amino-groups and the protonated amino-group are both stabilized by solvation, and the monoprotonated species is not favoured particularly by the formation of the intramolecular hydrogen bond. However, it is likely that the monoprotonated diamine is largely in the cyclic, intramolecularly hydrogen-bonded form. For example, it has been estimated that $N,N,N',N',2,2$-hexamethyl-1,3-propanediamine is cyclized to the extent of 77% in aqueous solution (Hine and Li, 1975).

$$\begin{array}{c} (CH_2)_p \\ Me{-}\overset{+}{N}{-}H\cdots N{-}Me \\ (CH_2)_q \end{array}$$

[50] $p = 3, q = 3$
[51] $p = 4, q = 3$
[52] $p = 5, q = 3$
[53] $p = 4, q = 4$

$$\text{^{+}H}_3\text{N(CH}_2)_4\text{NH}_3\text{^{+}} + \text{H}_2\text{O} \xrightleftharpoons{K_{a1}} \text{H}_2\text{N(CH}_2)_4\text{NH}_3\text{^{+}} + \text{H}_3\text{O}^+ \qquad (19)$$

$$\text{H}_2\text{N(CH}_2)_4\text{NH}_3\text{^{+}} + \text{H}_2\text{O} \xrightleftharpoons{K_{a2}} \text{H}_2\text{N(CH}_2)_4\text{NH}_2 + \text{H}_3\text{O}^+ \qquad (20)$$

The pK_{a2}-values for dissociation of the diprotonated diamines [50] to [53] in aqueous solution are shown in Table 14. These values are lower than for the corresponding acyclic diamines and the values decrease in the order to be expected if the monoprotonated species form intramolecular hydrogen bonds. The strongest intramolecular hydrogen bond is expected for the seven-membered ring as in [53].

Table 14 Dissociation of monocyclic diamines.

Diamine	$pK_{a2}{}^a$	$\Delta pK_a{}^b$
[50]	5.5	-1.64
[51]	3.1	-1.11
[52]	1.5	-0.89
[53]	0.4	$+0.48$

[a] In aqueous solution; [b] $\Delta pK_a = \log_{10} K$, where K is the equilibrium constant for the equilibrium in (21) in $[^2\text{H}_6]\text{Me}_2\text{SO}$ solvent.

Dissociation of the monoprotonated species does not occur in a range convenient for measurement in aqueous solution, except in the case of [50] for which $pK_{a2} = 11.9$ was found; for the remainder, a lower limit of $pK_{a2} > 12$ was estimated. Relative values of pK_{a2} for [50]–[53] were estimated by measuring the position of the equilibrium in (21) in $[^2\text{H}_6]\text{Me}_2\text{SO}$ from the nmr spectra. The results are shown in Table 14 as $\Delta pK_a = \log_{10} K$, where K is the value of the equilibrium constant for (21).

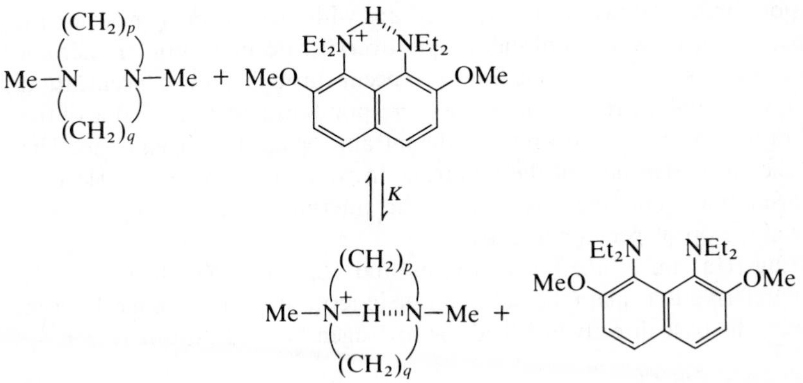

An aqueous pK_a-value of 16.1 has been measured for 1,8-bis(dimethyl-amino)-2,7-dimethoxynaphthalene from studies in 35% (v/v) Me_2SO-H_2O (Hibbert and Simpson, 1987a). The values of ΔpK_a in Table 14 imply that [50], [51] and [52] are more acidic and [53] is less acidic than 1,8-bis(dimethyl-amino)-2,7-dimethoxynaphthalene. Correspondingly, the amines derived from [50], [51] and [52] are less basic and that derived from [53] is more basic than 1,8-bis(dimethylamino)-2,7-dimethoxynaphthalene. It follows that the amine 1,6-dimethyl-1,6-diazacyclodecane [53] is an extremely strong base in aqueous solution.

The basicity of the monocyclic amine derived from [53] does not compare to that of the conjugate base of the bicyclic amine [49], but [53] is a much stronger base than acyclic diamines. The higher basicity of 1,6-dimethyl-1,6-diazacyclodecane compared to acyclic amines is thought to be a result of steric inhibition of solvation of the amine lone pairs. Solvation of the monoprotonated amine is also sterically hindered, but stabilization occurs by formation of an intramolecular hydrogen bond. Thus, the acid–base behaviour of [50]–[53] in solution is claimed (Alder et al., 1988a) to resemble the behaviour of acyclic diamines in the gas phase in which intramolecular hydrogen bonding plays an important role.

Mechanisms of proton transfer

In the late 1960s, two mechanisms had been suggested to account for the reduced values of the rate coefficients for proton transfer from intramolecu-larly hydrogen-bonded acids to bases. One, favoured by Eigen (Eigen, 1964; Eigen et al., 1964), involved rapid opening of the intramolecular hydrogen bond to give a low concentration of a non-hydrogen-bonded species (22). Under these conditions the reaction is first order in base catalyst (B) and the expressions in (23) apply for the rate coefficients in the forward and reverse directions (k_f and k_r, respectively). It was assumed that the open species formed an intermolecular hydrogen bond with the solvent and reacted at the diffusion-limited rate with the base, perhaps with proton transfer occurring down a chain of water molecules. The overall rate of proton transfer is reduced because the reaction is forced to occur through a low concentration intermediate and the reduction in rate corresponds exactly to the value of the equilibrium constant for opening of the intramolecular hydrogen bond. The rate, therefore, depends on the difference in strength of the intramolecular hydrogen bond and the strength of the intermolecular hydrogen bond between the open species and solvent.

Eyring (Haslam et al., 1965a; Eyring and Haslam, 1966; Jensen et al., 1966) preferred to explain the reduced rate in terms of a one-step mechanism in which the base directly attacked the hydrogen-bonded proton as in (24).

$$
\begin{bmatrix} -X \\ \vert \\ H \\ \vdots \\ -Y \end{bmatrix}
\underset{k_{-1}}{\overset{k_1}{\rightleftharpoons}}
\begin{bmatrix} -X^{\overset{\displaystyle H}{\diagup}} \\ \\ -Y \end{bmatrix}
\underset{k_{-2}[BH^+]}{\overset{k_2[B]}{\rightleftharpoons}}
\begin{bmatrix} -X^- \\ \\ -Y \end{bmatrix}
\tag{22}
$$

$$
k_f = (k_1/k_{-1})\,k_2 \; ; \; k_r = k_{-2}
\tag{23}
$$

The reduction in rate may then arise from a number of factors. Partial breakage of the hydrogen bond has occurred at the transition state and contributes to the activation energy of the proton-transfer reaction. In most cases the intramolecular hydrogen bond in the ground state will be bent and the non-linear reaction coordinate will hinder proton transfer (Menger, 1985). The reaction will also involve intimate contact between the hydrogen-bonded acid and the base, and this is likely to hinder the reaction because the most rapid proton transfers in aqueous solution are thought to proceed through a chain of hydrogen-bonded solvent molecules (Eigen, 1964).

$$
\begin{bmatrix} -X \\ \vert \\ H \\ \vdots \\ -Y \end{bmatrix} + B
\longrightarrow
\begin{bmatrix} \begin{bmatrix} -X \\ \\ H\cdots B \\ \\ -Y \end{bmatrix} \end{bmatrix}^{\ddagger}
\longrightarrow
\begin{bmatrix} -X^- \\ \\ -Y \end{bmatrix} + BH^+
\tag{24}
$$

Interest in the problem was stimulated by a suggestion (Kresge, 1973, 1975) that a choice between the two mechanisms may provide information applicable to proton-transfer reactions in general, as to whether solvation changes accompany transfer of the proton, precede it, or even lag behind (Kurz and Kurz, 1972; Albery, 1975). The two-step mechanism (22), involves solvent reorganization in the formation of the open intermediate which is intermolecularly hydrogen bonded to the solvent. Proton transfer then occurs. It was also recognized that many of the proton transfers in enzymic reactions occurred through hydrogen-bonded species, and, there-fore, information was needed from studies on model systems to help understand this more complex problem.

The early experiments concerned with proton transfer from hydrogen-bonded acids did not provide information which permitted a choice between the two mechanisms (Kresge, 1973). These experiments included the measurement of kinetic isotope effects (Haslam *et al.*, 1965b; Eyring and Haslam, 1966; Haslam and Eyring, 1967), activation parameters (Haslam *et al.*, 1965a), the effect of different solvents (Jensen *et al.*, 1966) and substi-tuent effects in the intramolecularly hydrogen-bonded acid (Miles *et al.*,

1966; Fueno *et al.*, 1973). The reactions were studied using the temperature-jump technique.

The first experiments to provide definite evidence involved an investigation of the effect of substituents in the base catalyst on the rate of reaction (Bernasconi and Terrier, 1975; Hibbert and Awwal, 1976, 1977). Eigen plots or Brønsted plots were constructed for proton removal by general bases from an intramolecularly hydrogen-bonded acid. The acids used in this first work were a substituted salicylate ion [54] and an ammonium centre with an intramolecular hydrogen bond to a nitro-group [55]. For the mechanism in (22), it would be predicted that for a thermodynamically favourable reaction in the forward direction the proton-transfer step would occur at the diffusion-controlled limit with a rate coefficient of *ca.* 1×10^{10} dm^3 mol^{-1} s^{-1}. It follows from (23) that the rate coefficient will be independent of the strength of the catalysing base (B) and hence independent of a substituent change in the base. Thus, the value of the Brønsted exponent β for the rate coefficients in the forward direction is predicted to be zero and a value of α of unity is predicted for the reverse reaction, since the reverse rate coefficients will be linearly related to the equilibrium constant for the reaction. This will apply up to a pK-value for the catalysing base such that the proton-transfer step is no longer thermodynamically favourable. This will occur at the point where $\Delta pK = \log(k_1/k_{-1})$, and a break in the Eigen plot should then be observed (Hibbert, 1986).

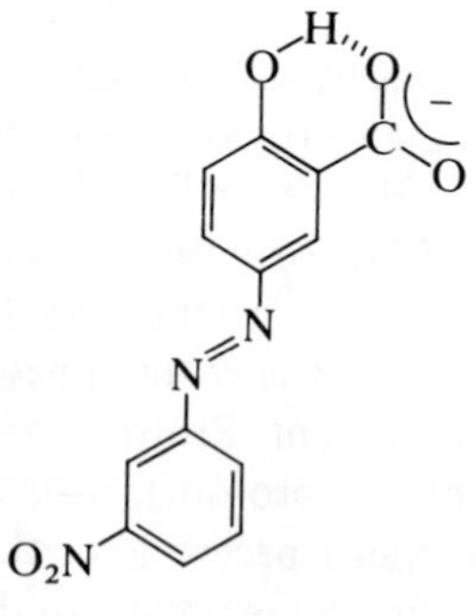

[54]

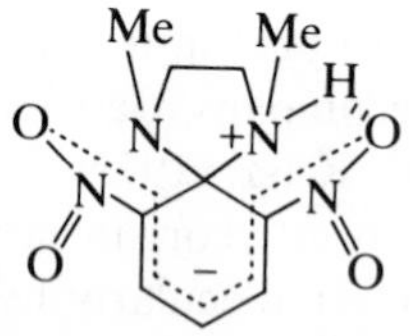

[55]

In contrast, for a proton transfer from a hydrogen-bonded acid occurring by the mechanism in (24), the predicted dependence of k_f and k_r on the strength of the catalysing base is quite different. In this case, when the pK_{BH^+} of the base and the pK-value of the hydrogen-bonded acid are closely matched, that is at $\Delta pK \simeq 0$, it would be predicted that the proton in the transition state will be roughly half-transferred and Brønsted exponents α and β of around 0.5 should be observed.

Kinetic data obtained by the temperature-jump technique for the sali-cylate ion [54] in the region $\Delta pK = -0.9$ to $+0.8$ gave values of $\beta = 0.0 \pm 0.1$ and $\alpha = 1.0 \pm 0.1$. For [55], data were obtained in the region $\Delta pK = -7.0$ to -1.9, giving $\beta = 0.0$ and $\alpha = 1.0$. These observations were taken as definite evidence for the operation of the mechanism in (22) rather than that in (24). Similar behaviour has subsequently been found for the reaction of the substituted naphthylammonium ions [56] and [57] (Barnett and Hibbert, 1984). These reactions are sufficiently slow such that, in some cases, the stopped flow method and even conventional spectrophotometric methods were used. Data for [57] were also obtained in the region of the break in the Eigen plot by using an isotope-exchange method (Kresge and Powell, 1981). The break was found to occur at $\Delta pK \simeq 6$. However, this does not imply a value for k_1/k_{-1} of $ca.$ $1 \times 10^{-6} \, \text{mol dm}^{-3}$, since it is likely that the proton-transfer step from the open form of these naphthylammo-nium ions occurs below the diffusion limit because of severe steric hindrance. It is, however, excellent evidence for the operation of mechanism (22). Thus, for three different classes of intramolecular hydrogen bonds, [54], [55] and [56], firm evidence has been obtained that the preferred mechanism consists of rapid opening of the hydrogen bond to give a non-intramolecularly hydrogen-bonded species present in low concentration and from which the proton is removed by a base.

[56]   [57]

In recent years, evidence has been found that both mechanisms of proton transfer can occur for certain intramolecularly hydrogen-bonded acids. Also, new kinetic behaviour has been obtained which allows a much more detailed examination of the reaction steps in (22). Kinetic data for the second ionization of substituted phenylazoresorcinols in the presence of hydroxide ions (25) were some of the first to be obtained for an intramolecu-larly hydrogen-bonded acid. The reciprocal relaxation time (τ^{-1}) for the approach to equilibrium in a temperature-jump experiment was measured at different hydroxide-ion concentrations. A linear dependence of τ^{-1} on $[\text{OH}^-]$ was obtained of the form of (26) (Eigen and Kruse, 1963; Inskeep et $al.$, 1968; Rose and Stuehr, 1971). However, careful measurements at lower hydroxide-ion concentrations (Perlmutter-Hayman and Shinar, 1975; Perl-mutter-Hayman et $al.$, 1976; Yoshida and Fujimoto, 1977) revealed that the variation of τ^{-1} with hydroxide-ion concentration was more complex; as the hydroxide-ion concentration was increased, the value of τ^{-1} first decreased and subsequently increased in a hook-shaped dependence. In some cases the complex dependence was not detectable and in other cases only just observable, and it is easy to see why it was overlooked in the first measurements.

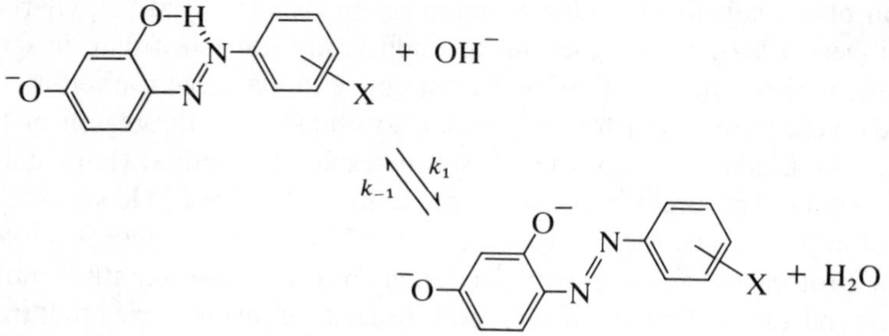

$$\tau^{-1} = k_1[OH^-] + k_{-1} \tag{26}$$

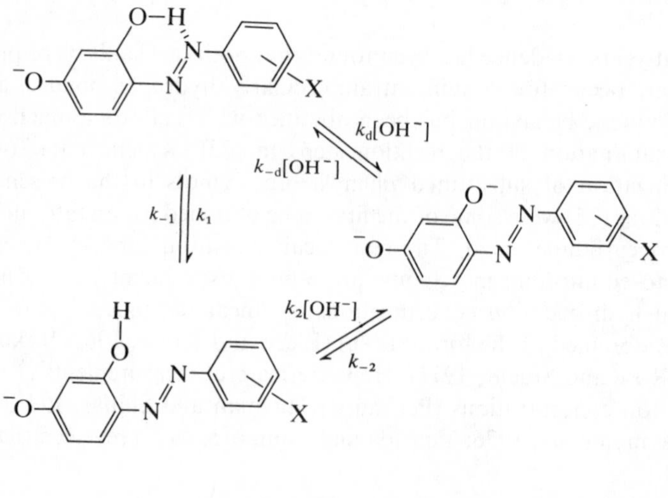

The complex kinetic dependence on hydroxide-ion concentration was explained by the mechanism in (27). Proton removal from the phenylazo-resorcinol monoanion by the hydroxide ion to give the dianion occurs by two different routes. One route is first order with respect to the hydroxide ion with rate coefficients k_d and k_{-d}, and is assumed to consist of a direct attack by the hydroxide ion on the hydrogen-bonded proton. The other route leads to a complex dependence of the rate of approach to equilibrium on the hydroxide-ion concentration, and involves prior opening of the hydrogen bond (rate coefficients k_1 and k_{-1}) followed by proton removal (rate coefficients k_2 and k_{-2}). Equation (28) is derived from the mechanism in (27) on the assumption that the proton transfer from the open species occurs more rapidly than the closing of the hydrogen bond ($k_2[OH^-]$ $> k_{-1}$) and that the open form is present in low concentration. A good fit to the experimental data can be obtained using (28), and values for the rate coefficients k_d, k_{-d} and k_1 and the ratio $k_2/k_{-2}k_{-1}$ are obtained from the best fit, using a separately measured value for the equilibrium constant (K) for the overall reaction.

$$\tau^{-1} = (k_1 + k_d[OH^-])(1 + K[OH^-]) \qquad (28)$$

The hooked-shaped dependence of τ^{-1} on the hydroxide-ion concentration has now been obtained for reaction of the monoanions of substituted 4-phenylazoresorcinols [58]–[62] (Perlmutter-Hayman et al., 1976; Briffett et al., 1988), 2-phenylazoresorcinol [63] (Hibbert and Sellens, 1988), 4,6-bis(phenylazo)resorcinols [64] and [65] (Hibbert and Simpson, 1983, 1985) and 2,4-bis(phenylazo)resorcinol [66] (Hibbert and Simpson, 1985). The 4-phenylazoresorcinols and the bis(phenylazo)resorcinols were studied in aqueous solution, and 2-phenylazoresorcinol was studied in 95% (v/v) $Me_2SO–H_2O$. The values of the rate coefficients which give rise to this complex dependence are very finely balanced; a slight change in substituent or other modification of the structure or reaction conditions can lead to a quite different kinetic behaviour (see later). For compounds [58]–[66], which show the complex dependence, the values of the rate coefficients are given in Table 15.

One of the surprising results that emerges from the data in Table 15, is the low value of the rate coefficient, typically, $k_1 \simeq 1 \times 10^3 \text{ s}^{-1}$, for opening of the intramolecular hydrogen bond in these compounds. The dependence of τ^{-1} on $[OH^-]$ requires that the proton-transfer step occurs rapidly compared to closing of the hydrogen bond; that is to say $k_2[OH^-] > k_{-1}$ over the range of hydroxide-ion concentrations studied. Since proton removal from the open form of the phenylazoresorcinol monoanions is a normal proton transfer (Eigen, 1964) and is thermodynamically favourable, it is likely to occur at the diffusion-controlled limit with a value of k_2 of ca. $1 \times 10^{10} \text{ dm}^3 \text{ mol}^{-1} \text{ s}^{-1}$. Hence, from the inequality $k_2[OH^-] > k_{-1}$, it follows that the value of k_{-1} must be lower than ca. 1×10^6 to $1 \times 10^7 \text{ s}^{-1}$. Thus with $k_1 \simeq 1 \times 10^3 \text{ s}^{-1}$, the equilibrium constant for opening the hydrogen bond must have a value greater than ca. 1×10^{-3} to 1×10^{-4}. A value for the equilibrium constant for opening of the intramolecular hydrogen bond in this range would have been anticipated, but the rate coefficients for opening and closing of the hydrogen bonds are unexpectedly low.

To provide further information about this process, the kinetics of the rotational equilibrium (29), involving 2-phenylazoresorcinol, were investigated by nmr spectroscopy (Hibbert and Sellens, 1988a). The equilibrium

Table 15 Rate coefficients for reaction of phenylazoresorcinols according to (27).

Phenylazoresorcinol	pK	K /dm^3 mol^{-1}	$10^{-6}k_d$ /dm^3 mol^{-1} s^{-1}	$10^{-3}k_1$ /s^{-1}	Reference
[58] R^1 = H, R^2 = H	12.0	63	0.98	7.7	a
[59] R^1 = SO$_3^-$, R^2 = H	11.8	96	0.96	2.0	a
[60] R^1 = NO$_2$, R^2 = H	11.7	130	0.95	0.11	a
[61] R^1 = H, R^2 = NO$_2$	11.5	186	1.56	2.3	a
[62] R^1 = Cl, R^2 = H	—	1311	1.1	0.45	b
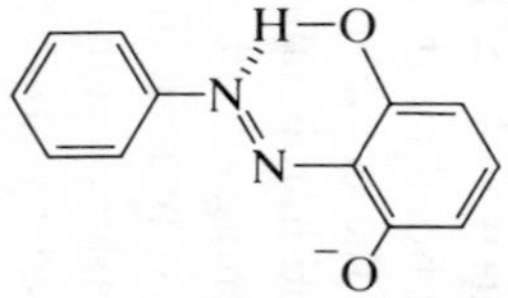 [63]	14	58.3	1.0	4.2	

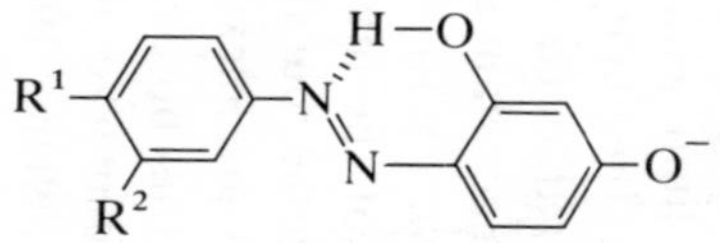

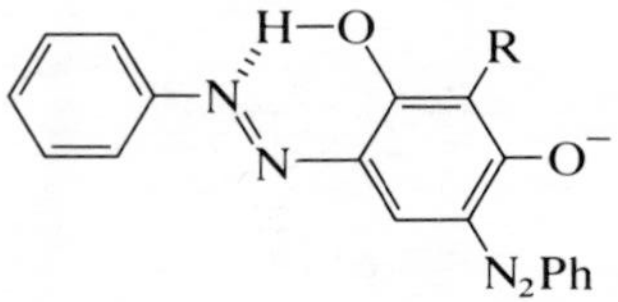

[64] R = H

[65] R = CH₃

[66]

11.3	4600	19	20	d
12.2	209	0.45	12	e
13.6	24.5	0.40	16	d

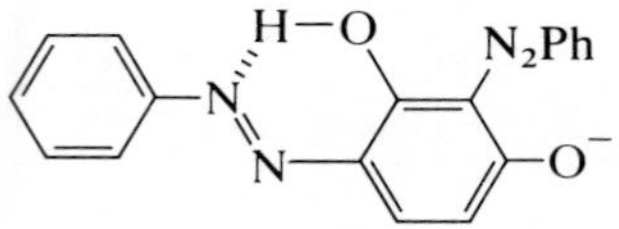

$R = H$ [64]
$R = CH_3$ [65]

[a] Perlmutter-Hayman et al. (1976), aqueous solution, 25°C; [b] Briffett et al. (1988), 70% (v/v) Me₂SO–H₂O, 15°C; [c] Hibbert and Sellens (1986); 95% (v/v) Me₂SO–H₂O, 15°C; [d] Hibbert and Simpson (1985); 20% (v/v) dioxan–H₂O, 5°C. [e] Hibbert and Simpson (1985); aqueous solution, 15°C.

between the open and closed forms of the phenylazoresorcinol monoanions cannot be studied in the same way because equilibrium lies too heavily in favour of the closed form. However, the equilibrium in (29) is an identity reaction, with an equilibrium constant of unity, and nmr spectroscopy is particulary useful in these cases. At temperatures below 300 K, the ^{1}H-nmr spectrum of 2-phenylazoresorcinol in [^{2}H$_8$]toluene shows distinct signals for the two hydroxyl groups ($\delta = 13.01$ and 7.49 ppm for the hydrogen-bonded and non-hydrogen bonded groups, respectively). Above this temperature, a collapsed peak is observed with $\delta = 9.98$ ppm. Line-broadening measurements at various temperatures were used to obtain activation parameters and a rate coefficient for rotation $k = 1.9 \pm 0.4 \times 10^3 \, s^{-1}$ at 298 K. This value is below that for bond rotation in 2-acetylresorcinol [67] and in 2-methoxycarbonyl resorcinol [68] in diethyl ether and dichlorofluoromethane, respectively (Koelle and Forsén, 1974). The results for 4,6-di-t-butyl-2-phenylazoresorcinol [69] and 2-phenylazo-1,3,5-trihydroxybenzene [70] in [^{2}H$_8$]toluene are slower still at $k = 2.2 \times 10^2$ and $0.44 \times 10^2 \, s^{-1}$, respectively.

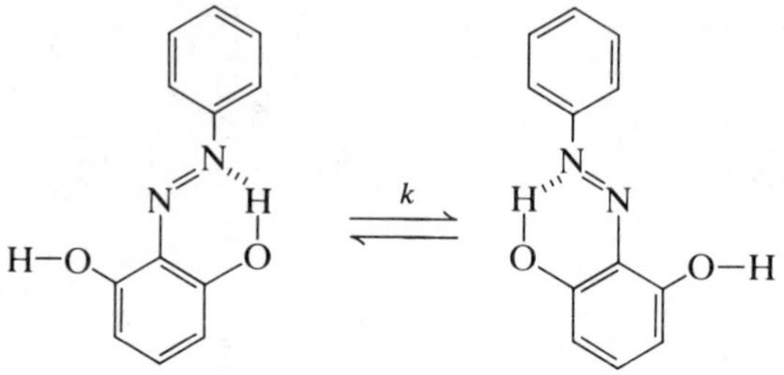

(29)

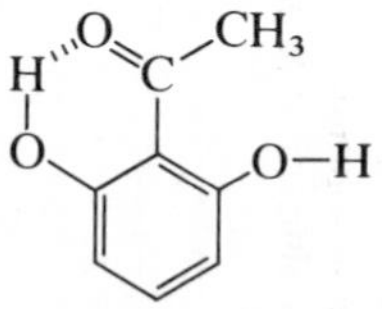

[67]

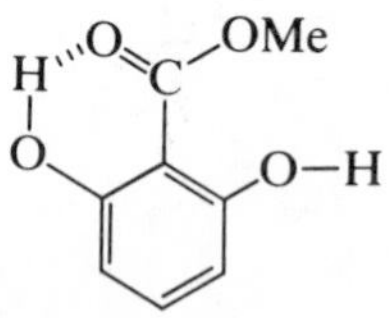

[68]

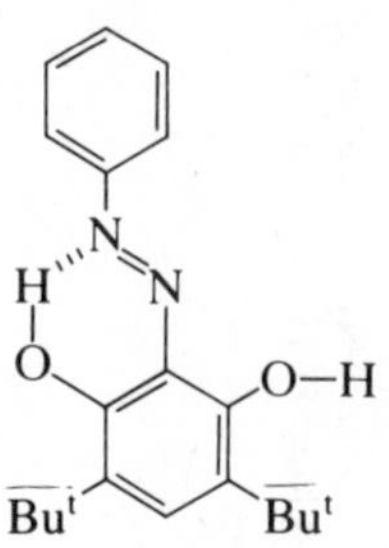

[69]

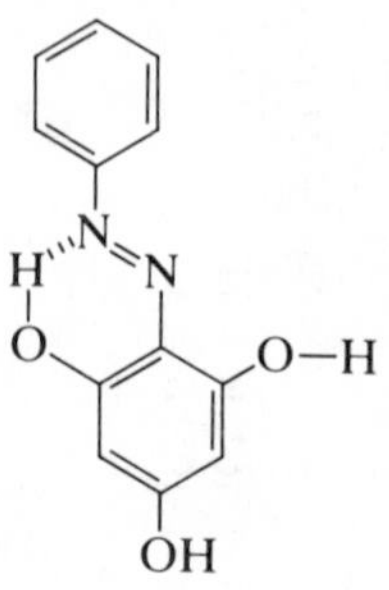

[70]

Further information about the rate of opening and closing of the intramolecular hydrogen bond in 4,6-bis(phenylazo)resorcinol monoanions was obtained from temperature-jump investigations of the proton-transfer behaviour in phenol buffers in 70% (v/v) Me_2SO-H_2O (Briffett et al., 1985; Briffett and Hibbert, 1988). In aqueous solution in the presence of hydroxide ions, the reaction of the monoanion of 4,6-bis(phenylazoresorcinol) to give the dianion occurs according to mechanism (27) and a hook-shaped dependence of τ^{-1} on $[OH^-]$ is observed. In 70% Me_2SO-H_2O containing phenol buffers, the hydroxide ion makes a negligible contribution to the rate, and buffer catalysis can be studied. The observed dependence of τ^{-1} on buffer concentration was curvilinear (Fig. 15). This implies a change in the rate-limiting step with increasing buffer concentration, and mechanism (30) was proposed to explain the results. If it is assumed that the open species is present in low concentration, the rate expression (31) is derived in which K is the overall equilibrium constant for the reaction and r is the buffer ratio ($r = [B^-]/[BH]$).

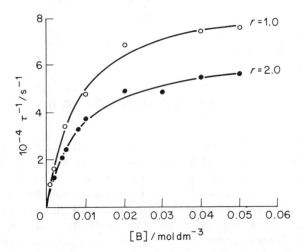

Fig. 15 Kinetic data for proton removal from the 4,6-bis(phenylazo)resorcinol monoanion to the 2-methylcresolate ion in 70% (v/v) Me_2SO-H_2O.

Mechanism (30) is similar to that on the lower route of (27) except that the proton-transfer step only becomes rapid at high buffer concentrations. At low buffer concentrations, the condition $k_2[B^-] < k_{-1}$ is satisfied, and the proton transfer is rate limiting; $\tau^{-1} = (k_1/k_{-1})k_2[B^-] + k_{-2}[BH]$. At high buffer concentrations the condition $k_2[B^-] \gg k_{-1}$ applies, and opening of the hydrogen bond becomes rate limiting; $\tau^{-1} = k_1 + (k_{-1}k_{-2}/k_2)([BH]/[B^-])$.

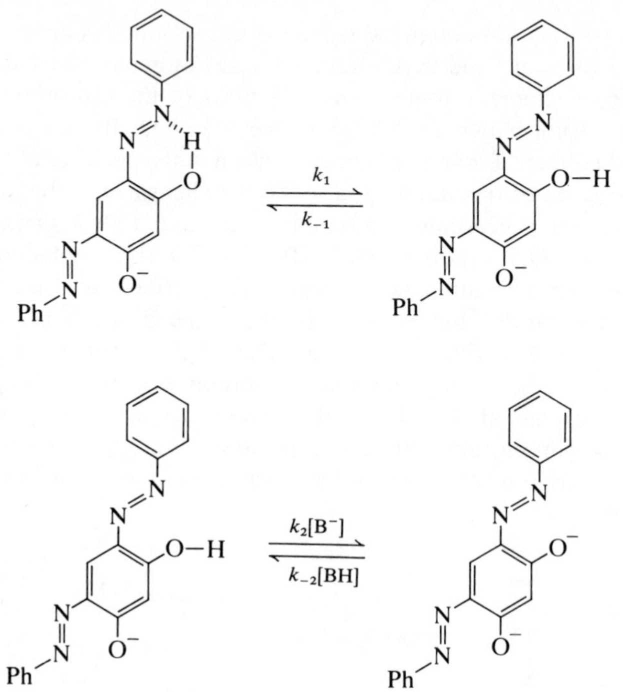

$$(30)$$

$$\tau^{-1} = k_1(1 + 1/Kr)\,[\text{B}^-]/(k_{-1}/k_2 + [\text{B}^-]) \qquad (31)$$

The first evidence that two step proton transfer from a hydrogen-bonded acid could occur consisted of Eigen plots for proton removal by buffer bases. The demonstration of a change in the rate-limiting step as in (30) and (31) provides even more clear evidence and permits the calculation of the rate coefficients and equilibrium constant for opening of the hydrogen bond.

The solid lines in Fig. 15, which refer to the data for reaction of the 4,6-bis(phenylazo)resorcinol monoanion in 2-methylphenol buffers at two buffer ratios are plots of (31) using best-fit values of k_1 and k_{-1}/k_2 and the separately measured value of the equilibrium constant ($K = 0.84$) for the reaction. The data at a buffer ratio $r = 1.0$ were fitted using $k_1 = 4.1 \times 10^4\,\text{s}^{-1}$ and $k_{-1}/k_2 = 7.9 \times 10^{-3}\,\text{mol}\,\text{dm}^{-3}$, and at $r = 2.0$ the best-fit values were $k_1 = 4.2 \times 10^4\,\text{s}^{-1}$ and $k_{-1}/k_2 = 8.4 \times 10^{-3}\,\text{mol}\,\text{dm}^{-3}$. Studies were made in three different phenol buffers and at several buffer ratios in each buffer. Similar values of the rate coefficient k_1, which refers to the unimolecular opening of the intramolecular hydrogen bond, were obtained at different buffer ratios and in different buffers. The results are shown in Table 16.

Table 16 Rate coefficients for proton transfer from the 4,6-bis(phenylazo)resorcinol monoanion to buffer species as described in (30).[a]

Buffer	K	$10^{-4}k_1$ /s^{-1}	$10^3 k_{-1}/k_2$ /dm^3 mol^{-1}
C$_6$H$_5$OH	0.28 ± 0.03	3.6 ± 0.6	5.0 ± 0.2
2-MeC$_6$H$_4$OH	0.84 ± 0.02	3.9 ± 0.6	7.5 ± 1.1
2,6-(Pr^i)$_2$C$_6$H$_3$OH	3.4 ± 0.3	3.9 ± 0.3	9.1 ± 0.2

[a] 70% (v/v) Me$_2$SO–H$_2$O at 15°C.

The use of 70% (v/v) Me$_2$SO–H$_2$O has permitted detailed studies of the buffer-catalysed reaction which was not possible in aqueous solution, and quite different kinetic behaviour was, thereby, uncovered. Because of the different effects of the change in solvent from a wholly aqueous solution to 70% (v/v) Me$_2$SO–H$_2$O on the values of the ionic product of water (Bernasconi and Terrier, 1975; Sorkhabi et al., 1978) and the acid-dissociation constant of phenol (Bernasconi and Terrier, 1975; Halle et al., 1970), the hydroxide-ion concentration in a 1:1 phenol/phenolate buffer is much lower in the mixed solvent than in pure water. The value of the equilibrium constant for the reaction is also affected by the change in solvent. One drawback in using Me$_2$SO–H$_2$O mixtures appears to be the greater susceptibility of reactions to electrolyte effects. Buffer association (Halle et al., 1970; Bernasconi and Terrier, 1975; Hibbert and Robbins, 1978) and specific salt effects (Bernasconi et al., 1985; Hibbert and Sellens, 1988b) are found in this solvent. These effects may reduce the accuracy of the data in Table 16, but the overall kinetic behaviour is not affected.

The results in Table 16 permit the calculation of values for the equilibrium constant and rate coefficients for opening and closing of the intramolecular hydrogen bond in the monoanion of 4,6-bis(phenylazo)resorcinol. The rate coefficient for opening ($k_1 \simeq 4 \times 10^4$ s^{-1}) is obtained from the data at high buffer concentrations where this process is rate limiting. The results for k_{-1}/k_2 allow the calculation of a k_{-1}-value of ca. 4×10^7 s^{-1} for closing of the hydrogen bond, assuming that proton removal from the open species is diffusion limited and occurs with a rate coefficient $k_2 \simeq 5 \times 10^9$ dm^3 mol^{-1} s^{-1}. Thus, the equilibrium constant for opening the hydrogen bond has a value of ca. 1×10^{-3}.

The Gibbs free energy reaction profiles in Fig. 16 have been calculated from the results in Table 16 and the mechanism in (30) and refer to reaction in a 1:1 2-methylphenol buffer at buffer concentrations of 0.001 and 0.1 mol dm^{-3} (Fig. 16(a) and (b), respectively). TS(1) is the transition state for opening of the intramolecular hydrogen bond and TS(2) is the transition

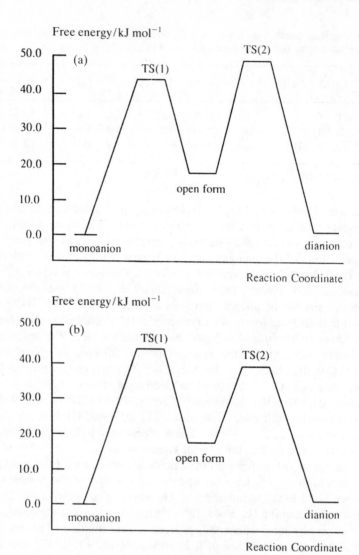

Fig. 16 Free energy/reaction coordinate diagram for proton transfer from the 4,6-bis(phenylazo)resorcinol monoanion to give the dianion in the presence of 2-methylphenol buffers at a 1:1 buffer ratio and at buffer concentrations of (a) 0.001 and (b) $0.10 \, \text{mol}^{-1} \, \text{dm}^{-3}$.

state for proton removal from the open intermediate. At the lower buffer concentration, the proton-transfer step occurs more slowly than closing of the hydrogen bond, and proton transfer is rate limiting. At the higher concentration, proton transfer occurs more rapidly than hydrogen-bond closure, and opening of the hydrogen bond is rate limiting.

To probe further the structure of the transition states and open intermediate, substituent effects were investigated (Briffett and Hibbert, 1988) by introducing 4-chloro-, 4-methoxyl, and 4-methyl groups into the 4,6-bis-(phenylazo)resorcinol. The value of the overall equilibrium constant for the reaction (K) was increased by an electron-withdrawing substituent (4-Cl) and reduced by electron-releasing groups. However, the derived values of k_1 and k_{-1}/k_2 were insensitive to these substituents and showed little change within experimental error. Similarly, a change in solvent to 70% (v/v) Me_2SO-D_2O showed that the value of k_1 was unchanged within experimental error by the isotopic substitution, although the value of k_{-1}/k_2 was increased slightly. The solvent isotope effect is composed of a number of contributions and, at present, it is not possible to analyse the individual effects of isotope substitution on the rate coefficients and equilibrium constant for opening of the intramolecular hydrogen bond.

Kinetic behaviour similar to that in Fig. 15 has also recently been found for other types of intramolecularly hydrogen-bonded acids. Proton transfer from the monoanions of 5,8-dihydroxy-1,4-naphthoquinone [71] and its 2-methyl derivative [72] to buffer species was studied in 50 and 70% (v/v) Me_2SO-H_2O using the temperature-jump technique (Hibbert and Spiers, 1988). A curvilinear dependence of τ^{-1} on buffer concentration was found and analysis of the results according to an equation similar to (31) gave values for the rate coefficients for opening and closing of the intramolecular hydrogen bond for [71] in 70% (v/v) Me_2SO-H_2O of $k_1 \simeq 1 \times 10^5 \text{ s}^{-1}$ and $k_{-1} \simeq 5 \times 10^8 \text{ s}^{-1}$, respectively. Therefore, the equilibrium constant for hydrogen-bond opening had a value of $ca.$ 2×10^{-4}. The results for the monoanions of [71] and [72] were quite similar, despite the interesting

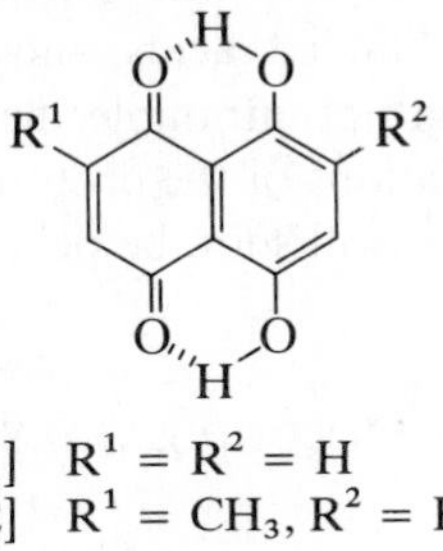

[71] $R^1 = R^2 = H$
[72] $R^1 = CH_3, R^2 = H$

observation (de la Vega, 1982; de la Vega $et\ al.$, 1982) that intramolecular proton transfer along the hydrogen bonds in [71] and [72] show important differences. On the basis of the appearance of the ^{1}H-nmr spectrum and independent theoretical calculations, it was concluded that, for [71], the intramolecular proton movement along the hydrogen bond occurred rapidly on the nmr timescale by a tunnelling mechanism but that for [72] the equilibrium in (32) occurred slowly.

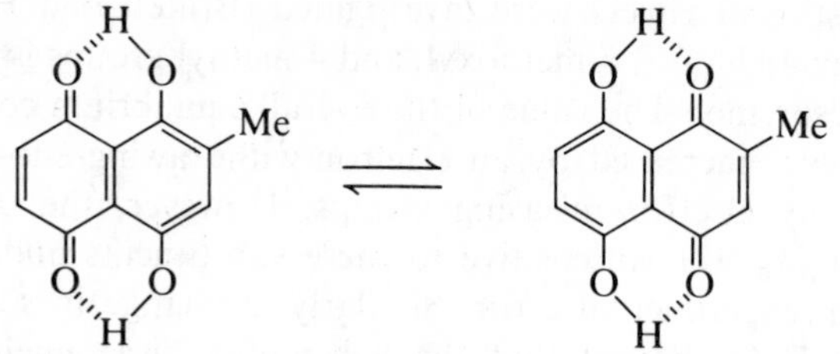

(32)

An indication has been obtained that the opening of the salicylate hydrogen bond may become partially rate limiting in proton transfer (33) from substituted salicylate ions to hydroxide ions and buffer species in 50% (v/v) $Me_2SO–H_2O$ (Hibbert and Spiers, 1989a). Temperature-jump measurements of the equilibration between the salicylate ion and its dissociated species lead to curved plots of τ^{-1} against buffer concentration and against hydroxide-ion concentration. Analysis of the results in terms of the mechanism in (33) gave the approximate values $k_1 = 5 \times 10^5 \, s^{-1}$ and $k_{-1} = 3 \times 10^9 \, s^{-1}$.

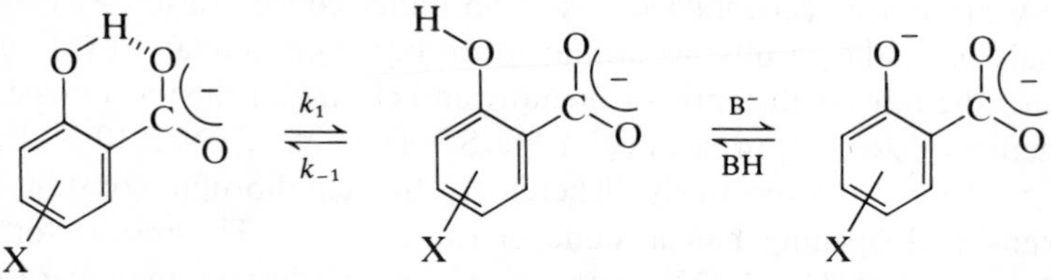

(33)

Our detailed knowledge of the mechanism of proton transfer from intramolecularly hydrogen-bonded acids and our understanding of the factors that affect the strength of intramolecular hydrogen bonds have now reached a stage when correlations of the proton-transfer behaviour with the nature of the hydrogen bond will soon be possible.

INTRAMOLECULAR HYDROGEN BOND CATALYSIS

The study of intramolecular catalysis, which deals with the way in which reactive and catalytic groups interact within the same molecule, is an important way of testing some of the hypotheses that have been put forward to account for the unusual features of enzymic reactions. The intramolecular model mimics the reactions that occur within the enzyme–substrate complex.

Intramolecular reactions often differ from their intermolecular counterparts in the exceptionally high rates that are observed and some reactions can occur intramolecularly that are impossible between separate molecules. Because of the importance of intramolecular catalysis, the subject has been reviewed frequently, particularly with reference to its connection with enzymic catalysis (Page, 1973, 1984; Fife, 1975; Jencks, 1975; Kirby, 1980; Fersht, 1985; Menger, 1985). The present coverage is limited to examples of intramolecular catalysis that owe some of their efficiency to intramolecular hydrogen bonding. The role that hydrogen bonds play in enzymic reactions is discussed in Section 5.

The most efficient examples of intramolecular reactions are those that occur by nucleophilic attack in ring-closing reactions (Kirby, 1980). Although acid–base catalysis of an intramolecular reaction and intramolecular nucleophilic reactions are often closely related, intramolecular acid–base catalysis is usually characterized by relatively low efficiency. Exceptions to this generalization are those examples of acid–base catalysis in which formation of an intramolecular hydrogen bond occurs at some stage in the reaction, and this is often found to lead to increased efficiency. In most of these examples, the salicylate ion is present as the leaving group (Kirby, 1980), but, recently, a few examples with other leaving groups have been discovered. A variety of reactions which involve displacement of the salicylate ion have been studied and, in many cases, these reactions occur at enhanced rates compared with the displacement of other leaving groups. Evidence is available that the enhanced rate is due to partial formation of the salicylate hydrogen bond in the transition state, which is thereby stabilized. This can be looked upon as intramolecular acid catalysis by the carboxyl group of salicylic acid, although, as will be seen, there are important differences in comparison with the usual intramolecular acid catalysis. The different classes of reaction will be discussed in turn.

Acetal hydrolysis

The interest in intramolecular catalysis of the hydrolysis of acetals arises because of its relevance to the mechanism of the lysozyme-catalysed cleavage of polysaccharides. One mechanism (Dunn and Bruice, 1973; Fife, 1975; Fersht, 1985; Kirby, 1987) that has been postulated is shown in (34); residue Glu-35 is thought to act as a general acid catalyst in the cleavage of the glycosidic bond to generate a carboxonium ion. The role of residue Asp-52 may be to provide electrostatic stabilization for the forming carboxonium ion or to assist through simultaneous general acid (Glu-35) and nucleophilic catalysis (Asp-52).

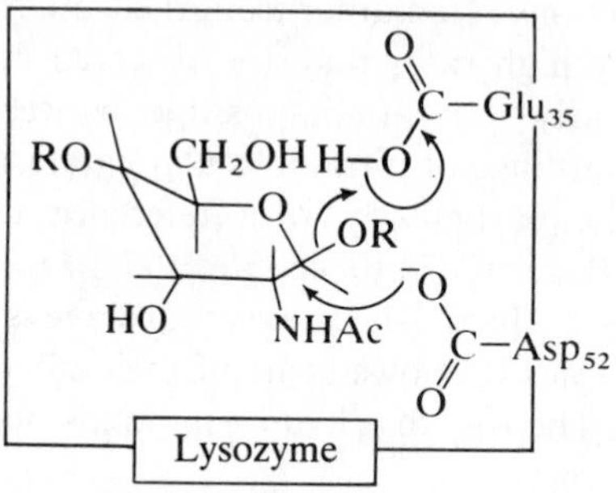

(34)

Suitably substituted acetals have been shown to hydrolyse rapidly by a mechanism that involves intramolecular general acid catalysis similar to that proposed for Glu-35 in (34). The largest effects have been found for acetals with the salicylate ion as the leaving group. For example, the spontaneous hydrolysis (35) of 2-methoxymethoxybenzoic acid [73] occurs 300-fold more rapidly than the same reaction of 4-methoxymethoxybenzoic acid [74] and *ca.* 600-fold more rapidly than the reaction of 2-methoxymethoxybenzoic acid methyl ester [75] (Capon *et al.*, 1969; Dunn and Bruice, 1970). The

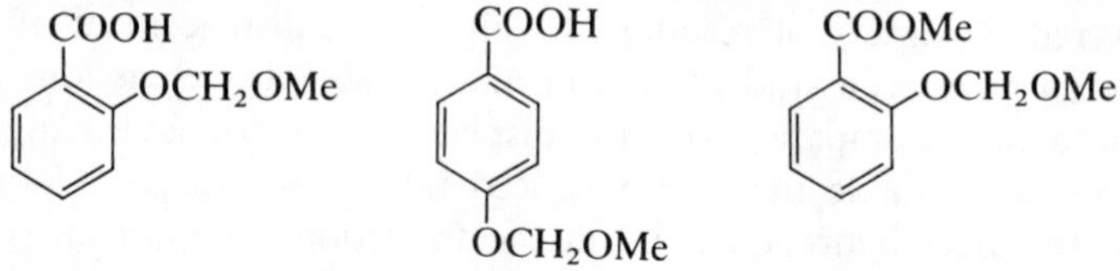

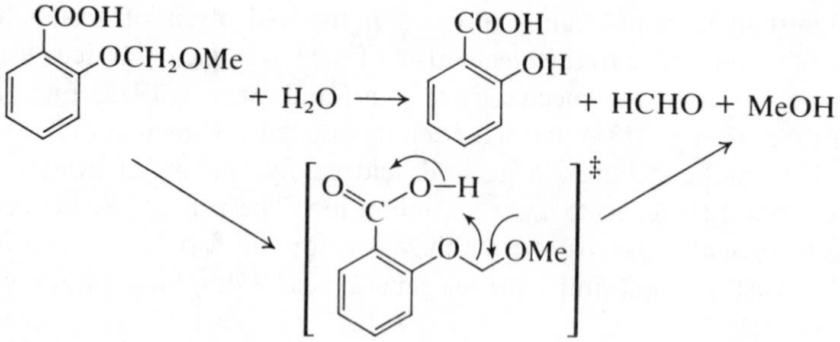

spontaneous hydrolysis involves displacement by solvent molecules and, in the case of 2-methoxymethoxybenzoic acid, the enhanced rate is due to intramolecular acid catalysis as in (35). There seems to be general agreement that mechanism (35) is followed, and specific acid catalysis of the reaction of the 2-methoxymethoxybenzoate anion (Dunn and Bruice, 1973), which is kinetically equivalent, has been ruled out. The reactions of [73], [74] and [75] show specific acid catalysis at low pH-values.

To calculate an effective molarity for catalysis by the carboxyl group in mechanism (35), it is necessary to compare the value of the rate coefficient with that for a similar reaction which is catalysed intermolecularly by a general acid. Unfortunately, for [73], [74] and [75], catalysis by general acids is not detectable. However, intramolecular catalysis has been observed in the reaction of [76] and the reaction of [78] is subject to intermolecular general acid catalysis in buffer solutions. The spontaneous hydrolysis of [76] in aqueous solution occurs with a rate coefficient k_0 that is 1.4×10^4-fold higher than that for [77] (Fife and Anderson, 1971). The value of k_0, which refers to intramolecular acid catalysis by the carboxyl group in [76], is a factor of 580 greater than the second-order rate coefficient for the intermolecular formic acid-catalysed hydrolysis of [78]. Actually, the two reactions were studied under different conditions, and, if some allowance is made for this and for the difference in the pK_a-values of formic acid and the carboxyl group in [76], the ratio of rate coefficients is closer to *ca.* 7×10^3 mol dm^{-3} (Kirby, 1980). This value of the effective molarity is much higher than the maximum value, *ca.* 80 mol dm^{-3}, normally anticipated for intramolecular general acid catalysis (Kirby, 1980) and illustrates the advantage that is conferred by the intramolecular hydrogen bond. Efficient hydrogen-bond catalysis appears to be characterized by much higher effective molarities than intramolecular general acid catalysis. A weak hydrogen bond is present in the ground state and the strengthening that occurs at the transition state, as the leaving group departs and charge is developed, leads to stabilization of the transition state and to enhanced rates of reaction.

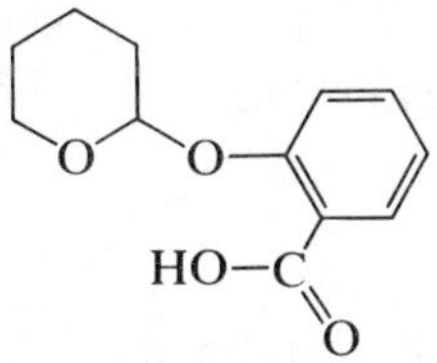

[76]

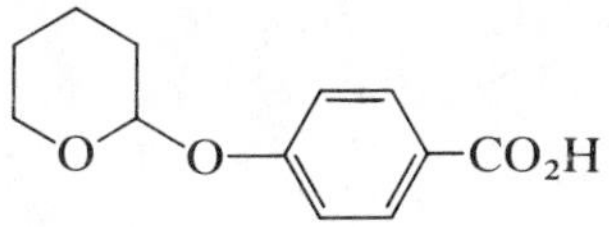

[77]

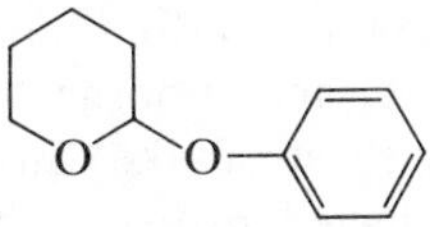

[78]

Interestingly, Fife has attempted to mimic even more closely the postulated mechanism for lysozyme catalysis by incorporating a nucleophilic centre in addition to the general acid, and this model, the anion of benzaldehyde disalicyl acetal is shown in (36) (Anderson and Fife, 1973). This species was estimated to be 3×10^9-fold more reactive than the corresponding dimethylester which does not possess the catalytic carboxyl group in ring A or the catalytic carboxylate group in ring B. However, the estimated effect of the carboxylate group in ring B in providing electrostatic stabilization of the developing carbonium ion was only a factor of 50. It was also demonstrated (Fife and Przystas, 1977) that [79] was 100-fold more reactive than [80] and this was attributed to electrostatic stabilization of the developing carbonium ion which can operate for [79] but not for [80]. The conditions for the observation of efficient intramolecular hydrogen-bond and electrostatic catalysis in the hydrolysis of acetals now appear to be well defined (Fife and Przystas, 1979).

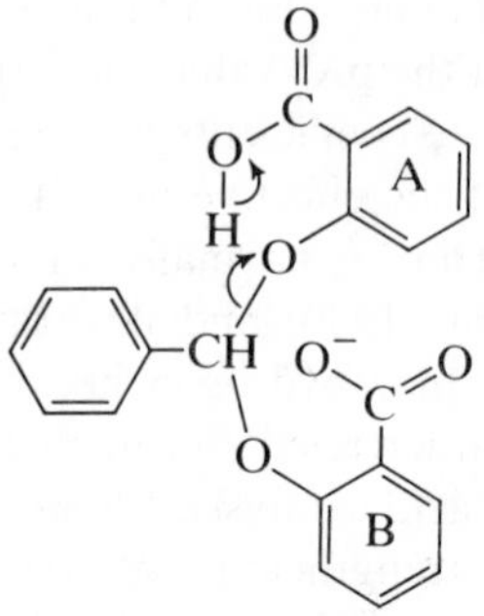

(36)

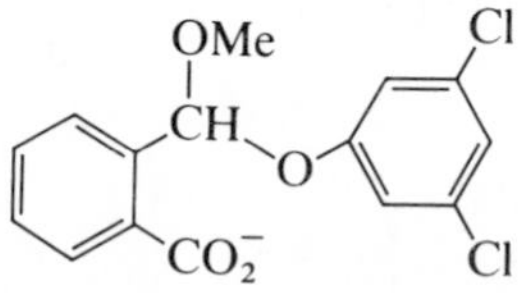

[79]

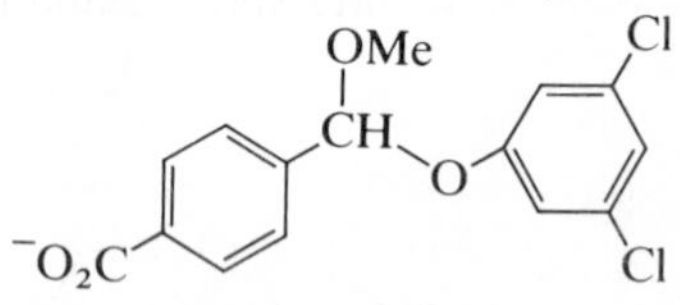

[80]

One of the requirements for efficient hydrogen-bond catalysis is a strong hydrogen bond which is stabilized by conjugation in a ring. For example, the intramolecular hydrogen bond in the leaving group of [81] is stabilized in this way and [81] reacts about two-fold more rapidly than 2-methoxymethoxybenzoic acid [73] (Kirby and Osborne, 1989). However, the catalytic ability of the carboxyl group is weak in the case of [82] in which the leaving group possesses a non-conjugated hydrogen bond (Kirby and Osborne, 1989).

[81]   [82]

Details of the transition-state structure for intramolecular hydrogen-bond catalysis with salicylate leaving groups (35) have been probed by investigating the effect of 4- and 5-substituents on the rate of hydrolysis of 2-methoxymethoxybenzoic acid (Craze and Kirby, 1974). In this way the substituent effects on the leaving tendency and on the intramolecular acid catalysis were separated. The effect on the leaving group was entirely normal, but it was found that a change in the pK_a-value of the carboxyl group brought about by substituents had little effect on the magnitude of the intramolecular catalysis. The same result was found for substituted benzaldehyde acetals with salicylate leaving groups (Buffet and Lamaty, 1976) for which effective molarities of *ca.* 7×10^4 mol dm^{-3} were calculated and shown to be largely independent of the substituent. This appears to be a characteristic of salicylate leaving groups and has also been observed with the salicylate ion as the leaving group in ester hydrolysis (Fersht and Kirby, 1968; Bromilow and Kirby, 1972) and in the hydrolysis of ethers (Barber and Kirby, 1987). In the hydrolysis of esters, the small effect of substituents implies that proton transfer from the carboxyl group to the leaving group is scarcely advanced at the transition state so that little charge has developed on the carboxyl group. Alternatively, it could be interpreted as meaning that the strength of the hydrogen bond present in the transition state for intramolecular hydrogen-bond catalysis is little influenced by the electronic effect of substituents.

There have been a few studies of acetals with hydrogen-bonded leaving groups other than the salicylate ion which could hydrolyse by a mechanism involving hydrogen-bond catalysis. To assess the contribution of intramolecular catalysis, the rate coefficients (k_0) for spontaneous solvolysis of [84] (Capon *et al.*, 1969), [85] (Kirby and Percy, 1987, 1989) and [86] (Hibbert and Spiers, 1989b) can be compared with the rate coefficient for reaction of the unsubstituted acetal [83]. A kinetic term in the rate law for the hydrolysis of [83] corresponding to spontaneous solvolysis is undetectable in comparison with the specific acid-catalysed rate (Capon *et al.*, 1969). However, this in itself illustrates the assistance provided by the intramolecular catalytic groups in [84], [85] and [86] for which spontaneous solvolysis is detectable. An estimate of the k_0-value for [83] can be obtained by extrapolation of the

k_0-values for acetals with better leaving groups to 1-naphthol as the leaving group. The results of one such estimate at 65°C (Hibbert and Spiers, 1989b) are shown for [83] along with the observed k_0-values for [84], [85] and [86]. Roughly, it appears that the effects are in the ratio 1900:1000:40 for [84], [85] and [86], respectively, relative to [83]. To assess accurately the catalytic effect of the groups in [84], [85] and [86] it would also be necessary to estimate the electronic effects of the substituents on the leaving group abilities in comparison with the 1-naphthol leaving group.

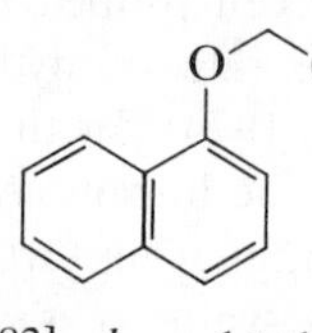

[83] $k_0 = 1 \times 10^{-7} \, \text{s}^{-1}$
(estimated value; see text)

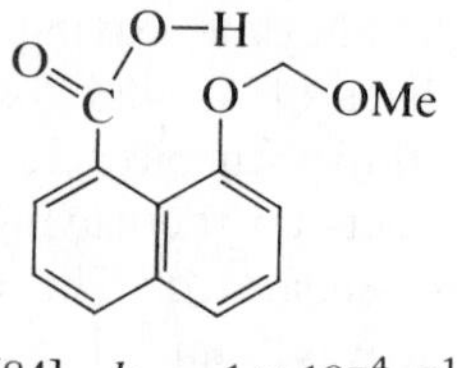

[84] $k_0 = 1 \times 10^{-4} \, \text{s}^{-1}$

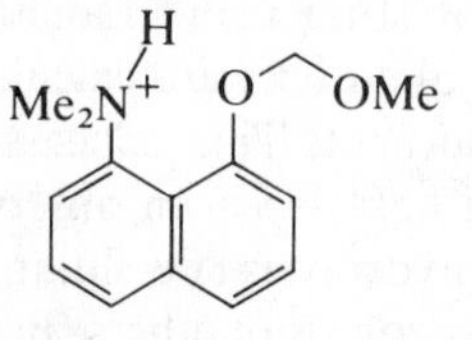

[85] $k_0 = 1.9 \times 10^{-4} \, \text{s}^{-1}$

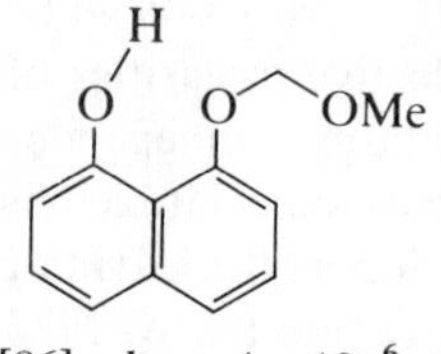

[86] $k_0 = 4 \times 10^{-6} \, \text{s}^{-1}$

In another estimate (Kirby and Percy, 1989), the carboxyl group in 1-methoxymethoxy-8-naphthoic acid and the dimethylammonium group in the 1-methoxymethoxy-8-N,N-dimethylnaphthylammonium ion are estimated to lead to rate increases by intramolecular catalysis of < $ca.$ 900 and 1.9×10^3 compared to the value of $ca.$ 1×10^4 calculated for the intramolecular catalytic effect of the carboxyl group in 2-methoxymethoxybenzoic acid. The salicylate ion remains the most efficient leaving group thus far discovered that can take part in hydrogen-bond catalysis of the hydrolysis of acetals.

Hydrolysis of ethers and esters, and related reactions

A preliminary study of the effect of a salicylate leaving group on the rate of hydrolysis of an ether has been published (Barber and Kirby, 1987). Hydrolysis of the salicyl 1-arylethyl ether (37) occurs 900-fold more rapidly than reaction of the corresponding ether with the carboxyl group replaced by the carbomethoxy group. The reactions are catalysed by the hydronium ion but, in addition, for reaction (37) a term in the rate expression corresponding to spontaneous hydrolysis is detectable. This is considered to arise from intramolecular hydrogen-bond catalysis as in (38).

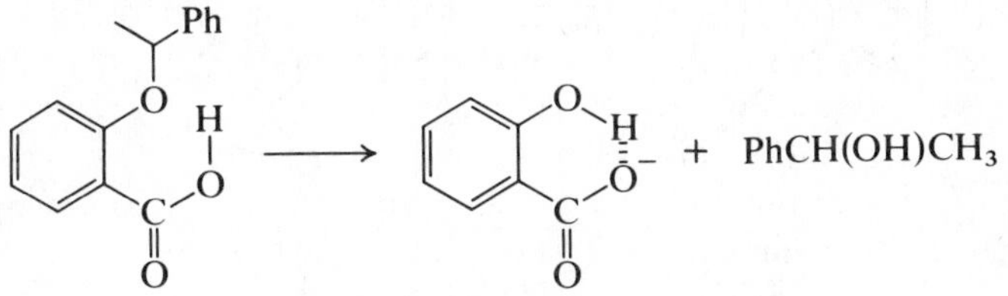

$$PhCH(OH)CH_3 \qquad (37)$$

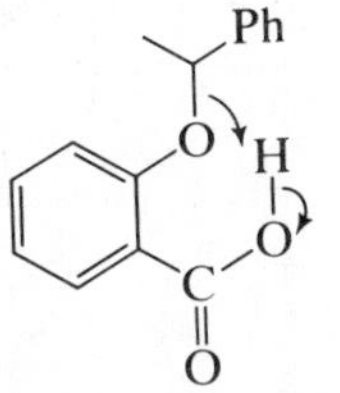

$$(38)$$

   The hydrolysis of acetyl salicylate is one of the classic examples of intramolecular general base catalysis of the attack of solvent by the ionized carboxylate group, (39). The rate coefficient for hydrolysis according to (39) is a factor of 50 times higher than the rate coefficient for reaction of phenyl acetate with solvent. For catalysis by the carboxylate group, an effective molarity of $13\,mol\,dm^{-3}$ was calculated (Fersht and Kirby, 1968) by comparing the rate coefficient for the intramolecular reaction with that for the intermolecularly general base-catalysed reaction of acetyl salicylate with solvent. These rate ratios are typically those found for intramolecular general base catalysis. The hydrolysis of the salicyl phosphate dianion (Bromilow and Kirby, 1972) is thought to occur by the completely different mechanism shown in (40). The reaction proceeds by rapid proton transfer to give the less favoured tautomer of salicyl phosphate dianion which then collapses in a reaction involving hydrogen-bond catalysis of departure of the leaving salicylate ion. In comparison with the rates of reaction of phosphate esters, which do not possess the catalytic carboxyl group, the kinetic advantage conferred by the latter is calculated to be of the order of 10^{10} (Bromilow and Kirby, 1972).

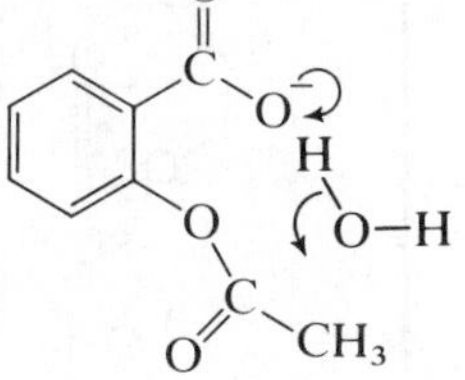

$$(39)$$

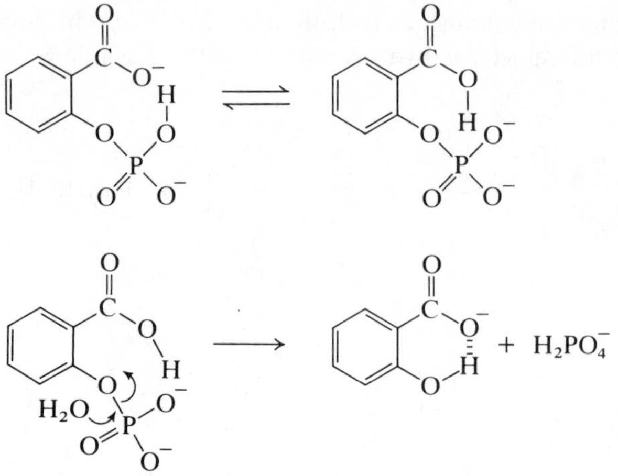

(40)

Evidence is available that the hydrolysis of salicyl sulphate is also subject to hydrogen-bond catalysis as in (41) (Hopkins *et al.*, 1983). Participation by the carboxyl group in the hydrolysis of salicyl sulphate was first identified by Benkovic (Benkovic, 1966), but details of the transition-state structure are now more closely defined. The hydrolysis of phosphate and sulphate esters are characterized by very similar substituent effects. These were analysed into the separate contributions transmitted to the reaction centre through the phenolic group and through the carboxyl group according to a procedure devised by Jaffé (Jaffé, 1954). For both phosphate and sulphate ester hydrolysis, the substituent effects were compatible with little charge development at the carboxyl group in the transition states and with little proton transfer from the carboxyl group to the phenolic oxygen. As previously observed for intramolecular catalysis of the hydrolysis of acetals, this appears to be a characteristic of salicylate leaving groups. However, the substituent effect could again be interpreted in terms of there being little change in the strength of the hydrogen bond in the transition state when the substituent is changed.

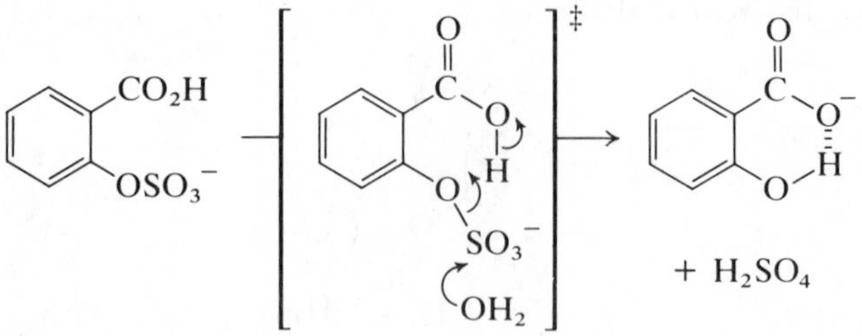

(41)

The mechanism of hydrolysis of 2-carboxyphenylsulphamic acid (42) might be expected to follow that for the hydrolysis of salicyl sulphate, but actually it is thought to proceed by classic intramolecular acid catalysis rather than by hydrogen-bond catalysis. Evidence for a substantial degree of proton transfer from the carboxyl group in the transition state has been obtained (Hopkins and Williams, 1982).

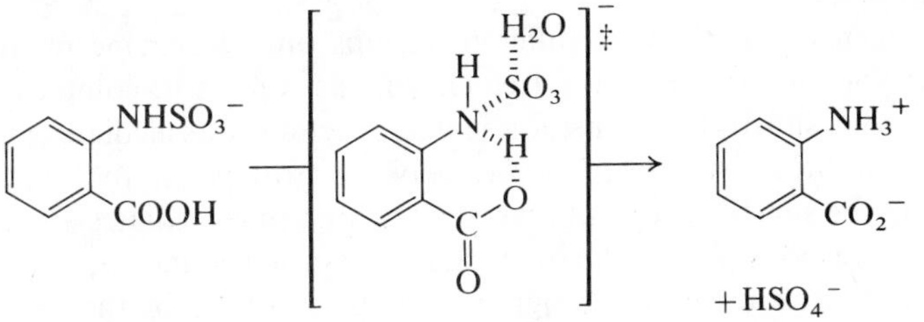

Kinetic studies of the acyl transfer (43) gave rate coefficients that were only a factor of ca. 10 higher than for the corresponding reaction of 1-acetoxynaphthalene (Hibbert and Malana, 1990). For acyl transfers between phenolate ions of roughly similar basicity, the evidence suggests that a concerted mechanism operates (Ba-Saif et al., 1987, 1989), but in the case of 1-hydroxy-8-acetoxynaphthalene, it appears that little assistance to departure of the leaving group is provided by development of the intramolecular hydrogen bond. The leaving group, the 1,8-dihydroxynaphthalene mono-anion, was found to give about a 40-fold enhancement in the rate of the acetal hydrolysis of 1-methoxymethoxy-8-hydroxynaphthalene in comparison to 1-methoxymethoxynaphthalene (see the earlier section on acetal solvolysis).

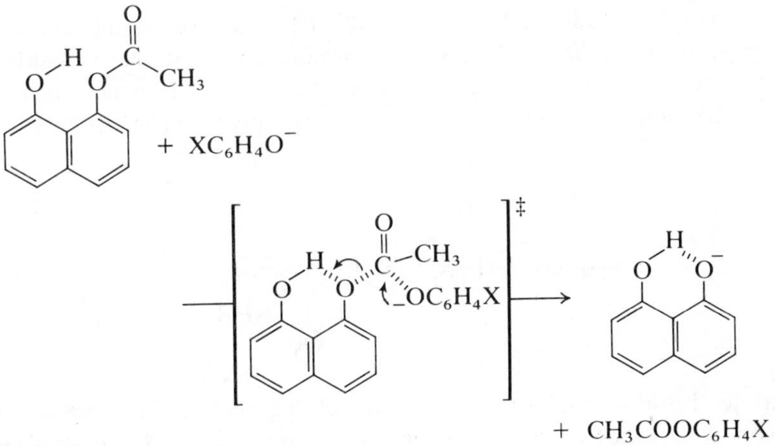

5   Enzyme catalysis and hydrogen bonding

The reactions that occur when a substrate is bound to an enzyme can be looked upon as multifunctional intramolecular catalysis within a macro-molecular complex. The purpose of the present section is to illustrate some of the varied and important roles that hydrogen bonds play in this form of intramolecular catalysis.

Hydrogen bonds between different groups on the enzyme maintain the structure of the protein in the free enzyme as well as in complexes of the enzyme with substrates, transition states, reaction intermediates and products. In addition, hydrogen bonds between groups on the enzyme and groups in the substrate, transition state, reaction intermediates and products are partly responsible for the high catalytic efficiency and the specificity of enzymes. For example, the recognition of the substrate by the enzyme may occur by a number of hydrogen-bond interactions. Stabilization of developing charge in a transition state by formation of a hydrogen bond to a group on the enzyme will lead to catalysis if the hydrogen bond is weaker or not present in the enzyme–substrate complex. In recent years, information about the involvement of hydrogen bonds in the mechanism of several enzyme-catalysed reactions has become available through crystallographic studies and kinetic measurements. The use of genetic engineering to change residues in the enzyme and to investigate the effect on enzyme structure and enzyme kinetics in comparison to the wild-type enzyme has provided particularly detailed and accurate information on enzyme mechanisms and the contribution of hydrogen bonding.

CHYMOTRYPSIN

Crystallographic studies (Blow, 1976) of the structure of the enzyme, enzyme-substrate complexes and enzyme–product complexes have identified a common feature in catalysis by the serine protease enzymes such as α-chymotrypsin. This is the well-known charge-relay system (44), in which

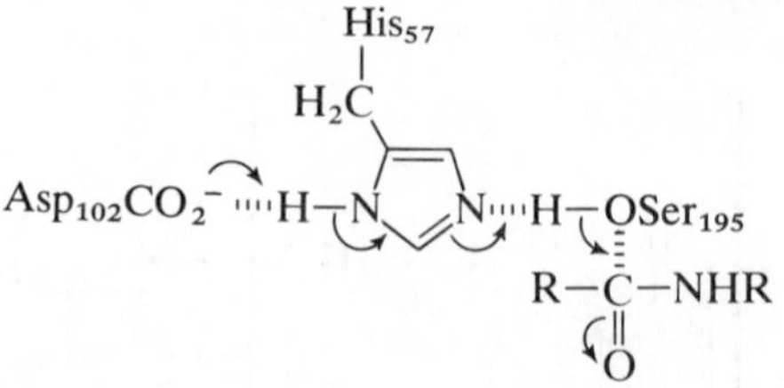

$$\text{Asp}_{102}\text{CO}_2^- \cdots \text{H--N} \underset{}{\overset{}{\diagup}} \text{N} \cdots \text{H--OSer}_{195} \qquad (44)$$

aspartate, histidine and serine residues act together in bringing about nucleophilic attack by the serine hydroxyl group on a peptide carbonyl. In the enzyme, the aspartate, histidine and serine groups are held by two hydrogen bonds. Model systems which attempt to mimic parts of this catalytic triad have been important in showing that the proposed enzyme mechanism is chemically reasonable (Fife, 1975).

THERMOLYSIN

Two other enzymic reactions for which detailed information is available about the involvement of hydrogen bonds in the catalytic process are the closely related zinc-containing enzymes carboxypeptidase A and thermolysin. The proposed mechanism (Hangauer *et al.*, 1984) for the thermolysin-catalysed cleavage of peptides (45) is based on the results of kinetic studies and of X-ray crystallography of the enzyme (Holmes and Matthews, 1982) and of a large number of enzyme–inhibitor complexes (Matthews, 1988). The inhibitors were chosen in an attempt to mimic possible interactions of the enzyme with substrates, products and with postulated reaction intermediates and transition states. In all these cases, the various species make a number of important hydrogen-bond interactions with groups on the enzyme. In (45), which shows the proposed mechanism for the attack of water at the peptide carbonyl group, the carbonyl group is hydrogen bonded to the imidazole ring of His-231 and is held as a ligand by the enzyme's zinc ion. In addition, the amide proton is hydrogen bonded to the carbonyl group of Ala-113. Attack of water at the peptide carbonyl may be assisted through general base catalysis by the carboxylate group of Glu-143.

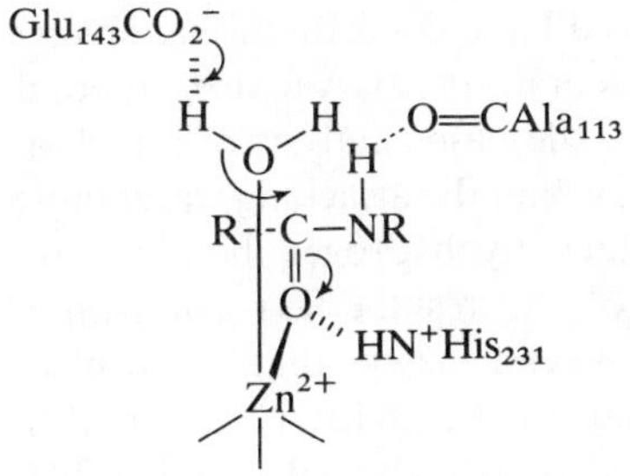

$$\begin{array}{c} Glu_{143}CO_2^- \\[2pt] \vdots \\ H \quad H \quad O{=}CAla_{113} \\ \quad O \quad H \\ R{-}C{-}NR \\ \quad O \\ \quad {}^{'''}HN^+His_{231} \\ Zn^{2+} \end{array} \qquad (45)$$

CARBOXYPEPTIDASE

The enzyme carboxypeptidase A is particularly amenable to structural investigation; crystal structures of the enzyme, of complexes of the enzyme with substrates, substrate analogues and inhibitors, and of transition-state analogues are available. To isolate an enzyme–substrate complex for a one-substrate enzyme reaction, or for an enzyme reaction where water is a reactant, substrates are chosen which show reduced reactivity in the enzymic reaction. In two-substrate enzymic reactions, the situation is more simple because it may be possible to study complexes of the enzyme with each of the substrates in turn.

From crystallographic studies, fast reaction kinetics and site-directed mutagenesis to produce mutant enzymes differing from the native enzyme in one or more specified residues, the mechanism in (46) has been proposed for the attack of water at the carbonyl carbon of benzoylglycylphenylalanine bound to carboxypeptidase (Christianson and Lipscomb, 1989).

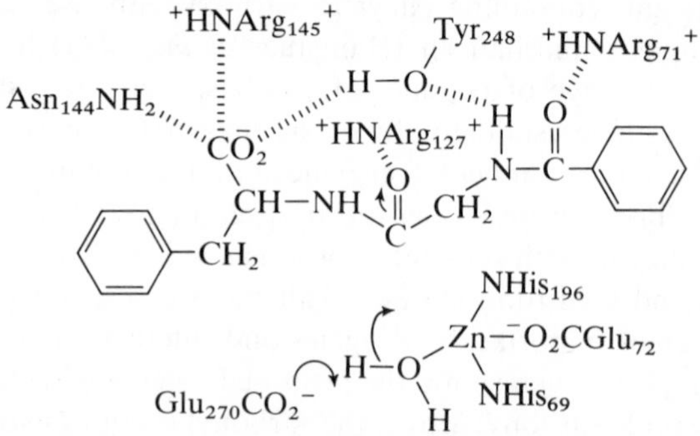

$$(46)$$

Hydrogen bonds are involved in this mechanism at several points. For example, the residue Tyr-248 is thought to donate a hydrogen bond to the carboxylate group of the bound substrate and to accept a hydrogen bond from the amide proton of the peptide. To accomplish this, it is necessary for the enzyme to undergo a large conformational change on substrate binding (Christianson and Lipscomb, 1989). A hydrogen bond between Glu-270 and bound water permits facile base catalysis of attack at the peptide carbonyl group. The zinc ion may bind the attacking water molecule and Arg-127 may assist nucleophilic attack by hydrogen bonding at the peptide carbonyl group. Replacement of the residue Tyr-248 with phenylalanine by site-directed mutagenesis gives an enzyme that binds substrate less effectively but otherwise shows similar kinetic ability (Gardell et al., 1985; Hilvert et al., 1986). Originally, it had been suggested that Tyr-248 may be involved as a general acid catalyst in assisting departure of the amine leaving group, but in view of the properties of the mutant enzyme it has now been suggested that the role of Tyr-248 is to bind substrate, and Glu-270 may act as a general acid catalyst in leaving group departure (Christianson and Lipscomb, 1989).

TYROSYL tRNA SYNTHETASE

The technique of site-directed mutagenesis has been successful in providing details of the mechanism of catalysis by tyrosyl tRNA synthetase. Quantitative information about the strength of interactions between hydrogen-bond donors and acceptors on the enzyme and substrates has been obtained by selective replacement of chosen enzyme residues and investigations of the effect on binding constants and rate coefficients. Tyrosyl tRNA synthetase (E) catalyses two steps in the synthesis of tRNA, shown in (47) and (48). Equation (47) involves the activation of tyrosine (Tyr) to generate enzyme-bound tyrosyl adenylate (E·Tyr-AMP) and pyrophosphate (P_2) from reaction with adenosine triphosphate (ATP) and, in (48), tRNA undergoes aminoacylation and adenosine monophosphate (AMP) is displaced.

$$E + Tyr + ATP \rightarrow E \cdot Tyr\text{-}AMP + P_2 \tag{47}$$

$$E \cdot Tyr\text{-}AMP + tRNA \rightarrow Tyr\text{-}tRNA + AMP + E \tag{48}$$

An outline mechanism for tyrosine activation has been proposed (Fersht, 1975; Fersht et al., 1975a,b; Ward and Fersht, 1988a) on the basis of conventional kinetic and binding studies, and this is shown in (49). For the aminoacylation step, some aspects of the reaction are still not known such as the point at which AMP is displaced, but the currently preferred mechanism (Fersht and Jakes, 1975; Ward and Fersht, 1988b) is that given in (50). This is compatible with the observed kinetics which show that two moles of tyrosine bind in each enzyme turnover during which one molecule of Tyr-tRNA appears.

$$
\begin{array}{c}
\quad\quad E \cdot ATP \\
ATP \nearrow \quad\quad \searrow Tyr \\
E \quad\quad\quad\quad\quad\quad E \cdot Tyr \cdot ATP \rightleftharpoons E \cdot Tyr\text{-}AMP \cdot P_2 \rightleftharpoons E \cdot Tyr\text{-}AMP + P_2 \\
Tyr \searrow \quad\quad \nearrow ATP \\
\quad\quad E \cdot Tyr
\end{array}
\tag{49}
$$

$$E \cdot Tyr\text{-}AMP + tRNA \rightleftharpoons E \cdot Tyr\text{-}AMP \cdot tRNA \rightleftharpoons E \cdot Tyr\text{-}tRNA \cdot AMP$$

$$E \cdot Tyr\text{-}tRNA \cdot AMP \rightleftharpoons E \cdot Tyr\text{-}tRNA + AMP \tag{50}$$

$$E \cdot Tyr\text{-}tRNA \overset{Tyr}{\rightleftharpoons} E \cdot Tyr\text{-}tRNA \cdot Tyr \rightleftharpoons E + Tyr\text{-}tRNA + Tyr$$

To provide information about the likely interactions of substrates and reaction intermediates with the enzyme, several important crystal structures have been analysed. The wild-type enzyme isolated from *Bacillus stearothermophilus* is a dimeric molecule (rmm *ca.* 90 000) with 419 amino-acid residues per subunit, and a crystal structure to 300 pm resolution has been published (Irwin *et al.*, 1976; Bhat *et al.*, 1982). A crystal-structure determination to 250 pm resolution (Brick and Blow, 1987) has been carried out for a complex between a mutant enzyme and the substrate tyrosine. The mutant contains 319 amino-acid residues, 1 to 317 and residues 418 and 419 of the wild-type enzyme, and is more suitable for precise structure determination. The mutant binds tyrosine and ATP and catalyses the first step of the enzyme reaction (47) with very similar values of the kinetic parameters compared to the wild-type enzyme (Waye *et al.*, 1983). The shortened enzyme does not bind tRNA and does not catalyse the aminoacyl transfer between bound tyrosyl adenylate and tRNA. Based on the structure determination, at least five hydrogen bonds are found to be involved in the binding of tyrosine by the enzyme, as shown in [87].

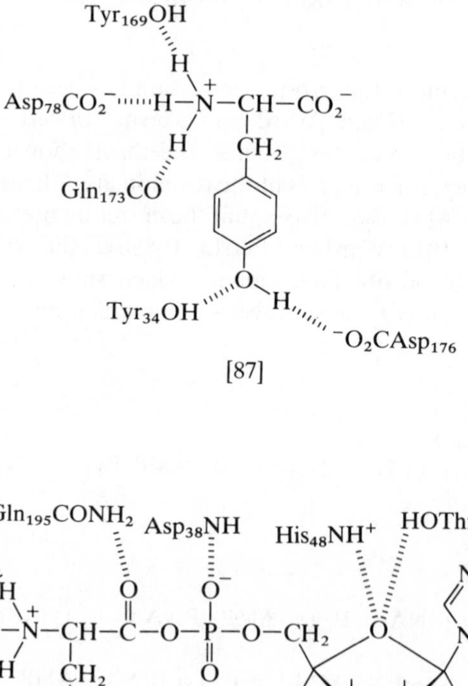

$$
\begin{array}{c}
\text{Tyr}_{169}\text{OH} \\[4pt]
\vdots \\
\text{H} \\
\quad \backslash\ \overset{+}{} \\
\text{Asp}_{78}\text{CO}_2^- \cdots \text{H} - \overset{+}{\text{N}} - \text{CH} - \text{CO}_2^- \\
\quad / \quad\ \ | \\
\quad \text{H} \quad\ \ \text{CH}_2 \\
\text{Gln}_{173}\text{CO} \\
\end{array}
$$

(aromatic ring) — O⋯H⋯ $^-\text{O}_2\text{CAsp}_{176}$ Tyr$_{34}$OH⋯

[87]

(structure with Gln$_{195}$CONH$_2$, Asp$_{38}$NH, His$_{48}$NH$^+$, HOThr$_{51}$, NH$_2$, Tyr$_{169}$OH, Asp$_{78}$CO$_2^-$⋯H—N$^+$—CH—C—O—P—O—CH$_2$, CH$_2$, Tyr$_{34}$OH⋯O—H⋯$^-$O$_2$CAsp$_{176}$, Gly$_{36}$C=O⋯H, HSCys$_{35}$, O⋯HNGly$_{192}$)

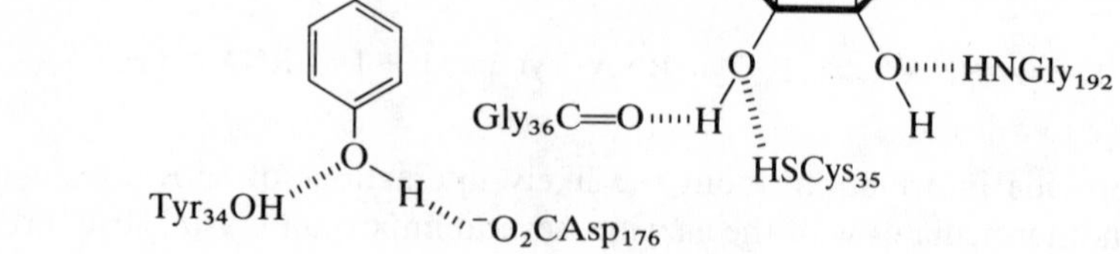

[88]

The structure of the complex between native enzyme and tyrosyl adenylate has been elucidated (Rubin and Blow, 1981). Tyrosyl adenylate is one of the products of the first step in the enzyme mechanism (47), and is the substrate for the second step of the reaction (48). Eleven possible hydrogen bonds are identified and the structure is given in [88]. Previously, the structure of the complex between the enzyme and the competitive inhibitor tyrosinyl adenylate had been clarified (Monteilhet and Blow, 1978).

The structure has been determined for a mutant enzyme in which the residue threonine-51 of the wild-type enzyme from *B. Stearothermophilus* is replaced by proline (Brown *et al.*, 1987). This is one of the differences between the wild-type enzymes isolated from *B. stearothermophilus* and from *Escherichia coli*. Interestingly, the mutant shows increased catalysis of the formation of tyrosyl adenylate and increased affinity for tyrosyl adenylate compared to the wild-type enzyme (Wilkinson *et al.*, 1984; Ho and Fersht, 1986). The precise reasons for the enhanced reactivity and binding ability of the mutant are not known (Ho and Fersht, 1986). However, on the basis of structural differences between the two enzymes, it has been suggested (Brown *et al.*, 1987) that the mutation brings about some solvation changes within the enzyme. The enhanced binding ability may arise because substrate binding to the wild-type enzyme is accompanied by displacement of solvent from the vicinity of Thr-51 which will make binding less favourable. In the mutant enzyme, the non-polar proline residue is not associated with the solvent and binding of the substrate is not disfavoured in this way.

Eight of the 11 hydrogen bonds between tyrosyl adenylate and enzyme found in the crystal structure of the complex [88] involve side-chain residues of the enzyme and can be mutated. The remaining three, formed with residues Gly-36, Gly-192 and Asp-38, are residues in the main chain and cannot be mutated without destroying a major part of the enzyme. Fersht and his research group, in collaboration with the research group of Blow, have begun a systematic replacement of the eight residues to investigate whether hydrogen bonds are present and to determine their role in the enzymic mechanism. The procedure involves the preparation of a mutant enzyme with the appropriate residue replaced or deleted using the techniques of site-directed mutagenesis. Detailed kinetic and binding studies and structural studies are then used to discover the effect of each mutation. Information about the change in strength of the hydrogen bonds as the reaction proceeds from the enzyme–substrate complex through the transition state to the reactive intermediate and the enzyme–product complex has been deduced.

The analysis (Fersht, 1988) of hydrogen-bond strengths and of the magnitude of their effect on catalysis can be illustrated with reference to the simplified scheme in (51)–(53). The first step is the formation of an enzyme–

substrate complex which is assumed to involve a hydrogen bond between site B on the substrate and group XH on the enzyme. The enzyme–substrate complex is transformed through the transition state ($\neq$) to the product complex (52), which then dissociates (53).

$$S{-}B\cdots HOH + HOH\cdots HX{-}E \rightleftharpoons S{-}B\cdots HX{-}E + HOH\cdots HOH \quad (51)$$

$$S{-}B\cdots HX{-}E \rightarrow [S{-}B\cdots HX{-}E]^{\ddagger} \rightarrow P{-}B\cdots HX{-}E \quad (52)$$

$$P{-}B\cdots HX{-}E \rightarrow P{-}B\cdots HOH + HOH\cdots HX{-}E \quad (53)$$

$$\Delta G_{bind} = G(S{-}B\cdots HX{-}E) + G(HOH\cdots HOH)$$
$$- G(S{-}B\cdots HOH) - G(H_2O\cdots HX{-}E) - G_R \quad (54)$$

The Gibbs free energy for enzyme–substrate binding (ΔG_{bind}) is given by (54) and consists of terms for the free energy of dissociation for each hydrogen bond and a term G_R for any other free energy changes on binding. Consider now the results of deletion of the group XH on the enzyme such that in the mutant the hydrogen-bond interaction between substrate group B and XH is simply removed, and the substrate group B is not solvated in the enzyme–substrate complex. The corresponding equation for substrate binding is (55) and the difference (ΔG_{app}) in free energy of binding for the wild-type (E) and mutant (E') enzymes is given by (56) in which ΔG_{reorg} is any change in Gibbs free energy brought about by the mutation other than that allowed for in the hydrogen-bond terms. The value of ΔG_{app} can be measured by determining the ratio of binding constants of the wild-type and mutant enzymes.

$$S{-}B\cdots HOH + H_2O/E' \rightleftharpoons S{-}B/E' + HOH\cdots HOH \quad (55)$$

$$\Delta G_{app} = \{G(E'/H_2O) - G(H_2O\cdots HX{-}E)\}$$
$$- \{G(S{-}B/E') - G(S{-}B\cdots HX{-}E)\} - \Delta G_{reorg} \quad (56)$$

In some cases, the magnitude of ΔG_{app} will be approximately the same as ΔG_{bind}, and in this case ΔG_{app} will represent the contribution made by the hydrogen-bond interaction S—B $\cdots$ HX—E to substrate binding. This may occur in the case of uncharged donors and acceptors, but, when a charged group is involved, it has been argued (Fersht, 1988) that measurement of ΔG_{app} is likely to overestimate ΔG_{bind}. Even in these cases, however, the use of site-directed mutagenesis with a detailed kinetic analysis of the reaction steps does provide information about the changes in the bond-dissociation energy of S—B $\cdots$ HX—E as the reaction proceeds. For example, it is possible to determine the changes in ΔG_{app} between successive intermediates and transition states on the reaction coordinate by measuring the changes in rate coefficients and equilibrium constants as a result of the mutation. If these changes between successive species are denoted by $\Delta\Delta G_{app}$ and if a bond S—B $\cdots$ HX—E is present and changes in strength through the various intermediates and transition states, (56) can be applied between any two states. The terms $G(E'/H_2O)$ and $G(H_2O \cdots HX—E)$ are constant for a particular enzyme mutation, and if it is assumed that $G(S—B/E')$ and ΔG_{reorg} are constant along the reaction coordinate, then (57) is obtained. This means that $\Delta\Delta G_{app}$ between the two successive states on the reaction coordinate measures the difference between the Gibbs free energies for dissociation of the hydrogen bond S—B $\cdots$ HX—E in the two states. This is turn measures the contribution of the hydrogen bond to the difference in stability of the two states.

$$\Delta\Delta G_{app} = \Delta G(S—B \cdots HX—E) \tag{57}$$

Table 17 Effects of hydrogen bonds on free energy differences as measured by ΔG_{app} in tyrosyl activation catalysed by mutations of tyrosyl tRNA synthetase (44).

| Intermediate or transition state | ΔG_{app}/kJ mol^{-1} | | |
	Mutation: Tyr-34 to Phe-34	Cys-35 to Gly-35	His-48 to Gly-48
E·Tyr	−2.2	+0.20	−1.6
E·Tyr·ATP	−2.0	+0.33	−3.5
[E·Tyr·ATP]$^+$	−2.2	−5.3	−6.8
E·Tyr·AMP·P$_2$	−2.9	−6.9	−6.8
E·TyrAMP	−3.7	−6.9	−7.7

The analysis described above has been used to estimate the contribution of hydrogen bonding at each stage along the reaction coordinate for reaction (47) catalysed by tyrosyl tRNA synthetase. Three examples (Wells and Fersht, 1986; Fersht, 1988) of the changes in free energy brought about by mutation are illustrated in Table 17. The species [E·Tyr·ATP]$^+$ in Table 17 is the transition state for formation of tyrosyl adenylate from the enzyme-bound species, tyrosine and ATP, and the other species in Table 17 are the reaction intermediates for the mechanism in (49). The mutations considered are Tyr-34 to Phe, Cys-35 to Gly, and His-48 to Gly residues. The roles of these residues in binding tyrosyl adenylate are shown in [88]; Tyr-34 is

hydrogen bonded to the tyrosine hydroxyl group, and Cys-35 and His-48 are
hydrogen bonded to the sugar moiety of tyrosyl adenylate. Whether or not
the measured values of ΔG_{app} reflect the true contribution of each particular
hydrogen bond to the relative stability of successive intermediates on the
reaction coordinate depends on whether the conditions for the application of
(56) apply and whether the assumptions in deducing (57) are satisfied. One
assumption that is necessary is that the hydrogen bond should be present in
both intermediates. Some of this information is available since structures of
the E·Tyr (Brick and Blow, 1987) and E·Tyr-AMP (Rubin and Blow, 1981)
intermediates are known, but structures of the other species are not yet
available.

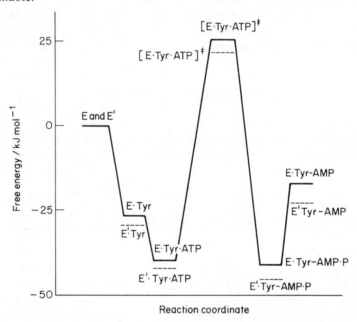

Fig. 17 Free energy/reaction coordinate diagram for tyrosine activation [see (49)],
with wild-type tyrosyl-tRNA synthetase (E) and the Tyr-34 to Phe mutant (E').

The values of ΔG_{app} for the mutation Tyr-34 to Phe show that the E·Tyr
complex is destabilized by $2.2\,\text{kJ mol}^{-1}$ on replacement of Tyr-34. This
mutation deletes a group which donates a hydrogen bond to the tyrosine
hydroxyl group of the substrate. The effect of the deletion shows that the
transition state and enzyme–tyrosyl adenylate complex are also stabilized by
the hydrogen bond in the wild-type enzyme and are destabilized in the
mutant. Stabilization of the enzyme–tyrosyl adenylate complex by the
hydrogen bond is greatest, implying that the hydrogen bond increases in

strength, in proceeding along the reaction coordinate. A free energy profile for the reaction involving the wild type and the mutant (Tyr-34 to Phe) is shown in Fig. 17 (Wells and Fersht, 1986). Similar profiles have been constructed for other mutations and the result is a clear picture of the role of the hydrogen bonds in catalysis.

In the case of the Tyr-34 to Phe mutation, the effect on substrate binding is quite small, and the binding constant for tyrosine is larger by about a factor of two for the wild-type compared to the mutant enzyme. However, it is found that deletion of a residue on the enzyme that interacts with a charged group on the substrate has a much larger effect on binding and catalytic coefficients (Fersht et al., 1985). Generally, for uncharged donors and acceptors, deletion of a residue weakens binding by ca. 2–6 kJ mol^{-1}, whereas deletion of a residue interacting with a charged donor or acceptor weakens binding by a further ca. 12 kJ mol^{-1} (Fersht et al., 1985). Data referring to the reaction in (47) are shown in Table 18 to illustrate these effects. In Table 18, the values determined for k_{cat} and K_m are given; k_{cat} refers to the rate coefficient for decomposition of the enzyme–substrate complex and K_m is given by the ratio $K_m = (k_{cat} + k_{-1})/k_1$ for the mechanism in (58) and (59). The ratio k_{cat}/K_m is the second-order rate coefficient for the reaction at low substrate (S) concentrations, $[S] \ll K_m$. The first four rows in Table 18 are data for the wild-type enzyme and the mutations previously considered in Table 17.

$$E + S \underset{k_{-1}}{\overset{k_1}{\rightleftharpoons}} ES \tag{58}$$

$$ES \xrightarrow{k_{cat}} product + E \tag{59}$$

The effect of deletion of residues which interact with charged groups on the substrate is illustrated by the results for the mutations Gln to Gly-195 and Tyr to Phe-169 given in Table 18. It is suggested (Fersht et al., 1985) that the residue Gln-195 may bind to the charged carboxylate group of tyrosine in the E·Tyr complex and to an ester carbonyl group in the E·Tyr-AMP complex. Evidence from the crystal structure shows that this is the case for the E·Tyr-AMP complex (Rubin and Blow, 1981), but the crystal structure of the E·Tyr species does not show a strong interaction of Gln-195 with the tyrosine carboxylate group (Brick and Blow, 1987). Deletion of Gln-195 may result in a much reduced value of k_{cat} because of a loss in binding energy in the transition state. The value of $K_m(Tyr)$, which is closely related to the dissociation constant of the E·Tyr complex, is also increased by the deletion because the stabilizing hydrogen bond is no longer present. The residue Tyr-169 is hydrogen bonded strongly to the ammonium group of tyrosine in the

Table 18 Values of k_{cat} and K_m for wild-type and mutant enzymes in tyrosine activation [see (47) and (49)].

Enzyme	k_{cat}/s^{-1}	$10^3 K_m(ATP)$ /mol dm^{-3}	$10^6 K_m(Tyr)$ /mol dm^{-3}	$k_{cat}/K_m(ATP)$ /dm^3 mol^{-1} s^{-1}	$k_{cat}/K_m(Tyr)$ /dm^3 mol^{-1} s^{-1}
Wild-type	8.35	1.08	2.23	7.73×10^3	3.74×10^6
Tyr to Phe-34	6.86	1.2	4.4	5.72×10^3	1.56×10^6
Cys to Gly-35	2.95	2.6	2.7	1.13×10^3	1.09×10^6
His to Gly-48	2.00	1.3	3.2	1.54×10^3	6.25×10^5
Gln to Gly-195	0.19	2.5	100	76	1.90×10^3
Tyr to Phe-169	6.05	1.25	1030	4.84×10^3	5.88×10^3
Thr to Ala-51	8.75	0.54	2.0	1.62×10^4	4.36×10^6
Thr to Pro-51	12.0	0.058		2.08×10^5	

E·Tyr complex and, on deletion, the binding of tyrosine to the enzyme is much less favourable thermodynamically.

In some cases, mutation can lead to enhanced catalytic ability of the enzyme. Results for the mutation Thr-51 to Pro-51 (Wilkinson *et al.*, 1984) have been mentioned previously. The results for this and for the mutation Thr-51 to Ala-51 (Fersht *et al.*, 1985) are also shown in Table 18. These mutations and that of Thr-51 to Cys-51 have been studied in some detail (Ho and Fersht, 1986). In each case it is found that the transition state is stabilized for formation of tyrosine adenylate from tyrosine and ATP within the enzyme; the mutant Thr-51 to Pro-51 increases the rate coefficient for the reaction by a factor of 20. However, the enzyme-bound tyrosine adenylate is also stabilized by the mutation and this results in a reduced rate of reaction of tyrosine adenylate with tRNA (48), the second step in the process catalysed by tyrosine tRNA synthetase. Overall, therefore, the mutants are poorer catalysts for the formation of aminoacyl tRNA. The enzyme from *E. coli* has the residue Pro-51 whereas Thr-51 is present in the enzyme from *B. stearothermophilus*. The enzyme from *E. coli* is more active than the latter enzyme in both the formation of tyrosine adenylate and in the aminoacylation of tRNA (Jones *et al.*, 1986b). It is therefore suggested (Ho and Fersht, 1986) that the enzyme from *E. coli* with Pro-51 must additionally have evolved ways of stabilizing the transition state for formation of tyrosine adenylate without the concomitant stabilization of tyrosine adenylate and reduction in the rate of aminoacylation of tRNA found for the Pro-51 mutant.

Even though hydrogen bonds between uncharged groups on the enzyme and substrate do not affect the values of k_{cat} and K_m as markedly as do interactions between charged residues, the effect on enzyme specificity can be quite large. For example, the effect of the mutation Tyr-34 to Phe-34 on the values of k_{cat} and K_m have been compared for the catalysed reactions of tyrosine and phenylalanine with ATP (Fersht *et al.*, 1985). The residue Tyr-34 makes a hydrogen bond with the hydroxyl group of tyrosine as in [87] but is unable to interact in this way with phenylalanine. The values of k_{cat}/K_m for reaction of tyrosine in the presence of wild-type and mutant enzymes were compared to the values for reaction of phenylalanine, and it was found that k_{cat}/K_m for tyrosine activation was 1.5×10^5-fold higher for the wild-type enzyme but only 1.0×10^4-fold higher for the mutant enzyme. Thus, the hydrogen bond formed between Tyr-34 and tyrosine in the wild-type enzyme favours reaction with tyrosine over phenylalanine and leads to a 15-fold specificity for tyrosine over phenylalanine. The combined effects of several hydrogen bonds, especially those involving charged residues, can lead to high degrees of specificity for the natural substrate. The magnitude of the free energy of binding for substrates, products, transition states and intermediates will be determined by many interactions, such as the hydro-phobic effect as well as by hydrogen bonds, but it is suggested (Fersht *et al.*, 1985) that hydrogen bonds may be of particular importance in determining the enzyme's preference for a particular reaction pathway involving one specific substrate.

One of the important consequences of studying catalysis by mutant enzymes in comparison with wild-type enzymes is the possibility of identify-ing residues involved in catalysis that are not apparent from crystal structure determinations. This has been usefully applied (Fersht *et al.*, 1988) to the tyrosine activation step in tyrosine tRNA synthetase (47) and (49). The residues Lys-82, Arg-86, Lys-230 and Lys-233 were replaced by alanine. Each mutation was studied in turn, and comparison with the wild-type enzyme revealed that each mutant was substantially less effective in cata-lysing formation of tyrosyl adenylate. Kinetic studies showed that these residues interact with the transition state for formation of tyrosyl adenylate and pyrophosphate from tyrosine and ATP and have relatively minor effects on the binding of tyrosine and tyrosyl adenylate. However, the crystal structures of the tyrosine–enzyme complex (Brick and Blow, 1987) and tyrosyl adenylate complex (Rubin and Blow, 1981) show that the residues Lys-82 and Arg-86 are on one side of the substrate-binding site and Lys-230 and Lys-233 are on the opposite side. It would be concluded from the crystal structures that not all four residues could be simultaneously involved in the catalytic process. Movement of one pair of residues close to the substrate moves the other pair of residues away. It is therefore concluded from the kinetic effects observed for the mutants that, in the wild-type enzyme, formation of the transition state for the reaction involves a conformational change to a structure which differs from the enzyme structure in the complex with tyrosine or tyrosine adenylate. The induced fit to the transition-state structure must allow interaction with all four residues simultaneously.

6 Summary

Although hydrogen bonding has been recognized as the single most import-ant intermolecular interaction for many years, it is only now that the subtleties of this kind of bonding are becoming appreciated. This review has advanced the hypothesis that there are three kinds of hydrogen bond: weak, strong and very strong. These are determined by the shape of the potential energy well, and the respective positions of hydrogen and deuterium within the well can be used to provide information about which well applies in a particular example. In future, it may be possible to "fine tune" such a potential energy well by varying electronic or steric factors within the system and to observe the effect on the properties of the hydrogen bond. Currently, the weak, strong and very strong categories serve mainly to explain the many puzzling features of hydrogen bonding. New techniques, such as the measurement of isotopic shifts in the nmr spectrum, may serve as probes to such studies.

The debate over the upper limit to the strength of hydrogen bonding appears to have been decided, with the bifluoride ion being the most likely contender, but the issue is still not finally resolved. Even so, the indications are that this is one of the few hydrogen bonds with a single minimum potential well and as such serves as the standard against which to measure other very strong hydrogen bonds. In the past decade, several unusually short hydrogen bonds have come to light and there is every expectation that more will be discoverd, especially involving $^+$N—H $\cdots$ N systems. These types of hydrogen bond have provided some of the more unexpected discoveries, in particular proton sponges and caged protons.

The role of hydrogen bonding in catalysis has been discussed, although mainly in terms of the salicylate ion as a leaving group. With its strong intramolecular hydrogen bond playing an essential part in the reaction mechanism, it seems likely that this will herald other systems where the role of a strong hydrogen bond may serve as the key step in a catalytic process.

Finally, the area where it is expected that hydrogen-bonding studies will receive most attention in the next decade is in studies of enzyme action. The recognition of a target molecule by an enzyme, the manoeuvering of this molecule to the active site, the orientation of the substrate at the site of action and the binding of transition states and intermediates are all determined by a programmed pattern of hydrogen bonds. Weak, strong and even very strong hydrogen bonds may all have a role to play, and it is here that the fine tuning of a potential well by steric and electronic changes near the hydrogen bond may allow the subtle action necessary for the specific catalysis by the enzyme. At present, this area is only just beginning to be explored, but promises to be the most exciting. It remains to be seen whether the role of hydrogen bonding can be established as clearly as it has been in nucleic-acid transcription.

Acknowledgements

We are grateful to all our research students who have helped to maintain our interest and enthusiasm for this area of chemistry.

References

Abu-Dari, K., Raymond, K. N. and Freyberg, D. P. (1979). *J. Am. Chem. Soc.* **101**, 3639

Albert, N. and Badger, R. M. (1958). *J. Chem. Phys.* **29**, 1193

Albery, W. J. (1975). *In* "Proton Tranfser Reactions" (eds E. F. Caldin and V. Gold). Chapman and Hall, London

Alder, R. W. (1989). *Chem. Rev.* **89**, 1215.

Alder, R. W. and Sessions, R. B. (1983). *In* "The Chemistry of Functional Groups. The Chemistry of Amino, Nitroso and Nitro Compounds" (ed. S. Patai). Wiley, London

Alder, R. W., Bowman, P. S., Steele, W. R. S. and Winterman, D. R. (1968). *J. Chem. Soc., Chem. Commun.* 723

Alder, R. W., Bryce, M. R., Goode, N. C., Miller, N. and Owen, J. (1981). *J. Chem. Soc., Perkin Trans. 1* 2840

Alder, R. W., Eastment, P., Hext, N. M., Moss, R. E., Orpen, A. G. and White, J. M. (1988a). *J. Chem. Soc., Chem. Commun.* 1528

Alder, R. W., Goode, N. C., Miller, N., Hibbert, F., Hunte, K. P. P. and Robbins, H. J. (1978). *J. Chem. Soc., Chem. Commun.* 79

Alder, R. W., Orpen, A. G. and Sessions, R. B. (1983a). *J. Chem. Soc., Chem. Commun.* 999

Alder, R. W., Moss, R. E. and Sessions, R. B. (1983b). *J. Chem. Soc., Chem. Commun.* 997

Alder, R. W., Moss, R. E. and Sessions, R. B. (1983c). *J. Chem. Soc., Chem. Commun.* 1000

Alder, R. W., Orpen, A. G. and White, J. M. (1988b). *Acta. Chem. Cryst.* **C44**, 287

Allen, G. and Dwek, R. A. (1966). *J. Chem. Soc. (B)* 161

Allen, L. C. (1975). *J. Am. Chem. Soc.* **97**, 6921

Almlöf, J. (1972). *Chem. Phys. Lett.* **17**, 49

Al-Rawi, J. M. A., Bloxsidge, J. P., O'Brien, C., Caddy, D. E., Elvidge, J. A., Jones, J. R. and Evans, E. A. (1979). *J. Chem. Soc., Perkin Trans. 2* 1593

Altman, L. A., Laungani, D., Gunnarsson, G., Wennerström, H. and Forsén, S. (1978). *J. Am. Chem. Soc.* **100**, 8264

Anderson, C. C. and Hassel, O. (1926). *J. Phys. Chem.* **123**, 151

Anderson, E. and Fife, T. H. (1973). *J. Am. Chem. Soc.* **95**, 6437

Appelman, E. H. and Kim, H. (1982). *J. Chem. Phys.* **76**, 1664

Appelman, E. H. and Thompson, R. C. (1984). *Angew. Chem., Int. Ed. Engl.* **27**, 392

Appelman, E. H., Wilson, W. W. and Kim, H. (1981). *Spectrochim. Acta* **A37**, 385

Arshadi, M. and Kebarle, P. (1970). *J. Phys. Chem.* **74**, 1483

Aue, D. H., Webb, H. M. and Bowers, M. T. (1973). *J. Am. Chem. Soc.* **95**, 2699

Aue, D. H., Webb, H. M. and Bowers, M. T. (1976). *J. Am. Chem. Soc.* **98**, 318

Ault, B. S. (1978). *J. Phys. Chem.* **82**, 844

Ault, B. S. (1979). *J. Phys. Chem.* **83**, 837

Ault, B. S. (1982). *Acc. Chem. Res.* **15**, 103

Awwal, A., Burt, R. and Kresge, A. J. (1981). *J. Chem. Soc., Perkin Trans. 2* 1566

Badger, R. M. and Bauer, S. H. (1937). *J. Chem. Phys.* **5**, 839

Barber, S. E. and Kirby, A. J. (1987). *J. Chem. Soc., Chem. Commun.* 1775

Barnett, G. H. and Hibbert, F. (1984). *J. Am. Chem. Soc.* **106**, 2080

Bartl, H. and Küppers, H. (1978). *In* "Nukleare Festkörperforschung am FRG: Ergebnisbericht 1977/78 der externen Arbeitsgruppen" (eds G. Heger and H. Weitzel), KfK 2719, pp. 7–9. Kernforschungszentrum, Karlsruhe

Bartl, H. and Küppers, H. (1980). *Z. Krist.* **152**, 161

Ba-Saif, S., Luthra, A. K. and Williams, A. (1987). *J. Am. Chem. Soc.* **109**, 6362

Ba-Saif, S., Luthra, A. K. and Williams, A. (1989). *J. Am. Chem. Soc.* **111**, 2647

Bassetti, M., Cerichelli, G. and Floris, B. (1988). *J. Chem. Res. (S)* 236
Bellamy, L. J. and Owen, A. J. (1969). *Spectrochim. Acta, Sect. A* **25**, 329
Benkovic, S. J. (1966). *J. Am. Chem. Soc.* **88**, 5511
Beno, M. A., Sundell, R. and Williams, J. M. (1984). *Croatica Chem. Acta* **57**, 695
Berglund, B. and Vaughan, R. W. (1980). *J. Chem. Phys.* **73**, 2037
Berglund, B., Lindgren, J. and Tegenfeldt, J. (1978). *J. Mol. Struct.* **43**, 169
Bernander, L. and Olofsson, G. (1972). *Tetrahedron* **28**, 3251
Bernasconi, C. F. and Terrier, F. (1975). *J. Am. Chem. Soc.* **97**, 7458
Bernasconi, C. F., Kanavarioti, A. and Killion, R. B. (1985). *J. Am. Chem. Soc.* **107**, 3612
Bertrand, J. A., Black, T. D., Eller, P. G., Helm, F. T. and Mahmood, R. (1976). *Inorg. Chem.* **15**, 2965
Bethell, D., McDonald, K. and Rao, K. S. (1977). *Tetrahedron Lett.* 1447
Bhat, T. N., Blow, D. M., Brick, P. and Nyborg, J. (1982). *J. Mol. Biol.* **158**, 699
Biagini-Cingi, M., Manotti-Lanfredi, A. M., Tiripicchio, A. and Tiripicchio-Camellini, M. (1977). *Acta Crystallogr., Sect. B* **33**, 3772
Biali, S. E. and Rappoport, Z. (1984). *J. Am. Chem. Soc.* **106**, 5641
Blinc, R. (1958). *Nature (London)* **182**, 1016
Blinc, R. and Zeks, B. (1974). "Soft Modes in Ferroelectrics and Antiferroelectrics". North-Holland, Amsterdam
Blow, D. M. (1976). *Acc. Chem. Res.* **9**, 145
Bondi, A. (1964). *J. Phys. Chem.* **68**, 441
Bonnet, B. and Mascherpa, G. (1980). *Inorg. Chem.* **19**, 785
Borodin, A. P. (1862). *Il Nuovo Cimento* **15**, 305
Bozorth, R. M. (1923). *J. Am. Chem. Soc.* **45**, 2128
Brick, P. and Blow, D. M. (1987). *J. Mol. Biol.* **194**, 287
Briffett, N. E. and Hibbert, F. (1988). *J. Chem. Soc., Perkin Trans. 2* 1041
Briffett, N. E., Hibbert, F. and Sellens, R. J. (1985). *J. Am. Chem. Soc.* **107**, 6712
Briffett, N. E., Hibbert, F. and Sellens, R. J. (1988). *J. Chem. Soc., Perkin Trans. 2* 2123
Bromilow, R. H. and Kirby, A. J. (1972). *J. Chem. Soc., Perkin Trans. 2* 149
Brown, K. A., Brick, P. and Blow, D. M. (1987). *Nature (London)* **326**, 416
Brown, S. J. and Clark, J. H. (1985). *J. Chem. Soc., Chem. Commun.* 672
Brunton, G. D. and Johnson, C. K. (1975). *J. Chem. Phys.* **62**, 3797
Buckingham, A. D. and Fan-Chen, L. (1981). *Int. Rev. Phys. Chem.* **1**, 253
Buemi, G. (1990). *J. Mol. Struct. (Theochem.)*, **209**, 89
Buemi, G. and Gandolfo, C. (1989). *J. Chem. Soc., Faraday Trans. 2* **85**, 215
Buffet, C. and Lamaty, G. (1976). *Recl. Trav. Chim. Pays-Bas* **95**, 1
Burdett, J. L. and Rogers, M. T. (1964). *J. Am. Chem. Soc.* **86**, 2105
Camerman, A., Mastropaolo, D. and Camerman, N. (1983). *J. Am. Chem. Soc.* **105**, 1584
Camilleri, P., Marby, C. A., Odell, B., Rzepa, H. S., Sheppard, R. N., Stewart, J. J. P. and Williams, D. J. (1989). *J. Chem. Soc., Chem. Commun.* 1722
Capon, B., Smith, M. C., Anderson, E., Dahm, R. H. and Sankey, G. H. (1969). *J. Chem. Soc (B)* 1038
Carlsen, L. and Duus, F. (1980). *J. Chem. Soc., Perkin Trans. 2* 1080
Cartwright, P. S., Gillard, R. D., Sillanpaa, E. R. T. and Valkonen, J. D. (1988). *Polyhedron* **7**, 2143
Chan, S. I., Lin, L., Clutter, D. and Dea, P. (1970). *Proc. Natl. Acad. Sci. USA* **65**, 816

Chiang, Y., Kresge, A. J. and More O'Ferrall, R. A. (1980). *J. Chem. Soc., Perkin Trans. 2* 1832

Christe, K. O. (1987). *J. Fluorine Chem.* **35**, 621

Christianson, D. W. and Lipscomb, W. N. (1989). *Acc. Chem. Res.* **22**, 62

Chunnilall, C J., Sherman, W. F. and Wilkinson, G. R. (1984). *J. Mol. Struct.* **115**, 205

Clair, R. L. and McMahon, T. B. (1979). *Can. J. Chem.* **57**, 473

Clark, D. R., Emsley, J. and Hibbert, F. (1988a). *J. Chem. Soc., Perkin Trans. 2* 919

Clark, D. R., Emsley, J. and Hibbert, F. (1988b). *J. Chem. Soc., Chem. Commun.* 1252

Clark, D. R., Emsley, J. and Hibbert, F. (1989). *J. Chem. Soc., Perkin Trans. 2* 1299

Clark, J. H., Kanippayoor, R. K. and Miller, J. M. (1981a). *J. Chem. Soc., Dalton Trans.* 1152

Clark, J. H., Emsley, J., Jones, D. J. and Overill, R. E. (1981b). *J. Chem. Soc., Dalton Trans.* 1219

Clary, D. C. and Connor, J. N. L. (1984). *J. Phys. Chem.* **88**, 2758

Cote, G. L. and Thompson, H.. W. (1951). *Proc. R. Soc. (A)* **210**, 206

Coulombeau, C. (1977). *J. Fluorine Chem* **9**, 483

Coulson, C. A. (1957). *Research* **10**, 149

Craze, G.-A. and Kirby, A. J. (1974). *J. Chem. Soc., Perkin Trans. 2* 61

Del Bene, J. E., Frisch, M. J. and Pople, J. A. (1985). *J. Phys. Chem.* **89**, 3669

Denne, W. A. and MacKay, M. F. (1971). *J. Cryst. Mol. Struct.* **1**, 311

Desmeules, P. J. and Allen, L. C. (1980). *J. Chem. Phys.* **72**, 4731

Dewar, D. H., Fergusson, J. E., Hentschel, P. R., Wilkins, C. J. and Williams, P. P. (1964). *J. Chem. Soc.* 688

Dunn, B. M. and Bruice, T. C. (1970). *J. Am. Chem. Soc.* **92**, 2410

Dunn, B. M. and Bruice, T. C. (1973). *Adv. Enzymol.* **37**, 1

Eckert, M. and Zundel, G. (1988). *J. Phys. Chem.* **92**, 7016

Eigen, M. (1964). *Angew. Chem. Int. Ed. Engl.* **3**, 1

Eigen, M. and Kruse, W. (1963). *Z. Naturforsch. (B)* **18**, 857

Eigen, M., Kruse, W., Maass, G. and de Maeyer, L. (1964). *Prog. React. Kin.* **2**, 285

Ellison, S. L. R. and Robinson, M. J. T. (1983). *J. Chem. Soc., Chem. Commun.* 745

Emsley, J. (1971). *J. Chem. Soc. (A)* 2702

Emsley, J. (1980). *Chem. Soc. Revs.* **9**, 91

Emsley, J. (1984). *Struct. Bonding (Berlin)* **57**, 147

Emsley, J. and Clark, J. H. (1974). *J. Chem. Soc., Dalton Trans.* 1125

Emsley, J. and Freeman, N. J. (1987). *J. Mol. Struct.* **161**, 193

Emsley, J. and Hoyte, O. P. A. (1976). *J. Chem. Soc., Dalton Trans.* 2219

Emsley, J., Hoyte, O. P. A. and Overill, R. E. (1977). *J. Chem. Soc., Perkin Trans. 2* 2079

Emsley, J., Hoyte, O. P. A. and Overill, R. E. (1978). *J. Am. Chem. Soc.* **100**, 3303

Emsley, J., Jones, D. J. and Kuroda, R. (1981). *J. Chem. Soc., Dalton Trans.* 2141

Emsley, J., Parker, R. J. and Overill, R. E. (1983). *J. Chem. Soc., Faraday Trans. II* **79**, 1347

Emsley, J., Freeman, N. J., Parker, R. J. and Overill, R. E. (1986a). *J. Chem. Soc., Perkin Trans. 2* 1479

Emsley, J., Freeman, N. J., Parker, R. J., Dawes, H. M. and Hursthouse, M. B. (1986b). *J. Chem. Soc., Perkin Trans. 1* 471

Emsley, J., Gold, V. and Szeto, W. T. A. (1986c). *J. Chem. Soc., Dalton Trans.* 2641

Emsley, J., Freeman, N. J., Bates, P. A. and Hursthouse, M. B. (1987). *J. Mol. Struct.* **161**, 181

Emsley, J., Freeman, N. J., Bates, P. A. and Hursthouse, M. B. (1988a). *J. Chem. Soc., Perkin Trans. 1* 297

Emsley, J., Ma, L. Y. Y., Bates, P. A. and Hursthouse, M. B. (1988b). *J. Mol. Struct.* **178**, 297

Emsley, J., Reza, N. M., Dawes, H. M., Hursthouse, M. B. and Kuroda, R. (1988c). *Phosphorus and Sulfur* **35**, 141

Emsley, J., Ma, L. Y. Y., Bates, P. A., Motevalli, M. and Hursthouse, M. B. (1989). *J. Chem. Soc., Perkin Trans. 2* 527

Emsley, J., Ma, L. Y. Y., Karlaulov, S. A., Motevalli, M. and Hursthouse, M. B. (1990a). *J. Mol. Struct.* **216**, 143

Emsley, J., Ma, L. Y. Y., Nyburg, S. C. and Parkins, A. W. (1990b). *J. Mol. Struct.*, **240**, 59

Endo, S., Chino, T., Tsuboi, S. and Koto, K. (1989). *Nature (London)* **340**, 452

Ernstbrunner, E. E. (1970). *J. Chem. Soc. A* 1558

Evans, D. F. (1982). *J. Chem. Soc., Chem. Commun.* 1226

Eyring, E. and Haslam, J. L. (1966). *J. Phys. Chem.* **70**, 293

Fenn, M. D. and Spinner, E. (1984). *J. Phys. Chem.* **88**, 3993

Fersht, A. R. (1975). *Biochemistry* **14**, 5

Fersht, A. R. (1985). "Enzyme Structure and Mechanism". Freeman, New York

Fersht, A. R. (1988). *Biochemistry* **27**, 1577

Fersht, A. R. and Jakes, R. (1975). *Biochemistry* **14**, 3350

Fersht, A. R. and Kirby, A. J. (1968). *J. Am. Chem. Soc.* **90**, 5826

Fersht, A. R., Mulvey, R. S. and Koch, G. L. E. (1975a). *Biochemistry* **14**, 13

Fersht, A. R., Ashford, J. S., Bruton, C. J., Jakes, R., Koch, G. L. E. and Hartley, B. S. (1975b). *Biochemistry* **14**, 1

Fersht, A. R., Shi, J.-P., Knill-Jones, J., Lowe, D. M., Wilkinson, A. J., Blow, D. M., Brick, P., Carter, P., Waye, M. M. Y. and Winter, G. (1985). *Nature (London)* **314**, 235

Fersht, A. R., Knill-Jones, J. W., Bedouelle, H. and Winter, G. (1988). *Biochemistry* **27**, 1581

Fife, T H. (1975). *Adv. Phys. Org. Chem.* **11**, 1

Fife, T. H. and Anderson, E. (1971). *J. Am. Chem. Soc.* **93**, 6610

Fife, T. H. and Przystas, T. J. (1977). *J. Am. Chem. Soc.* **99**, 6693

Fife, T. H. and Przystas, T. J. (1979). *J. Am. Chem. Soc.* **101**, 1202

Forsén, S. and Nilsson, M. (1960). *Acta Chem. Scand.* **14**, 1333

Freeman, N. J. (1987). PhD. thesis, University of London

Frevel, L. K. and Rinn, H. W. (1962). *Acta Crystallogr.* **15**, 286

Fueno, T., Kajimoto, O., Nishigaki, Y. and Yoshioka, T. (1973). *J. Chem. Soc., Perkin Trans. 2* 738

Fujiwara, F. Y. and Martin, J. S. (1971). *Can. J. Chem.* **49**, 3071

Fujiwara, F. Y. and Martin, J. S. (1974a). *J. Am. Chem. Soc.* **96**, 7625

Fujiwara, F. Y. and Martin, J. S. (1974b). *J. Am. Chem. Soc.* **96**, 7632

Fyfe, W. S. (1953). *J. Chem. Phys.* **21**, 2

Gardell, S. J., Craik, C. S., Hilvert, D., Urdea, M. S. and Rutter, W. J. (1985). *Nature (London)*, **317**, 551

Geraldes, C. F. G. C., Barros, M. T., Maycock, C. D. and Silva, M. I. (1990). *J. Mol. Struct.*, **238**, 335

Gilli, G., Bellucci, F., Ferretti, V. and Bertolasi, V. (1989). *J. Am. Chem. Soc.* **111**, 1023

Glocker, G. and Evans, G. E. (1942). *J. Chem. Phys.* **10**, 607

"Gmelin Handbook of Inorganic Chemistry" (1982). 8th edn, Fluorine Supplement, Vol. 3, System-number 5, Springer-Verlag, Berlin
Gold, V. (1963). *Proc. Chem. Soc.* 141
Gold, V. (1969). *Adv. Phys. Org. Chem.* 7, 259
Gold, V. and Grist, S. (1971). *J. Chem. Soc. (B)* 1665
Gold, V. and Grist, S. (1972). *J. Chem. Soc., Perkin Trans.* 2 89
Gold, V. and Lowe, B. M. (1967). *J. Chem. Soc. (A)* 936
Gold, V., Morris, K. P. and Wilcox, C. F. (1982). *J. Chem. Soc., Perkin Trans.* 2 1615
Grens, E., Grinvalde, A. and Stradins, J. (1975). *Spectrochim Acta, Sect. A* 31, 555
Grimsrud, E. P. and Kebarle, P. (1973). *J. Am. Chem. Soc.* 95, 7939
Guissani, Y. and Ratajczak, H. (1981). *Chem. Phys.* 62, 319
Gunnarsson, G., Wennerström, H., Egan, W. and Forsén, S. (1976). *Chem. Phys. Lett.* 38, 96
Hadzi, D. (1965). *Pure. Appl. Chem.* 11, 435
Hadzi, D. and Bratos, S. (1976). In "The Hydrogen Bond" (eds P. Schuster, G. Zundel and C. Sandorfy), Vol. II, Ch. 12. North-Holland, Amsterdam
Hadzi, D. and Orel, B. (1973). *J. Mol. Struct.* 18, 227
Halle, J. C., Gaboriaud, R. and Schaal, R. (1970). *Bull. Soc. Chim. Fr.* 2047
Hamilton, W. C. (1962). *Acta Cryst.* 15, 353
Hamilton, W. C. and Ibers, J. A. (1968). "Hydrogen Bonding in Solids". Benjamin, New York
Hangauer, D. G., Monzingo, A. F. and Matthews, B. W. (1984). *Biochemistry* 23, 5730
Hansen, P. (1983). *Ann. Rep. NMR Spectroscopy* 15, 105
Harmon, K. M., Madeira, S. L. and Carling, R. W. (1974). *Inorg. Chem.* 13, 1260
Harmon, K. M., Gennick, I. and Potvin, M. M. (1977). *Inorg. Chem.* 16, 2033
Harmon, K. M. and Lovelace, R. R. (1982). *J. Phys. Chem.* 86, 900
Harrell, S. A. and McDaniel, D. H. (1964). *J. Am. Chem. Soc.* 86, 4497
Haslam, J. L. and Eyring, E. M. (1967). *J. Phys. Chem.* 71, 4471
Haslam, J. L., Eyring, E. M., Epstein, W. W., Christiansen, G. A. and Miles, M. H. (1965a). *J. Am. Chem. Soc.* 87, 1
Haslam, J. L., Eyring, E. M., Epstein, W. W., Jensen, R. P. and Jaget, C. W. (1965b). *J. Am. Chem. Soc.* 87, 4247
Hassel, O. and Luzanski, H. (1932). *Z. Krist.* 83, 448
Haselbach, E., Henriksson, A., Jachimowicz, F. and Wirz, J. (1972). *Helv. Chim. Acta* 55, 1757
Helder, J., Birker, P. J. M. W. L., Verschoor, G. C. and Reedijk, J. (1984). *Inorg. Chim. Acta* 85, 169
Helmholz, L. and Rogers, M. T. (1939). *J. Am. Chem. Soc.* 61, 2590
Helmholz, L. and Rogers, M. T. (1940). *J. Am. Chem. Soc.* 62, 1533
Hibbert, F. (1973). *J. Chem. Soc., Chem. Commun.* 463
Hibbert, F. (1974). *J. Chem. Soc., Perkin Trans.* 2 1862
Hibbert, F. (1984). *Acc. Chem. Res.* 17, 115
Hibbert, F. (1986). *Adv. Phys. Org. Chem.* 22, 113
Hibbert, F. and Awwal, A. (1976). *J. Chem. Soc., Chem. Commun.* 995
Hibbert, F. and Awwal, A. (1978). *J. Chem. Soc., Perkin Trans.* 2 939
Hibbert, F. and Hunte, K. P. P. (1981). *J. Chem. Soc., Perkin Trans.* 2 1562
Hibbert, F. and Hunte, K. P. P. (1983). *J. Chem. Soc., Perkin Trans.* 2 1895
Hibbert, F. and Malana, M. A. (1990). *J. Chem. Soc., Perkin Trans.* 2, 711

Hibbert, F. and Phillips, S. C. (1989). Unpublished work
Hibbert, F. and Robbins, H. J. (1978). *J. Am. Chem. Soc.* **100**, 8239
Hibbert, F. and Robbins, H. J. (1980). *J. Chem. Soc., Chem. Commun.* 141
Hibbert, F. and Sellens, R. J. (1986). *J. Chem. Soc., Perkin Trans. 2* 1757
Hibbert, F. and Sellens, R. J. (1988a). *J. Chem. Soc., Perkin Trans. 2* 529
Hibbert, F. and Sellens, R. J. (1988b). *J. Chem. Res. (S)* 368
Hibbert, F. and Simpson, G. R. (1983). *J. Am. Chem. Soc.* **105**, 1063
Hibbert, F. and Simpson, G. R. (1985). *J. Chem. Soc., Perkin Trans. 2* 1247
Hibbert, F. and Simpson, G. R. (1987a). *J. Chem. Soc., Perkin Trans. 2* 243
Hibbert, F. and Simpson, G. R. (1987b). *J. Chem. Soc., Perkin Trans. 2* 613
Hibbert, F. and Spiers, K. J. (1988). *J. Chem. Soc., Perkin Trans. 2* 1309
Hibbert, F. and Spiers, K. J. (1989a). *J. Chem. Soc., Perkin Trans. 2* 67
Hibbert, F. and Spiers, K. J. (1989b). *J. Chem. Soc., Perkin Trans. 2* 377
Hilvert, D., Gardell, S. J., Rutter, W. J. and Kaiser, E. T. (1986). *J. Am. Chem. Soc.* **108**, 5298
Hine, J. and Li, W.-S. (1975). *J. Org. Chem.* **40**, 1795
Ho, C. K. and Fersht, A. R. (1986). *Biochemistry* **25**, 1891
Holmes, M. A. and Matthews, B. W. (1982). *J. Mol. Biol.* **160**, 623
Hopkins, A. R. and Williams, A. (1982). *J. Org. Chem.* **47**, 1745
Hopkins, A. R., Green, A. L. and Williams, A. (1983). *J. Chem. Soc., Perkin Trans. 2* 1279
Howard, J. A. K., Knox, S. A. R., Terrill, N. J. and Yates, M. I. (1989). *J. Chem. Soc., Chem. Commun.* 640
Hsu, B. and Schlemper, E. O. (1980). *Acta Crystallogr., Sect. B* **36**, 3017
Hsu, B., Schlemper, E. O. and Fiar, C. K. (1980). *Acta Crystallogr., Sect. B* **36**, 1387
Hussain, M. S. and Al-Hamoud, S. A. A. (1985). *J. Chem. Soc., Dalton Trans.* 749
Hussain, M. S., Schlemper, E. O. and Fiar, C. K. (1980). *Acta Crystallogr., Sect. B* **36**, 1104
Hussain, M. S., Schlemper, E. O. and Yelon, W. B. (1981). *Acta Crystallogr., Sect. B* **37**, 347
Ibers, J. A. (1964). *J. Chem. Phys.* **40** 402
Ichikawa, M. (1978a). *Chem. Phys. Lett.* **79**, 583
Ichikawa, M. (1978b). *Acta Crystallogr. Sect. B* **34**, 2074
Ichikawa, M. (1981). *J. Cryst. Mol. Struct.* **11**, 167
Iijima, K., Ohnogi, A. and Shibata, S. (1987). *J. Mol. Struct.* **156**, 111
Imashiro, F., Maeda, S., Takegoshi, K., Terao, T. and Saika, A. (1987). *J. Am. Chem. Soc.* **109**, 5213
Inskeep, W. H., Jones, D. L., Silfvast, W. T. and Eyring, E. M. (1968). *Proc. Natl. Acad. Sci. USA* **59**, 1027
Irwin, M. J., Nyborg, J., Reid, B. R. and Blow, D. M. (1976). *J. Mol. Biol.* **105**, 577
Iwasaki, F. F., Iwasaki, H. and Saito, Y. (1967). *Acta Crystallogr.* **23**, 64
Jaffe, H. H. (1954). *J. Am. Chem. Soc.* **76**, 4261
James, B. R. and Morris, R. H. (1980). *J. Chem. Soc., Chem. Commun.* 31
Janoschek, R. (1982). *Croat. Chim. Acta* **55**, 75
Janssen, C. L., Allen, W. D., Schaefer, H. F. and Bowman, J. M. (1986). *Chem. Phys. Lett.* **131**, 352
Jarret, R. M. and Saunders, M. (1985). *J. Am. Chem. Soc.* **107**, 2648
Jarret, R. M. and Saunders, M. (1986). *J. Am. Chem. Soc.* **108**, 7549
Jeffrey, G. A. and Yeon, Y. (1986). *Acta Crystallogr. Sect. B* **42**, 410

Jencks, W. P. (1975). *Adv. Enzymol.* **43**, 219
Jenkins, H. D. B. and Pratt, K. F. (1977). *J. Chem. Soc., Faraday Trans. 2* **73**, 812
Jensen, R. P., Eyring, E. M. and Walsh, W. M. (1966). *J. Phys. Chem.* **70**, 226
Joesten, M. D. and Schaad, L. J. (1974). "Hydrogen Bonding". Dekker, New York
Jones, D. J., Roziere, J. and Lehmann, M. S. (1986a). *J. Chem. Soc., Dalton Trans.* 651
Jones, M. D., Lowe, D. M., Borgford, T. and Fersht, A. R. (1986b). *Biochemistry* **25**, 1887
Jones, R. D. G. (1976a). *Acta Crystallogr., Sect. B.* **32**, 1807
Jones, R. D. G. (1976b). *Acta Crystallogr., Sect. B.* **32**, 2133
Joswig, W., Fuess, H. and Ferraris, G. (1982). *Acta Crystallogr., Sect. C.* **38**, 2798
Kamlet, M. J. and Taft, R. W. (1976). *J. Am. Chem. Soc.* **98**, 337, 2886
Karipides, A. and Miller, C. (1984). *J. Am. Chem. Soc.* **106**, 1494
Kassebaum, J. W. and Silverman, D. N. (1989). *J. Am. Chem. Soc.* **111**, 2691
Kawaguchi, K. and Hirota, E. (1986). *J. Chem. Phys.* **84**, 2953
Kehr, W. G., Breitinger, D. K. and Bauer, G. (1980). *Acta Crystallogr., Sect. B.* **36** 2545
Keil, F. and Ahlrichs, R. (1976). *J. Am. Chem. Soc.* **98**, 4787
Ketelaar, J. A. A. (1941). *J. Chem. Phys.* **9**, 775
Ketelaar, J. A. A. and Vedder, W. (1951). *J. Chem. Phys.* **19**, 654
Kirby, A. J. (1980). *Adv. Phys. Org. Chem.* **17**, 183
Kirby, A. J. (1987). *Crit. Rev. Biochem.* **22**, 283
Kirby, A. J. and Osborne, R. (1989). Results quoted in Kirby, A. J. and Percy, J. (1989).
Kirby, A. J. and Percy, J. (1987). *J. Chem. Soc., Chem. Commun.* 1774
Kirby, A. J. and Percy, J. (1989). *J. Chem. Soc., Perkin Trans. 2* 907
Koelle, U. and Forsén, S. (1974). *Acta Chem., Ser. A* **28**, 531
Koll, A., Rospenk, M., Sobczyk, L. and Glowiak, T. (1986). *Can. J. Chem* **64**, 1850
Kollman, P. and Allen, L. C. (1972). *Chem. Rev.* **72**, 283
Kol'tsov, A. I. and Kheifets, G. M. (1971). *Russ. Chem. Rev.* **40**, 773
Kreevoy, M. M. and Liang, T. (1980). *J. Am. Chem. Soc.* **102**, 3315
Kreevoy, M. M., Liang, T. and Chang, K.-C. (1977). *J. Am. Chem. Soc.* **99**, 5207
Kreevoy, M. M. and Ridl, B. A. (1981). *J. Phys. Chem.* **85**, 914
Kresge, A. J. (1973). *Chem. Soc. Rev.* **2**, 475
Kresge, A. J. (1975). *Acc. Chem. Res.* **8**, 354
Kresge, A. J. and Allred, A. L. (1963). *J. Am. Chem. Soc.* **85**, 1541
Kresge, A. J. and Chiang, Y. (1973). *J. Phys. Chem.* **77**, 822
Kresge, A. J. and Powell, M. F. (1981). *J. Am. Chem. Soc.* **103**, 972
Kruh, R., Fuwa, K. and McEver, T. E. (1956). *J. Am. Chem. Soc.* **78**, 4256
Küppers, H., Kvick, A. and Olovsson, I. (1981). *Acta Crystallogr., Sect. B* **37**, 1203
Kurz, J. L. and Kurz, L. C. (1972). *J. Am. Chem. Soc.* **94**, 445
Kurz, J. L., Myers, M. T. and Ratcliff, K. M. (1984). *J. Am. Chem. Soc.* **106**, 5631
Kvick, A., Koetzle, T. F. and Takusagawa, F. (1974). *J. Chem. Phys.* **60**, 3866
Lapachev, V. V., Mainagashev, I. Ya., Stekhova, S. A., Fedotov, M. A., Krivopalov, V. P. and Mamaev, V. P. (1985). *J. Chem. Soc., Chem. Commun.* 494
Larson, J. W. and McMahon, T. B. (1986). *J. Am. Chem. Soc.* **108**, 1719
Larson, J. W. and McMahon, T. B. (1987). *J. Phys. Chem.* **91**, 554
Larson, J. W. and McMahon, T. B. (1988). *J. Am. Chem. Soc.* **110**, 1087
Larsson, G. and Nahringbauer, I. (1988). *Acta Crystallogr., Sect. B.* **24**, 666
Lau, Y. K., Ikuta, S. and Kebarle, P. (1982). *J. Am. Chem. Soc.* **104**, 1462

Laughton, P. M. and Robertson, R. E. (1969). *In* "Solute–Solvent Interactions" (eds J. F. Coetzee and C. D. Ritchie), p. 399. Dekker, New York

Lazaar, K. I. and Bauer, S. H. (1983). *J. Phys. Chem.* **87**, 2411

Le Fèvre, R. J. W. and Welsh, H. (1949). *J. Chem. Soc.* 2330

Legon, A. C., Millen, D. J. and Rogers, S. C. (1980). *Proc. R. Soc. London, Ser. A* **370**, 213

Lehmann, M. S. and Larsen, F. K. (1971). *Acta Chem. Scand.* **13**, 3859

Leipert, T. K. (1977). *Org. Mag. Res.* **9**, 157

Lintvedt, R. L. and Holtzclaw Jr, H. F. (1966). *J. Am. Chem. Soc.* **88**, 2713

Ludman, C. J., Waddington, T. C., Pang, E. K. C. and Smith, J. A. S. (1977). *J. Chem. Soc., Faraday Trans. 2* **73**, 1003

Lukehart, C. M. and Zeile, J. V. (1976). *J. Am. Chem. Soc.* **98**, 2365

Lynton, H. and Siew, R. Y. (1973). *Can. J. Chem.* **51**, 227

Macdonald, A. L., Speakman, J. C. and Hadzi, D. (1972). *J. Chem. Soc., Perkin Trans. 2* 825

Mallet, J. W. (1881). *Am. Chem. J.* **3**, 189

Massa, W. and Herdtweck, E. (1983). *Acta Crystallogr. Sect. C* **39**, 509

Matthews B. W. (1988). *Acc. Chem. Res.* **21**, 333

McDonald, T. R. R. (1960). *Acta Crystallogr.* **13**, 113

McGaw, B. L. and Ibers, J. A. (1963). *J. Chem. Phys.* **39**, 2677

McMahon, T. B. and Kebarle, P. (1986). *J. Am. Chem. Soc.* **108**, 6502

McMahon, T. B. and Larson, J. W. (1982a). *J. Am. Chem. Soc.* **104**, 5848

McMahon, T. B. and Larson, J. W. (1982b). *J. Am. Chem. Soc.* **104**, 6255

McMahon, T. B. and Larson, J. W. (1983). *J. Am. Chem. Soc.* **105**, 2944

McMahon, T. B. and Larson, J. W. (1984a). *Can. J. Chem.* **62**, 675

McMahon, T. B. and Larson, J. W. (1984b). *Inorg. Chem.* **23**, 2029

McMahon, T. B. and Roy, M. (1985). *Can. J. Chem.* **63**, 708

Menger, F. M. (1985). *Acc. Chem. Res.* **18**, 128

Meot-Ner, M. (1984). *J. Am. Chem. Soc.* **106**, 1257

Meyer, K. H. and Hopff, H. (1921). *Chem. Ber.* **54**, 579

Meyer, K. H. and Schoeller, V. (1920). *Chem. Ber.* **53**, 1410

Mikenda, W. (1986). *J. Mol. Struct.* **147**, 1

Miles, M. H., Eyring, E. M., Epstein, W. W. and Anderson, M. T. (1966). *J. Phys. Chem.* **70**, 3490

Millefiori, S., Millefiori, A. and Granozzi, G. (1983). *J. Mol. Struct. (Theochem.)* **105**, 135

Millen, D. J., Legon, A. C. and Schrems, O. (1979). *J. Chem. Soc., Faraday Trans. 2* **75**, 592

Miller, P., Butler, R. A. and Lippincott, E. R. (1972). *J. Chem. Phys.* **57**, 5451

Misaki, S., Kashino, S. and Haisa, M. (1986). *Bull. Chem. Soc. Jpn* **59**, 1059

Misaki, S., Kashino, S. and Haisa, M. (1989a). *Acta Crystallogr., Sect. C* **45**, 62

Misaki, S., Kashino, S. and Haisa, M. (1989b). *Acta Crystallogr., Sect. C* **45**, 917

Monteilhet, C. and Blow, D. M. (1978). *J. Mol. Biol.* **122**, 407

Mootz, D. and Bartmann, F. (1988). *Angew. Chem., Int. Ed. Engl.* **27**, 391

Mootz, D. and Boenigk, D. (1986). *J. Am. Chem. Soc.* **108**, 6634

Mootz, D. and Boenigk, D. (1987). *Z. Anorg. Allgem. Chem.* **544**, 159

Mootz, D. and Oellers, E.-J. (1988). *Z. Anorg. Allgem. Chem.* **559**, 27

Mootz, D. and Poll, W. (1984a). *Z. Naturforsch.* **39b**, 290

Mootz, D. and Poll, W. (1984b). *Z. Naturforsch.* **39b**, 1300

Mootz, D. and Steffen, M. (1981). *Angew. Chem., Int. Ed. Engl.* **20**, 196

Mootz, D., Poll, W., Pawelke, G. and Appelman, E. H. (1988). *Angew. Chem., Int. Ed. Engl.* **27**, 392

More O'Ferrall, R. A. (1969). *J. Chem. Soc., Chem. Commun.* 114

More O'Ferrall, R. A. (1975). In "Proton Transfer Reactions" (eds E. F. Caldin and V. Gold). Chapman and Hall, London

Morokuma, K. (1977). *Acc. Chem. Res.* **10**, 294

Murray-Rust, P., Stallings, W. C., Monti, C. T., Preston, R. K. and Glusker, J. P. (1983). *J. Am. Chem. Soc.* **105**, 3206

Nadler, E. B. and Rappoport, Z. (1989). *J. Am. Chem. Soc.* **111**, 213

Nakamoto, K., Margoshes, M. and Rundle, R. E. (1955). *J. Am. Chem. Soc.* **77**, 6480

Neckel, A., Kuzmany, P. and Vinek, G. (1971). *Z. Naturforsch.* **26a**, 569

Nelmes, R. J. (1980). *Ferroelectrics* **24**, 237

Newman, R. and Badger, R. M. (1951). *J. Chem. Phys.* **19**, 1207

Ng, C. Y., Trevor, D. J., Tiedemann, P. W., Ceyer, S. T., Kronesbusch, P. L., Mahan, B. H. and Lee, Y. T. (1977). *J. Chem. Phys.* **67**, 4235

Noak, W. E. (1979). *Theor. Chim. Acta* **53**, 101

Noble, P. N. and Kortzeborn, R. N. (1970). *J. Chem. Phys.* **52**, 5375

Norrestam, R., Von Glehn, M. and Wachtmeister, C. A. (1974). *Acta Chem. Scand., Ser. B* **28**, 1149

Novak, A. (1974). *Structure and Bonding* **18**, 177

Odutola, J. A. and Dyke, T. R. (1980). *J. Chem. Phys.* **72**, 5062

Ogoshi, H. and Nakamoto, K. (1966). *J. Chem. Phys.* **45**, 3113

Ogoshi, H. and Yoshida, Z.-I. (1971). *Spectrochim. Acta, Sect. A* **27**, 165

Olah, G. A., Prakash, G. K. S. and Sommer, J. (1985). "Superacids", p. 48. Wiley-Interscience, New York

Ostlund, N. S. and Bellenger, L. W. (1975). *J. Am. Chem. Soc.* **97**, 1237

Olovsson, I. (1982). *Croat. Chim. Acta* **55**, 171

Olovsson, I. and Jönsson, P.-G. (1976). "The Hydrogen Bond" (eds P. Schuster, G. Zundel and C. Sandorfy), Vol. II, Ch. 8. North-Holland, Amsterdam

Owen, N. D. S. (1989). PhD thesis, York University

Page, M. I. (1973). *Chem. Soc. Rev.* **2**, 295

Page. M. I. (1984). "The Chemistry of Enzyme Action". Elsevier, Amsterdam

Pal, J., Murmann, K. and Schlemper, E. O. (1986). *Inorg. Chim. Acta* **115**, 153

Paratt, J. C. and Smith, J. A. S. (1975). *J. Chem. Soc., Faraday Trans. 2* **71**, 596

Pauling, L. (1933). *Z. Krist.* **85**, 380

Peinel, G. (1979). *Chem. Phys. Lett.* **65**, 324

Perlmutter-Hayman, B. and Shinar, R. (1975). *Int. J. Chem. Kinet.* **7**, 453

Perlmutter-Hayman, B., Sarfaty, R. and Shinar, R. (1976). *Int. J. Chem. Kinet.* **8**, 741

Perrin, C. L. and Thoburn, J. D. (1989). *J. Am. Chem. Soc.* **111**, 8010

Perrin, D. D. (1965). "Dissociation Constants of Organic Bases in Aqueous Solution", Butterworths, London; supplement (1972)

Peterson, S. W. and Levy, H. V. (1952). *J. Chem. Phys.* **20**, 704

Pfeffer, P., Valentine, K. and Parrish, F. (1978). *J. Am. Chem. Soc.* **100**, 1265

Piaggio, P., Tubino, R. and Dellepiane, C. (1983). *J. Mol. Struct.* **96**, 277

Pimentel, G. C. and McClellan, A. L. (1960). "The Hydrogen Bond". Freeman, San Francisco

Pimentel, G. C. and McClellan, A. L. (1971). *Ann. Rev. Phys. Chem.* **22**, 347

Pitzer, K. S. and Westrum, E. F. (1947). *J. Chem. Phys.* **15**, 526

Power, L. F. and Jones, R. D. G. (1971). *Acta Crystallogr., Sect. B* **27**, 181

Pratt, J. C. and Smith, J. A. S. (1975). *J. Chem. Soc., Faraday Trans. 2* **71**, 596

Pyzalka, D., Pyzalka, R. and Borowiak, T. (1983). *J. Cryst. Spectr. Res.* **13**, 211
Raban, M. and Yamamoto, G. (1977). *J. Org. Chem.* **42**, 2549
Rappoport, Z., Nugield, D. A. and Biali, S. E. (1988). *J. Org. Chem.* **53**, 4814
Regitz, M. and Schäfer, A. (1981). *Liebigs Ann. Chem.* 1172
Robertson, J. M. and Ubbelohde, A. R. (1939). *Proc. R. Soc. London, Ser. A* **170**, 222
Robinson, M. J. T., Rosen, K. M. and Workman, J. D. B. (1977). *Tetrahedron* **33**, 1655
Rohlfing, C. M., Allen, L. C. and Ditchfield, R. (1983). *J. Chem. Phys.* **79**, 4958
Rose, M. C. and Stuehr, J. E. (1971). *J. Am. Chem. Soc.* **93**, 4350
Rothschild, W. G. (1976). *In* "The Hydrogen Bond" (eds P. Schuster, G. Zundel and C. Sandorfy), Vol. II, Ch. 16. North-Holland, Amsterdam
Rubin, J. and Blow, A. R. (1981). *J. Mol. Biol.* **145**, 489
Sardella, D. J., Heinert, D. H. and Shapiro, B. L. (1969). *J. Org. Chem.* **34**, 2817
Saupe, T., Krieger, C. and Staab, H. A. (1986). *Angew. Chem., Int. Ed. Engl.* **25**, 451
Schaad, L. J. (1974). "Hydrogen Bonding" (eds M. D. Joesten and L. J. Schaad), Ch. 2. Dekker, New York
Schaefer, W. B. and Marsh, R. E. (1984). *J. Chem. Soc., Chem. Commun.* 1555
Schlemper, E. O., Hamilton, W. C. and La Placa, S. T. (1971). *J. Chem. Phys.* **54**, 3990
Schlemper, E. O. (1986). *Acta Crystallogr., Sect. C* **42**, 755
Schowen, R. L. (1972). *Prog. Phys. Org. Chem.* **9**, 197
Schultz, A. J., Srinivasan, R. G., Teller, J. M., Williams, J. M. and Lukehart, C. M. (1984). *J. Am. Chem. Soc.* **106**, 999
Schuster, P. (1976). *In* "The Hydrogen Bond" (eds P. Schuster, G. Zundel and C. Sandorfy), Vol. II, p. 427. North-Holland, Amsterdam
Semmingsen, D. (1974). *Acta Chem. Scand., Ser. B* **28** 169
Semmingsen, D. (1977). *Acta Chem. Scand., Ser. B* **31**, 114
Shapet'ko, N. N. (1973). *Org. Mag. Res.* **5**, 215
Shapet'ko, N. N., Bogachev, Yu. S., Berestova, S. S. and Lukovkin, G. M. (1975). *Org. Mag. Res.* **7**, 540
Shapet'ko, N. N., Bogachev, Y. S., Radushnova, I. L. and Shigorin, D. N. (1976). *Doklady Acad. Nauk, SSSR* **231**, 1085
Silverman, D. N. (1981). *J. Am. Chem. Soc.* **103**, 6242
Singh, I. and Calvo, C. (1975). *Can. J. Chem.* **53**, 1046
Singh, T. R. and Wood, J. L. (1969). *J. Chem. Phys.* **50**, 3572
Smit, P. H., Derissen, J. L. and van Duijneveldt, F. Y. (1979). *Mol. Phys.* **37**, 501
Somorjai, R. L. and Hornig, D. F. (1962). *J. Chem. Phys.* **36**, 1980
Soriano, J., Shamir, J., Netzer, A. and Marcus, Y. (1969). *Inorg. Nuclear Chem. Lett.* **5**, 209
Sorkhabi, H. A., Halle, J.-C. and Terrier, F. (1978). *J. Chem. Res. (S)* 108
Spackman, M. A. (1986). *J. Chem. Phys.* **85**, 6587
Speakman, J. C. (1972). *Struct. Bonding (Berlin)* **12**, 141
Speakman, J. C. and Currie, M. (1970). *J. Chem. Soc. (A)* 1923
Spinner, E. (1974). *Aust. J. Chem.* **27**, 1149
Spinner, E. (1977). *Aust. J. Chem.* **30**, 1167
Spinner, E. (1980a). *Aust. J. Chem.* **33**, 933
Spinner, E. (1980b). *J. Chem. Soc., Perkin Trans. 2* 395
Spinner, E. (1983). *J. Am. Chem. Soc.* **105**, 756
Staab, H. A. and Saupe, T. (1988). *Angew. Chem., Int. Ed. Engl.* **27**, 865

Staab, H. A., Hone, M. and Krieger, C. (1988a). *Tetrahedron Lett.* **29**, 1905
Staab, H. A., Krieger, C. and Hone, M. (1988b). *Tetrahedron Lett.* **29**, 5629
Staab, H. A., Saupe, T. and Krieger, C. (1983). *Angew. Chem., Int. Ed. Engl.* **22**, 731
Stepisnik, J. and Hadzi, D. (1972). *J. Mol. Struct.* **13**, 307
Størgard, A., Strich, J., Almlöf, J. and Roos, B. (1975). *Chem. Phys.* **8**, 405
Stomberg, R. (1981). *Acta Chem. Scand., Sect. A* **35**, 389
Stomberg, R. (1982). *Acta Chem. Scand., Sect. A* **36**, 101
Swain, C. G., Swain, M. S., Powell, A. L. and Alunni, S. (1983). *J. Am. Chem. Soc.* **105**, 502
Szarek, W. A., Hay, G. W. and Perlmutter, M. M. (1982). *J. Chem. Soc., Chem. Commun.* 1253
Szentivanyi, H. and Stomberg, R. (1984). *Acta Chem. Scand., Sect. A* **38**, 101
Taft, R. W., Abraham, M. H., Doherty, R. M. and Kamlet, M. J. (1985). *J. Am. Chem. Soc.* **107**, 3105
Takahashi, L. H., Radhakrishnan, R., Rosenfield, R. E., Meyer, E. F. and Trainor, D. A. (1989). *J. Am. Chem. Soc.* **111**, 3368
Takusagawa, F. and Koetzle, T. F. (1979). *Acta Crystallogr., Sect. B* **35**, 2126
Taylor, N. F. (1988). "Fluorinated Carbohydrates". American Chemical Society, Washington
Tayyari, S. F., Zeegers-Huyskens, Th. and Wood, J. L. (1979a). *Spectrochim. Acta, Sect. A* **35**, 1265
Tayyari, S. F., Zeegers-Huyskens, Th. and Wood, J. L. (1979b). *Spectrochim, Acta, Sect. A* **35**, 1289
Tichy, K., Rüegg, A. and Benes, J. (1980). *Acta Crystallogr. Sect. B* **36**, 1028
Tiedemann, P. W., Anderson, S. L., Ceyer, S. T., Hirooka, T., Ng, C. Y., Mahan, B. H. and Lee, Y. T. (1979). *J. Chem. Phys.* **71**, 605
Toros, Z. R. and Prodic, B. K. (1976). *Acta Crystallogr., Sect. B* **32**, 1096
Truter, M. R. and Vickery, B. L. (1972). *J. Chem. Soc., Dalton Trans.* 395
Ubbelohde, A. R. (1939). *Proc. Roy. Soc. London, Ser. A* **173**, 417
Ubbelohde, A. R. and Woodward, I. (1942). *Proc. Roy. Soc. London, Ser. A* **179**, 399
Umeyama, H. and Morokuma, K. (1977). *J. Am. Chem. Soc.* **99**, 1316; and refs therein
van Duijneveldt, F. B. and Murrell, J. N. (1967). *J. Chem. Phys.* **46**, 1759
de la Vega, J. R. (1982). *Acc. Chem. Res.* **15**, 185
de la Vega, J. R., Busch, J. H., Schauble, J. H., Kunze, K. L. and Haggert, B. E. (1982). *J. Am. Chem. Soc.* **104**, 3295
Vogt, H. H. and Gompper, R. (1981). *Chem. Ber.* **114**, 2884
Vinogradov, S. N. and Linnell, R. H. (1971). "Hydrogen Bonding" Van Nostrand-Reinhold, New York
Waddington, T. C. (1958). *Trans. Faraday Soc.* **54**, 25
Walters, E. A. and Long, F. A. (1972). *J. Phys. Chem.* **76**, 362
Ward, W. H. J. and Ferscht, A. R. (1988a). *Biochemistry* **27**, 1041
Ward, W. H. J. and Ferscht, A. R. (1988b). *Biochemistry* **27**, 5525
Waugh, J. S., Floyd, B. H. and Yost, D. M. (1953). *J. Phys. Chem.* **57**, 846
Waye, M. M. Y., Winter, G., Wilkinson, A. J. and Fersht, A. R. (1983). *EMBO J.* **2**, 1827
Wells, T. N. C. and Fersht, A. R. (1986). *Biochemistry* **25**, 1881
White, J. M., Alder, R. W. and Orpen, A. G. (1988a). *Acta Chem. Cryst., Sect. C* **44**, 662

White, J. M., Alder, R. W. and Orpen, A. G. (1988b). *Acta Chem. Cryst., Sect. C* **44**, 664
White, J. M., Alder, R. W. and Orpen, A. G. (1988c). *Acta Chem. Cryst., Sect. C* **44**, 872
White, J. M., Alder, R. W. and Orpen, A. G. (1988d). *Acta Chem. Cryst., Sect. C* **44**, 1465
White, J. M., Alder, R. W. and Orpen, A. G. (1988e). *Acta Chem. Cryst., Sect. C* **44**, 1467
Wierzchowski, K. L., Shugar, D. and Katriski, A. R. (1963). *J. Am. Chem. Soc.* **85**, 827
Wierzchowski, K. L. and Shugar, D. (1965). *Spectrochim. Acta* **21**, 943
Wilkinson, A. J., Fersht, A. R., Blow, D. M., Carter, P. and Winter, G. (1984). *Nature (London)* **307**, 187
Williams, D. E., Dumke, W. L. and Rundle, R. E. (1962). *Acta Cryst.* **15**, 627
Williams, J. M. and Schneemeyer, L. F. (1973). *J. Am. Chem. Soc.* **95**, 5780
Withers, S. G., Street, I. P. and Armstrong, C. R. (1986). *Biochemistry* **25**, 6021
Withers, S. G., Street, I. P. and Percival, M. D. (1988). See Ch. 5 of Taylor (1988)
Withers, S. G., Street, I. P. and Rupitz, K. (1989). *Biochemistry* **28**, 1581
Yamdagni, R. and Kebarle, P. (1971). *J. Am. Chem. Soc.* **93**, 7139
Yamdagni, R. and Kebarle, P. (1973). *J. Am. Chem. Soc.* **95**, 3504
Yoffe, S. T., Fedin, E. I., Petrovskii, M. I. and Kabachnik, M. I. (1966). *Tetrahedron* 2661
Yoshida, N. and Fujimoto, M. (1977). *Chem. Lett.* 1301
Zirnstein, M. A. and Staab, H. A. (1987). *Angew. Chem., Int. Ed. Engl.* **26**, 460
Zundel, G. and Fritsch, J. (1986). *In* "Chemical Physics of Solvation" (eds R. R. Dogonadze, E. Kálmán, A. A. Kornyshev and J. Ulstrop), Chs 2 and 3. Elsevier, Amsterdam
Zundel, G. (1988). *J. Mol. Struct.* **177**, 43
Zundel, G. and Eckert, M. (1989). *J. Mol. Struct. (Theochem.)* **200**, 73

Subject Index

Author Index

Numbers in italic refer to the pages on which references are listed at the end of each article

Cumulative Index of Authors

413

Cumulative Index of Titles